AF717854

Modular Forms, a Computational Approach

Modular Forms, a Computational Approach

William Stein

with an appendix by
Paul E. Gunnells

Graduate Studies in Mathematics

Volume 79

American Mathematical Society
Providence, Rhode Island

Editorial Board

David Cox
Walter Craig
Nikolai Ivanov
Steven G. Krantz
David Saltman (Chair)

This edition is published by the American Mathematical Society under license from the author.

2000 *Mathematics Subject Classification.* Primary 11F11, 11Y16, 11F67, 11F55, 11F75.

For additional information and updates on this book, visit
www.ams.org/bookpages/gsm-79

Library of Congress Cataloging-in-Publication Data

Stein, William, 1974–
Modular forms, a computational approach / William Stein ; with an appendix by Paul E. Gunnells.
p. cm. — (Graduate studies in mathematics ; v. 79)
Includes bibliograpical references and index.
ISBN-13: 978-0-8218-3960-7 (alk. paper)
1. Forms, Modular—Data processing—Textbooks. 2. Algebraic spaces—Data processing—Textbooks. I. Title.

QA243.S74 2007
512.7′3—dc22 2006047950

Copying and reprinting. Individual readers of this publication, and nonprofit libraries acting for them, are permitted to make fair use of the material, such as to copy a chapter for use in teaching or research. Permission is granted to quote brief passages from this publication in reviews, provided the customary acknowledgment of the source is given.

Republication, systematic copying, or multiple reproduction of any material in this publication is permitted only under license from the American Mathematical Society. Requests for such permission should be addressed to the Acquisitions Department, American Mathematical Society, 201 Charles Street, Providence, Rhode Island 02904-2294, USA. Requests can also be made by e-mail to reprint-permission@ams.org.

© 2007 by the author. All rights reserved.
© 2007 for the Appendix by Paul E. Gunnells. All rights reserved.
Printed in the United States of America.

∞ The paper used in this book is acid-free and falls within the guidelines established to ensure permanence and durability.
Visit the AMS home page at http://www.ams.org/

10 9 8 7 6 5 4 3 2 1 12 11 10 09 08 07

To my grandmother, Annette Maurer

Contents

Preface

This is a graduate-level textbook about algorithms for computing with modular forms. It is nontraditional in that the primary focus is not on underlying theory; instead, it answers the question *"how do you use a computer to explicitly compute spaces of modular forms?"*

This book emerged from notes for a course the author taught at Harvard University in 2004, a course at UC San Diego in 2005, and a course at the University of Washington in 2006.

The author has spent years trying to find good practical ways to compute with classical modular forms for congruence subgroups of $\mathrm{SL}_2(\mathbb{Z})$ and has implemented most of these algorithms several times, first in C++ [**Ste99b**], then in MAGMA [**BCP97**], and as part of the free open source computer algebra system SAGE (see [**Ste06**]). Much of this work has involved turning formulas and constructions buried in obscure research papers into precise computational recipes then testing these and eliminating inaccuracies.

The author is aware of no other textbooks on computing with modular forms, the closest work being Cremona's book [**Cre97a**], which is about computing with elliptic curves, and Cohen's book [**Coh93**] about algebraic number theory.

In this book we focus on how to compute *in practice* the spaces $M_k(N, \varepsilon)$ of modular forms, where $k \geq 2$ is an integer and ε is a Dirichlet character of modulus N (the appendix treats modular forms for higher rank groups). We spend the most effort explaining the general algorithms that appear so far to be the best (in practice!) for such computations. We will not discuss in any detail computing with quaternion algebras, half-integral weight forms, weight 1 forms, forms for noncongruence subgroups or groups other than GL_2, Hilbert and Siegel modular forms, trace formulas, p-adic modular forms, and modular abelian varieties, all of which are topics for additional books. We also rarely analyze the complexity of the algorithms, but instead settle for occasional remarks about their practical efficiency.

For most of this book we assume the reader has some prior exposure to modular forms (e.g., [**DS05**]), though we recall many of the basic definitions. We cite standard books for proofs of the fundamental results about modular forms that we will use. The reader should also be familiar with basic algebraic number theory, linear algebra, complex analysis (at the level of [**Ahl78**]), and algorithms (e.g., know what an algorithm is and what big oh notation means). In some of the examples and applications we assume that the reader knows about elliptic curves at the level of [**Sil92**].

Chapter 1 is foundational for the rest of this book. It introduces congruence subgroups of $\mathrm{SL}_2(\mathbb{Z})$ and modular forms as functions on the complex upper half plane. We discuss q-expansions, which provide an important computational handle on modular forms. We also study an algorithm for computing with congruence subgroups. The chapter ends with a list of applications of modular forms throughout mathematics.

In Chapter 2 we discuss level 1 modular forms in much more detail. In particular, we introduce Eisenstein series and the cusp form Δ and describe their q-expansions and basic properties. Then we prove a structure theorem for level 1 modular forms and use it to deduce dimension formulas and give an algorithm for explicitly computing a basis. We next introduce Hecke operators on level 1 modular forms, prove several results about them, and deduce multiplicativity of the Ramanujan τ function as an application. We also discuss explicit computation of Hecke operators. In Section 2.6 we make some brief remarks on recent work on asymptotically fast computation of values of τ. Finally, we describe computation of constant terms of Eisenstein series using an analytic algorithm. We generalize many of the constructions in this chapter to higher level in subsequent chapters.

In Chapter 3 we turn to modular forms of higher level but restrict for simplicity to weight 2 since much is clearer in this case. (We remove the weight restriction later in Chapter 8.) We describe a geometric way of viewing cuspidal modular forms as differentials on modular curves, which leads to modular symbols, which are an explicit way to present a certain homology group. This chapter closes with methods for explicitly computing cusp forms of weight 2 using modular symbols, which we generalize in Chapter 9.

In Chapter 4 we introduce Dirichlet characters, which are important both in explicit construction of Eisenstein series (in Chapter 5) and in decomposing spaces of modular forms as direct sums of simpler spaces. The main focus of this chapter is a detailed study of how to explicitly represent and compute with Dirichlet characters.

Chapter 5 is about how to explicitly construct the Eisenstein subspace of modular forms. First we define generalized Bernoulli numbers attached to a Dirichlet character and an integer then explain a new analytic algorithm for computing them (which generalizes the algorithm in Chapter 2). Finally we give without proof an explicit description of a basis of Eisenstein series, explain how to compute it, and give some examples.

Chapter 6 records a wide range of dimension formulas for spaces of modular forms, along with a few remarks about where they come from and how to compute them.

Chapter 7 is about linear algebra over exact fields, mainly the rational numbers. This chapter can be read independently of the others and does not require any background in modular forms. Nonetheless, this chapter occupies a central position in this book, because the algorithms in this chapter are of crucial importance to any actual implementation of algorithms for computing with modular forms.

Chapter 8 is the most important chapter in this book; it generalizes Chapter 3 to higher weight and general level. The modular symbols formulation described here is central to general algorithms for computing with modular forms.

Chapter 9 applies the algorithms from Chapter 8 to the problem of computing with modular forms. First we discuss decomposing spaces of modular forms using Dirichlet characters, and then explain how to compute a basis of Hecke eigenforms for each subspace using several approaches. We also discuss congruences between modular forms and bounds needed to provably generate the Hecke algebra.

Chapter 10 is about computing analytic invariants of modular forms. It discusses tricks for speeding convergence of certain infinite series and sketches how to compute every elliptic curve over $\mathbb{Q}$ with given conductor.

Chapter 11 contains detailed solutions to most of the exercises in this book. (Many of these were written by students in a course taught at the University of Washington.)

Appendix A deals with computational techniques for working with generalizations of modular forms to more general groups than $\mathrm{SL}_2(\mathbb{Z})$, such as $\mathrm{SL}_n(\mathbb{Z})$ for $n \geq 3$. Some of this material requires more prerequisites than the rest of the book. Nonetheless, seeing a natural generalization of the material in the rest of this book helps to clarify the key ideas. The topics in the appendix are directly related to the main themes of this book: modular symbols, Manin symbols, cohomology of subgroups of $SL_2(\mathbb{Z})$ with various coefficients, explicit computation of modular forms, etc.

Software. We use SAGE, Software for Algebra and Geometry Experimentation (see [**Ste06**]), to illustrate how to do many of the examples. SAGE is completely free and packages together a wide range of open source mathematics software for doing much more than just computing with modular forms. SAGE can be downloaded and run on your computer or can be used via a web browser over the Internet. The reader is encouraged to experiment with many of the objects in this book using SAGE. We do not describe the basics of using SAGE in this book; the reader should read the SAGE tutorial (and other documentation) available at the SAGE website [**Ste06**]. All examples in this book have been automatically tested and should work exactly as indicated in SAGE version at least 1.5.

Acknowledgements. David Joyner and Gabor Wiese carefully read the book and provided a huge number of helpful comments.

John Cremona and Kevin Buzzard both made many helpful remarks that were important in the development of the algorithms in this book. Much of the mathematics (and some of the writing) in Chapter 10 is joint work with Helena Verrill.

Noam Elkies made remarks about Chapters 1 and 2. Sándor Kovács provided interesting comments on Chapter 1. Allan Steel provided helpful feedback on Chapter 7. Jordi Quer made useful remarks about Chapter 4 and Chapter 6.

The students in the courses that I taught on this material at Harvard, San Diego, and Washington provided substantial feedback: in particular, Abhinav Kumar made numerous observations about computing widths of cusps (see Section 1.4.1) and Thomas James Barnet-Lamb made helpful remarks about how to represent Dirichlet characters. James Merryfield made helpful remarks about complex analytic issues and about convergence in Stirling's formula. Robert Bradshaw, Andrew Crites (who wrote Exercise 7.5), Michael Goff, Dustin Moody, and Koopa Koo wrote most of the solutions included in Chapter 11 and found numerous typos throughout the book. Dustin Moody also carefully read through the book and provided feedback.

H. Stark suggested using Stirling's formula in Section 2.7.1, and Mark Watkins and Lynn Walling made comments on Chapter 3.

Parts of Chapter 1 follow Serre's beautiful introduction to modular forms [**Ser73**, Ch. VII] closely, though we adjust the notation, definitions, and order of presentation to be consistent with the rest of this book.

I would like to acknowledge the partial support of NSF Grant DMS 05-55776. Gunnells was supported in part by NSF Grants DMS 02-45580 and DMS 04-01525.

Notation and Conventions. We denote canonical isomorphisms by $\cong$ and noncanonical isomorphisms by $\approx$. If V is a vector space and s denotes some sort of construction involving V, we let V_s denote the corresponding subspace and V^s the quotient space. E.g., if ι is an involution of V, then V_+ is $\mathrm{Ker}(\iota - 1)$ and $V^+ = V/\mathrm{Im}(\iota - 1)$. If A is a finite abelian group, then A_{tor} denotes the torsion subgroup and A/tor denotes the quotient A/A_{tor}. We denote right group actions using exponential notation. Everywhere in this book, N is a positive integer and k is an integer.

If N is an integer, a *divisor* t of N is a *positive* integer such that N/t is an integer.

Chapter 1

Modular Forms

This chapter introduces modular forms and congruence subgroups, which are central objects in this book. We first introduce the upper half plane and the group $\mathrm{SL}_2(\mathbb{Z})$ then recall some definitions from complex analysis. Next we define modular forms of level 1 followed by modular forms of general level. In Section 1.4 we discuss congruence subgroups and explain a simple way to compute generators for them and determine element membership. Section 1.5 lists applications of modular forms.

We assume familiarity with basic number theory, group theory, and complex analysis. For a deeper understanding of modular forms, the reader is urged to consult the standard books in the field, e.g., [**Lan95, Ser73, DI95, Miy89, Shi94, Kob84**]. See also [**DS05**], which is an excellent first introduction to the theoretical foundations of modular forms.

1.1. Basic Definitions

The group

$$\mathrm{SL}_2(\mathbb{R}) = \left\{ \begin{pmatrix} a & b \\ c & d \end{pmatrix} : ad - bc = 1 \text{ and } a, b, c, d \in \mathbb{R} \right\}$$

acts on the *complex upper half plane*

$$\mathfrak{h} = \{z \in \mathbb{C} : \mathrm{Im}(z) > 0\}$$

by *linear fractional transformations*, as follows. If $\gamma = \left(\begin{smallmatrix} a & b \\ c & d \end{smallmatrix}\right) \in \mathrm{SL}_2(\mathbb{R})$, then for any $z \in \mathfrak{h}$ we let

$$\gamma(z) = \frac{az + b}{cz + d} \in \mathfrak{h}. \tag{1.1.1}$$

Since the determinant of γ is 1, we have

$$\left(\frac{d}{dz}\gamma\right)(z) = \frac{1}{(cz+d)^2}.$$

Definition 1.1 (Modular Group). The *modular group* is the group of all matrices $\left(\begin{smallmatrix} a & b \\ c & d \end{smallmatrix}\right)$ with $a, b, c, d \in \mathbb{Z}$ and $ad - bc = 1$.

For example, the matrices

$$S = \begin{pmatrix} 0 & -1 \\ 1 & 0 \end{pmatrix} \quad \text{and} \quad T = \begin{pmatrix} 1 & 1 \\ 0 & 1 \end{pmatrix} \tag{1.1.2}$$

are both elements of $\mathrm{SL}_2(\mathbb{Z})$; the matrix S induces the function $z \mapsto -1/z$ on $\mathfrak{h}$, and T induces the function $z \mapsto z + 1$.

Theorem 1.2. *The group* $\mathrm{SL}_2(\mathbb{Z})$ *is generated by* S *and* T.

Proof. See e.g. [**Ser73**, §VII.1]. □

In SAGE we compute the group $\mathrm{SL}_2(\mathbb{Z})$ and its generators as follows:

```
sage: G = SL(2,ZZ); G
Modular Group SL(2,Z)
sage: S, T = G.gens()
sage: S
[ 0 -1]
[ 1  0]
sage: T
[1 1]
[0 1]
```

Definition 1.3 (Holomorphic and Meromorphic). Let R be an open subset of $\mathbb{C}$. A function $f : R \to \mathbb{C}$ is *holomorphic* if f is complex differentiable at every point $z \in R$, i.e., for each $z \in R$ the limit

$$f'(z) = \lim_{h \to 0} \frac{f(z+h) - f(z)}{h}$$

exists, where h may approach 0 along any path. A function $f : R \to \mathbb{C} \cup \{\infty\}$ is *meromorphic* if it is holomorphic except (possibly) at a discrete set S of points in R, and at each $\alpha \in S$ there is a positive integer n such that $(z - \alpha)^n f(z)$ is holomorphic at α.

The function $f(z) = e^z$ is a holomorphic function on $\mathbb{C}$; in contrast, $1/(z - i)$ is meromorphic on $\mathbb{C}$ but not holomorphic since it has a pole at i. The function $e^{-1/z}$ is not even meromorphic on $\mathbb{C}$.

Modular forms are holomorphic functions on $\mathfrak{h}$ that transform in a particular way under a certain subgroup of $\mathrm{SL}_2(\mathbb{Z})$. Before defining general modular forms, we define modular forms of level 1.

1.2. Modular Forms of Level 1

Definition 1.4 (Weakly Modular Function). A *weakly modular function* of *weight* $k \in \mathbb{Z}$ is a meromorphic function f on $\mathfrak{h}$ such that for all $\gamma = \left(\begin{smallmatrix} a & b \\ c & d \end{smallmatrix}\right) \in \mathrm{SL}_2(\mathbb{Z})$ and all $z \in \mathfrak{h}$ we have

$$f(z) = (cz + d)^{-k} f(\gamma(z)). \tag{1.2.1}$$

The constant functions are weakly modular of weight 0. There are no nonzero weakly modular functions of odd weight (see Exercise 1.4), and it is not obvious that there are any weakly modular functions of even weight $k \geq 2$ (but there are, as we will see!). The product of two weakly modular functions of weights k_1 and k_2 is a weakly modular function of weight $k_1 + k_2$ (see Exercise 1.3).

When k is even, (1.2.1) has a possibly more conceptual interpretation; namely (1.2.1) is the same as

$$f(\gamma(z))(d(\gamma(z)))^{k/2} = f(z)(dz)^{k/2}.$$

Thus (1.2.1) simply says that the weight k "differential form" $f(z)(dz)^{k/2}$ is fixed under the action of every element of $\mathrm{SL}_2(\mathbb{Z})$.

By Theorem 1.2, the group $\mathrm{SL}_2(\mathbb{Z})$ is generated by the matrices S and T of (1.1.2), so to show that a meromorphic function f on $\mathfrak{h}$ is a weakly modular function, all we have to do is show that for all $z \in \mathfrak{h}$ we have

$$f(z+1) = f(z) \qquad \text{and} \qquad f(-1/z) = z^k f(z). \tag{1.2.2}$$

Suppose f is a weakly modular function of weight k. A *Fourier expansion* of f, if it exists, is a representation of f as $f(z) = \sum_{n=m}^{\infty} a_n e^{2\pi i n z}$, for all $z \in \mathfrak{h}$. Let $q = q(z) = e^{2\pi i z}$, which we view as a holomorphic function on $\mathbb{C}$. Let D' be the open unit disk with the origin removed, and note that q defines a map $\mathfrak{h} \to D'$. By (1.2.2) we have $f(z+1) = f(z)$, so there is a function $F : D' \to \mathbb{C}$ such that $F(q(z)) = f(z)$. This function F is a complex-valued function on D', but it may or may not be well behaved at 0.

Suppose that F is well behaved at 0, in the sense that for some $m \in \mathbb{Z}$ and all q in a neighborhood of 0 we have the equality

$$F(q) = \sum_{n=m}^{\infty} a_n q^n. \tag{1.2.3}$$

If this is the case, we say that f is *meromorphic at* ∞. If, moreover, $m \geq 0$, we say that f is *holomorphic at* ∞. We also call (1.2.3) the *q-expansion* of f about ∞.

Definition 1.5 (Modular Function). A *modular function* of *weight* k is a weakly modular function of weight k that is meromorphic at ∞.

Definition 1.6 (Modular Form). A *modular form* of *weight* k (and *level* 1) is a modular function of weight k that is holomorphic on $\mathfrak{h}$ and at ∞.

If f is a modular form, then there are numbers a_n such that for all $z \in \mathfrak{h}$,

$$f(z) = \sum_{n=0}^{\infty} a_n q^n. \tag{1.2.4}$$

Proposition 1.7. *The above series converges for all* $z \in \mathfrak{h}$.

Proof. The function $f(q)$ is holomorphic on D, so its Taylor series converges absolutely in D. $\square$

Since $e^{2\pi i z} \to 0$ as $z \to i\infty$, we set $f(\infty) = a_0$.

Definition 1.8 (Cusp Form). A *cusp form* of *weight* k (and *level* 1) is a modular form of weight k such that $f(\infty) = 0$, i.e., $a_0 = 0$.

Let $\mathbb{C}[[q]]$ be the ring of all *formal power series* in q. If $k = 2$, then $dq = 2\pi i q dz$, so $dz = \frac{1}{2\pi i}\frac{dq}{q}$. If $f(q)$ is a cusp form of weight 2, then

$$2\pi i f(z) dz = f(q)\frac{dq}{q} = \frac{f(q)}{q} dq \in \mathbb{C}[[q]]dq.$$

Thus the differential $2\pi i f(z)dz$ is holomorphic at ∞, since q is a local parameter at ∞.

1.3. Modular Forms of Any Level

In this section we define spaces of modular forms of arbitrary level.

Definition 1.9 (Congruence Subgroup). A *congruence subgroup* of $\mathrm{SL}_2(\mathbb{Z})$ is any subgroup of $\mathrm{SL}_2(\mathbb{Z})$ that contains

$$\Gamma(N) = \mathrm{Ker}(\mathrm{SL}_2(\mathbb{Z}) \to \mathrm{SL}_2(\mathbb{Z}/N\mathbb{Z}))$$

for some positive integer N. The smallest such N is the *level of* Γ.

The most important congruence subgroups in this book are

$$\Gamma_1(N) = \left\{ \begin{pmatrix} a & b \\ c & d \end{pmatrix} \in \mathrm{SL}_2(\mathbb{Z}) : \begin{pmatrix} a & b \\ c & d \end{pmatrix} \equiv \begin{pmatrix} 1 & * \\ 0 & 1 \end{pmatrix} \pmod{N} \right\}$$

and

$$\Gamma_0(N) = \left\{ \begin{pmatrix} a & b \\ c & d \end{pmatrix} \in \mathrm{SL}_2(\mathbf{Z}) : \begin{pmatrix} a & b \\ c & d \end{pmatrix} \equiv \begin{pmatrix} * & * \\ 0 & * \end{pmatrix} \pmod{N} \right\},$$

where $*$ means any element. Both groups have level N (see Exercise 1.6).

Let k be an integer. Define the *weight k right action* of $\mathrm{GL}_2(\mathbf{Q})$ on the set of all functions $f : \mathfrak{h} \to \mathbf{C}$ as follows. If $\gamma = \left(\begin{smallmatrix} a & b \\ c & d \end{smallmatrix}\right) \in \mathrm{GL}_2(\mathbf{Q})$, let

$$(f^{[\gamma]_k})(z) = \det(\gamma)^{k-1}(cz+d)^{-k} f(\gamma(z)). \tag{1.3.1}$$

Proposition 1.10. *Formula* (1.3.1) *defines a right action of* $\mathrm{GL}_2(\mathbf{Z})$ *on the set of all functions* $f : \mathfrak{h} \to \mathbf{C}$*; in particular,*

$$f^{[\gamma_1\gamma_2]_k} = (f^{[\gamma_1]_k})^{[\gamma_2]_k}.$$

Proof. See Exercise 1.7. □

Definition 1.11 (Weakly Modular Function)**.** A *weakly modular function* of weight k for a congruence subgroup Γ is a meromorphic function $f : \mathfrak{h} \to \mathbf{C}$ such that $f^{[\gamma]_k} = f$ for all $\gamma \in \Gamma$.

A central object in the theory of modular forms is the *set of cusps*

$$\mathbf{P}^1(\mathbf{Q}) = \mathbf{Q} \cup \{\infty\}.$$

An element $\gamma = \left(\begin{smallmatrix} a & b \\ c & d \end{smallmatrix}\right) \in \mathrm{SL}_2(\mathbf{Z})$ acts on $\mathbf{P}^1(\mathbf{Q})$ by

$$\gamma(z) = \begin{cases} \frac{az+b}{cz+d} & \text{if } z \neq \infty, \\ \frac{a}{c} & \text{if } z = \infty. \end{cases}$$

Also, note that if the denominator c or $cz+d$ is 0 above, then

$$\gamma(z) = \infty \in \mathbf{P}^1(\mathbf{Q}).$$

The set of *cusps for a congruence subgroup* Γ is the set $C(\Gamma)$ of Γ-orbits of $\mathbf{P}^1(\mathbf{Q})$. (We will often identify elements of $C(\Gamma)$ with a representative element from the orbit.) For example, the lemma below asserts that if $\Gamma = \mathrm{SL}_2(\mathbf{Z})$, then there is exactly one orbit, so $C(\mathrm{SL}_2(\mathbf{Z})) = \{[\infty]\}$.

Lemma 1.12. *For any cusps* $\alpha, \beta \in \mathbf{P}^1(\mathbf{Q})$ *there exists* $\gamma \in \mathrm{SL}_2(\mathbf{Z})$ *such that* $\gamma(\alpha) = \beta$.

Proof. This is Exercise 1.8. □

Proposition 1.13. *For any congruence subgroup* Γ*, the set* $C(\Gamma)$ *of cusps is finite.*

Proof. This is Exercise 1.9. □

See [**DS05**, §3.8] and Algorithm 8.12 below for more discussion of cusps and results relevant to their enumeration.

In order to define modular forms for general congruence subgroups, we next explain what it means for a function to be holomorphic on the *extended upper half plane*

$$\mathfrak{h}^* = \mathfrak{h} \cup \mathbb{P}^1(\mathbb{Q}).$$

See [**Shi94**, §1.3–1.5] for a detailed description of the correct topology to consider on $\mathfrak{h}^*$. In particular, a basis of neighborhoods for $\alpha \in \mathbb{Q}$ is given by the sets $\{\alpha\} \cup D$, where D is an open disc in $\mathfrak{h}$ that is tangent to the real line at α.

Recall from Section 1.2 that a weakly modular function f on $\mathrm{SL}_2(\mathbb{Z})$ is holomorphic at ∞ if its q-expansion is of the form $\sum_{n=0}^{\infty} a_n q^n$.

In order to make sense of holomorphicity of a weakly modular function f for an arbitrary congruence subgroup Γ at any $\alpha \in \mathbb{Q}$, we first prove a lemma.

Lemma 1.14. *If $f : \mathfrak{h} \to \mathbb{C}$ is a weakly modular function of weight k for a congruence subgroup Γ and if $\delta \in \mathrm{SL}_2(\mathbb{Z})$, then $f^{[\delta]_k}$ is a weakly modular function for $\delta^{-1}\Gamma\delta$.*

Proof. If $s = \delta^{-1}\gamma\delta \in \delta^{-1}\Gamma\delta$, then

$$(f^{[\delta]_k})^{[s]_k} = f^{[\delta s]_k} = f^{[\delta\delta^{-1}\gamma\delta]_k} = f^{[\gamma\delta]_k} = f^{[\delta]_k}.$$

□

Fix a weakly modular function f of weight k for a congruence subgroup Γ, and suppose $\alpha \in \mathbb{Q}$. In Section 1.2 we constructed the q-expansion of f by using that $f(z) = f(z+1)$, which held since $T = \begin{pmatrix} 1 & 1 \\ 0 & 1 \end{pmatrix} \in \mathrm{SL}_2(\mathbb{Z})$. There are congruence subgroups Γ such that $T \notin \Gamma$. Moreover, even if we are interested only in modular forms for $\Gamma_1(N)$, where we have $T \in \Gamma_1(N)$ for all N, we will still have to consider q-expansions at infinity for modular forms on groups $\delta^{-1}\Gamma_1(N)\delta$, and these need not contain T. Fortunately, $T^N = \left(\begin{smallmatrix} 1 & N \\ 0 & 1 \end{smallmatrix}\right) \in \Gamma(N)$, so a congruence subgroup of level N contains T^N. Thus we have $f(z+H) = f(H)$ for some positive integer H, e.g., $H = N$ always works, but there may be a smaller choice of H. The minimal choice of $H > 0$ such that $\left(\begin{smallmatrix} 1 & H \\ 0 & 1 \end{smallmatrix}\right) \in \delta^{-1}\Gamma\delta$, where $\delta(\infty) = \alpha$, is called the *width of the cusp* α relative to the group Γ (see Section 1.4.1). When f is meromorphic at infinity, we obtain a Fourier expansion

$$f(z) = \sum_{n=m}^{\infty} a_n q^{n/H} \tag{1.3.2}$$

in powers of the function $q^{1/H} = e^{2\pi iz/H}$. We say that f is holomorphic at ∞ if in (1.3.2) we have $m \geq 0$.

What about the other cusps $\alpha \in \mathbb{P}^1(\mathbb{Q})$? By Lemma 1.12 there is a $\gamma \in \mathrm{SL}_2(\mathbb{Z})$ such that $\gamma(\infty) = \alpha$. We declare f to be *holomorphic at the cusp* α if the weakly modular function $f^{[\gamma]_k}$ is holomorphic at ∞.

Definition 1.15 (Modular Form). A *modular form* of integer *weight* k for a congruence subgroup Γ is a weakly modular function $f : \mathfrak{h} \to \mathbb{C}$ that is holomorphic on $\mathfrak{h}^*$. We let $M_k(\Gamma)$ denote the space of weight k modular forms of weight k for Γ.

Proposition 1.16. *If a weakly modular function f is holomorphic at a set of representative elements for $C(\Gamma)$, then it is holomorphic at every element of $\mathbb{P}^1(\mathbb{Q})$.*

Proof. Let $c_1, \ldots, c_n \in \mathbb{P}^1(\mathbb{Q})$ be representatives for the set of cusps for Γ. If $\alpha \in \mathbb{P}^1(\mathbb{Q})$, then there is $\gamma \in \Gamma$ such that $\alpha = \gamma(c_i)$ for some i. By hypothesis f is holomorphic at c_i, so if $\delta \in \mathrm{SL}_2(\mathbb{Z})$ is such that $\delta(\infty) = c_i$, then $f^{[\delta]_k}$ is holomorphic at ∞. Since f is a weakly modular function for Γ,

$$f^{[\delta]_k} = (f^{[\gamma]_k})^{[\delta]_k} = f^{[\gamma\delta]_k}. \tag{1.3.3}$$

But $\gamma(\delta(\infty)) = \gamma(c_i) = \alpha$, so (1.3.3) implies that f is holomorphic at α. □

1.4. Remarks on Congruence Subgroups

Recall that a congruence subgroup is a subgroup of $\mathrm{SL}_2(\mathbb{Z})$ that contains $\Gamma(N)$ for some N. Any congruence subgroup has finite index in $\mathrm{SL}_2(\mathbb{Z})$, since $\Gamma(N)$ does. What about the converse: is every finite index subgroup of $\mathrm{SL}_2(\mathbb{Z})$ a congruence subgroup? This is the *congruence subgroup problem*. One can ask about the congruence subgroup problem with $\mathrm{SL}_2(\mathbb{Z})$ replaced by many similar groups. If p is a prime, then one can prove that every finite index subgroup of $\mathrm{SL}_2(\mathbb{Z}[1/p])$ is a congruence subgroup (i.e., contains the kernel of reduction modulo some integer coprime to p), and for any $n > 2$, all finite index subgroups of $\mathrm{SL}_n(\mathbb{Z})$ are congruence subgroups (see [**Hum80**]). However, there are numerous finite index subgroups of $\mathrm{SL}_2(\mathbb{Z})$ that are not congruence subgroups. The paper [**Hsu96**] contains an *algorithm* to decide if certain finite index subgroups are congruence subgroups and gives an example of a subgroup of index 12 that is not a congruence subgroup.

One can consider modular forms even for noncongruence subgroups. See, e.g., [**Tho89**] and the papers it references for work on this topic. We will not consider such modular forms further in this book. Note that modular symbols (which we define later in this book) *are* computable for noncongruence subgroups.

Finding coset representatives for $\Gamma_0(N)$, $\Gamma_1(N)$ and $\Gamma(N)$ in $\mathrm{SL}_2(\mathbb{Z})$ is straightforward and will be discussed at length later in this book. To make the problem more explicit, note that you can quotient out by $\Gamma(N)$ first. Then the question amounts to finding coset representatives for a subgroup of $\mathrm{SL}_2(\mathbb{Z}/N\mathbb{Z})$ (and lifting), which is reasonably straightforward.

Given coset representatives for a finite index subgroup G of $\mathrm{SL}_2(\mathbb{Z})$, we can compute generators for G as follows. Let R be a set of coset representatives for G. Let $\sigma, \tau \in \mathrm{SL}_2(\mathbb{Z})$ be the matrices denoted by S and T in (1.1.2). Define maps $s, t : R \to G$ as follows. If $r \in R$, then there exists a unique $\alpha_r \in R$ such that $Gr\sigma = G\alpha_r$. Let $s(r) = r\sigma\alpha_r^{-1}$. Likewise, there is a unique β_r such that $Gr\tau = G\beta_r$ and we let $t(r) = r\tau\beta_r^{-1}$. Note that $s(r)$ and $t(r)$ are in G for all r. Then G is generated by $s(R) \cup t(R)$.

Proposition 1.17. *The above procedure computes generators for G.*

Proof. Without loss of generality, assume that $I = \left(\begin{smallmatrix} 1 & 0 \\ 0 & 1 \end{smallmatrix}\right)$ represents the coset of G. Let g be an element of G. Since σ and τ generate $\mathrm{SL}_2(\mathbb{Z})$, it is possible to write g as a product of powers of σ and τ. There is a procedure, which we explain below with an example in order to avoid cumbersome notation, which writes g as a product of elements of $s(R) \cup t(R)$ times a right coset representative $r \in R$. For example, if

$$g = \sigma\tau^2\sigma\tau,$$

then $g = I\sigma\tau^2\sigma\tau = s(I)y\tau^2\sigma\tau$ for some $y \in R$. Continuing,

$$s(I)y\tau^2\sigma\tau = s(I)(y\tau)\tau\sigma\tau = s(I)(t(y)z)\tau\sigma\tau$$

for some $z \in R$. Again,

$$s(I)(t(y)z)\tau\sigma\tau = s(I)t(y)(z\tau)\sigma\tau = \cdots .$$

The procedure illustrated above (with an example) makes sense for arbitrary g and, after carrying it out, writes g as a product of elements of $s(R) \cup t(R)$ times a right coset representative $r \in R$. But $g \in G$ and I is the right coset representative for G, so this right coset representative must be I. □

Remark 1.18. We could also apply the proof of Proposition 1.17 to write any element of G in terms of the given generators. Moreover, we could use it to write any element $\gamma \in \mathrm{SL}_2(\mathbb{Z})$ in the form gr, where $g \in G$ and $r \in R$, so we can decide whether or not $\gamma \in G$.

1.4.1. Computing Widths of Cusps. Let Γ be a congruence subgroup of level N. Suppose $\alpha \in C(\Gamma)$ is a cusp, and choose $\gamma \in \mathrm{SL}_2(\mathbb{Z})$ such that $\gamma(\infty) = \alpha$. Recall that the minimal h such that $\left(\begin{smallmatrix} 1 & h \\ 0 & 1 \end{smallmatrix}\right) \in \gamma^{-1}\Gamma\gamma$ is called the *width of the cusp* α for the group Γ. In this section we discuss how to compute h.

Algorithm 1.19 (Width of Cusp). *Given a congruence subgroup Γ of level N and a cusp α for Γ, this algorithm computes the width h of α. We assume that Γ is given by congruence conditions, e.g., $\Gamma = \Gamma_0(N)$ or $\Gamma_1(N)$.*

(1) [Find γ] Use the extended Euclidean algorithm to find $\gamma \in \mathrm{SL}_2(\mathbb{Z})$ such that $\gamma(\infty) = \alpha$, as follows. If $\alpha = \infty$, set $\gamma = 1$; otherwise, write $\alpha = a/b$, find c, d such that $ad - bc = 1$, and set $\gamma = \left(\begin{smallmatrix} a & b \\ c & d \end{smallmatrix}\right)$.

(2) [Compute Conjugate Matrix] Compute the following element of $\mathrm{Mat}_2(\mathbb{Z}[x])$:
$$\delta(x) = \gamma \begin{pmatrix} 1 & x \\ 0 & 1 \end{pmatrix} \gamma^{-1}.$$
Note that the entries of $\delta(x)$ are constant or linear in x.

(3) [Solve] The congruence conditions that define Γ give rise to four linear congruence conditions on x. Use techniques from elementary number theory (or enumeration) to find the smallest simultaneous positive solution h to these four equations.

Example 1.20. (1) Suppose $\alpha = 0$ and $\Gamma = \Gamma_0(N)$ or $\Gamma_1(N)$. Then $\gamma = \left(\begin{smallmatrix} 0 & -1 \\ 1 & 0 \end{smallmatrix}\right)$ has the property that $\gamma(\infty) = \alpha$. Next, the congruence condition is
$$\delta(x) = \gamma \begin{pmatrix} 1 & x \\ 0 & 1 \end{pmatrix} \gamma^{-1} = \begin{pmatrix} 1 & 0 \\ x & 1 \end{pmatrix} \equiv \begin{pmatrix} 1 & * \\ 0 & 1 \end{pmatrix} \pmod{N}.$$
Thus the smallest positive solution is $h = N$, so the width of 0 is N.

(2) Suppose $N = pq$ where p, q are distinct primes, and let $\alpha = 1/p$. Then $\gamma = \left(\begin{smallmatrix} 1 & 0 \\ p & 1 \end{smallmatrix}\right)$ sends ∞ to α. The congruence condition for $\Gamma_0(pq)$ is
$$\delta(x) = \gamma \begin{pmatrix} 1 & x \\ 0 & 1 \end{pmatrix} \gamma^{-1} = \begin{pmatrix} 1 - px & x \\ -p^2x & px + 1 \end{pmatrix} \equiv \begin{pmatrix} * & * \\ 0 & * \end{pmatrix} \pmod{pq}.$$
Since $p^2x \equiv 0 \pmod{pq}$, we see that $x = q$ is the smallest solution. Thus $1/p$ has width q, and symmetrically $1/q$ has width p.

Remark 1.21. For $\Gamma_0(N)$, once we enforce that the bottom left entry is 0 (mod N) and use that the determinant is 1, the coprimality from the other two congruences is automatic. So there is one congruence to solve in the $\Gamma_0(N)$ case. There are two congruences in the $\Gamma_1(N)$ case.

1.5. Applications of Modular Forms

The above definition of modular forms might leave the impression that modular forms occupy an obscure corner of complex analysis. This is *not* the case! Modular forms are highly geometric, arithmetic, and topological objects that are of extreme interest all over mathematics:

(1) **Fermat's last theorem:** Wiles' proof [**Wil95**] of Fermat's last theorem uses modular forms extensively. The work of Wiles et al. on modularity also massively extends computational methods for elliptic curves over $\mathbb{Q}$, because many elliptic curve algorithms, e.g., for computing L-functions, modular degrees, Heegner points, etc., require that the elliptic curve be modular.

(2) **Diophantine equations:** Wiles' proof of Fermat's last theorem has made available a wide array of new techniques for solving certain diophantine equations. Such work relies crucially on having access to tables or software for computing modular forms. See, e.g., [**Dar97, Mer99, Che05, SC03**]. (Wiles did not need a computer, because the relevant spaces of modular forms that arise in his proof have dimension 0!) Also, according to Siksek (personal communication) the paper [**BMS06**] would "have been entirely impossible to write without [the algorithms described in this book]."

(3) **Congruent number problem:** This ancient open problem is to determine which integers are the area of a right triangle with rational side lengths. There is a potential solution that uses modular forms (of weight 3/2) extensively (the solution is conditional on truth of the Birch and Swinnerton-Dyer conjecture, which is not yet known). See [**Kob84**].

(4) **Topology:** Topological modular forms are a major area of current research.

(5) **Construction of Ramanujan graphs:** Modular forms can be used to construct almost optimal expander graphs, which play a role in communications network theory.

(6) **Cryptography and Coding Theory:** Point counting on elliptic curves over finite fields is crucial to the construction of elliptic curve cryptosystems, and modular forms are relevant to efficient algorithms for point counting (see [**Elk98**]). Algebraic curves that are associated to modular forms are useful in constructing and studying certain error-correcting codes (see [**Ebe02**]).

(7) **The Birch and Swinnerton-Dyer conjecture:** This central open problem in arithmetic geometry relates arithmetic properties of elliptic curves (and abelian varieties) to special values of L-functions. Most deep results toward this conjecture use modular forms extensively (e.g., work of Kolyvagin, Gross-Zagier, and Kato). Also, modular forms are used to compute and prove results about special values of these L-functions. See [**Wil00**].

(8) **Serre's Conjecture on modularity of Galois representation:** Let $G_{\mathbb{Q}} = \mathrm{Gal}(\overline{\mathbb{Q}}/\mathbb{Q})$ be the Galois group of an algebraic closure of $\mathbb{Q}$. Serre conjectured and many people have (nearly!) proved that every continuous homomorphism $\rho : G_{\mathbb{Q}} \to \mathrm{GL}_2(\mathbb{F}_q)$, where $\mathbb{F}_q$ is a finite field and $\det(\rho(\text{complex conjugation})) = -1$, "arises" from a modular form. More precisely, for almost all primes p the coefficients a_p of a modular (eigen-)form $\sum a_n q^n$ are congruent to the traces of elements $\rho(\mathrm{Frob}_p)$, where Frob_p are certain special elements of $G_{\mathbb{Q}}$ called Frobenius elements. See [**RS01**] and [**DS05**, Ch. 9].

(9) **Generating functions for partitions:** The generating functions for various kinds of partitions of an integer can often be related to modular forms. Deep theorems about modular forms then translate into results about partitions. See work of Ramanujan, Gordon, Andres, and Ahlgren and Ono (e.g., [**AO01**]).

(10) **Lattices:** If $L \subset \mathbb{R}^n$ is an even unimodular lattice (the basis matrix has determinant ± 1 and $\lambda \cdot \lambda \in 2\mathbb{Z}$ for all $\lambda \in L$), then the theta series

$$\theta_L(q) = \sum_{\lambda \in L} q^{\lambda \cdot \lambda}$$

is a modular form of weight $n/2$. The coefficient of q^m is the number of lattice vectors with squared length m. Theorems and computational methods for modular forms translate into theorems and computational methods for lattices. For example, the 290 theorem of M. Bharghava and J. Hanke is a theorem about lattices, which asserts that an integer-valued quadratic form represents all positive integers if and only if it represents the integers up to 290; it is proved by doing many calculations with modular forms (both theoretical and with a computer).

1.6. Exercises

1.1 Suppose $\gamma = \left(\begin{smallmatrix} a & b \\ c & d \end{smallmatrix}\right) \in \mathrm{GL}_2(\mathbb{R})$ has positive determinant. Prove that if $z \in \mathbb{C}$ is a complex number with positive imaginary part, then the imaginary part of $\gamma(z) = (az+b)/(cz+d)$ is also positive.

1.2 Prove that every rational function (quotient of two polynomials) is a meromorphic function on $\mathbb{C}$.

1.3 Suppose f and g are weakly modular functions for a congruence subgroup Γ with $f \neq 0$.

(a) Prove that the product fg is a weakly modular function for Γ.
(b) Prove that $1/f$ is a weakly modular function for Γ.

(c) If f and g are modular functions, show that fg is a modular function for Γ.

(d) If f and g are modular forms, show that fg is a modular form for Γ.

1.4 Suppose f is a weakly modular function of odd weight k and level $\Gamma_0(N)$ for some N. Show that $f = 0$.

1.5 Prove that $\mathrm{SL}_2(\mathbb{Z}) = \Gamma_0(1) = \Gamma_1(1) = \Gamma(1)$.

1.6 (a) Prove that $\Gamma_1(N)$ is a group.

(b) Prove that $\Gamma_1(N)$ has finite index in $\mathrm{SL}_2(\mathbb{Z})$ (Hint: It contains the kernel of the homomorphism $\mathrm{SL}_2(\mathbb{Z}) \to \mathrm{SL}_2(\mathbb{Z}/N\mathbb{Z})$.)

(c) Prove that $\Gamma_0(N)$ has finite index in $\mathrm{SL}_2(\mathbb{Z})$.

(d) Prove that $\Gamma_0(N)$ and $\Gamma_1(N)$ have level N.

1.7 Let k be an integer, and for any function $f : \mathfrak{h}^* \to \mathbb{C}$ and $\gamma = \left(\begin{smallmatrix} a & b \\ c & d \end{smallmatrix}\right) \in \mathrm{GL}_2(\mathbb{Q})$, set $f^{[\gamma]_k}(z) = \det(\gamma)^{k-1} \cdot (cz+d)^{-k} \cdot f(\gamma(z))$. Prove that if $\gamma_1, \gamma_2 \in \mathrm{GL}_2(\mathbb{Z})$, then for all $z \in \mathfrak{h}^*$ we have

$$f^{[\gamma_1\gamma_2]_k}(z) = ((f^{[\gamma_1]_k})^{[\gamma_2]_k})(z).$$

1.8 Prove that for any $\alpha, \beta \in \mathbb{P}^1(\mathbb{Q})$, there exists $\gamma \in \mathrm{SL}_2(\mathbb{Z})$ such that $\gamma(\alpha) = \beta$.

1.9 Prove Proposition 1.13, which asserts that the set of cusps $C(\Gamma)$, for any congruence subgroup Γ, is finite.

1.10 Use Algorithm 1.19 to give an example of a group Γ and cusp α with width 2.

Chapter 2

Modular Forms of Level 1

In this chapter we study in detail the structure of level 1 modular forms, i.e., modular forms on $\mathrm{SL}_2(\mathbb{Z}) = \Gamma_0(1) = \Gamma_1(1)$. We assume some complex analysis (e.g., the residue theorem), linear algebra, and that the reader has read Chapter 1.

2.1. Examples of Modular Forms of Level 1

In this section we will finally see some examples of modular forms of level 1! We first introduce the Eisenstein series and then define Δ, which is a cusp form of weight 12. In Section 2.2 we prove the structure theorem, which says that all modular forms of level 1 are polynomials in Eisenstein series.

For an even integer $k \geq 4$, the *nonnormalized weight k Eisenstein series* is the function on the extended upper half plane $\mathfrak{h}^* = \mathfrak{h} \cup \mathbb{P}^1(\mathbb{Q})$ given by

$$G_k(z) = \sum_{m,n\in\mathbb{Z}}^{*} \frac{1}{(mz+n)^k}. \tag{2.1.1}$$

The star on top of the sum symbol means that for each z the sum is over all $m, n \in \mathbb{Z}$ such that $mz + n \neq 0$.

Proposition 2.1. *The function $G_k(z)$ is a modular form of weight k, i.e., $G_k \in M_k(\mathrm{SL}_2(\mathbb{Z}))$.*

Proof. See [**Ser73**, § VII.2.3] for a proof that $G_k(z)$ defines a holomorphic function on $\mathfrak{h}^*$. To see that G_k is modular, observe that

$$G_k(z+1) = \sum^* \frac{1}{(m(z+1)+n)^k} = \sum^* \frac{1}{(mz+(n+m))^k} = \sum^* \frac{1}{(mz+n)^k},$$

where for the last equality we use that the map $(m, n+m) \mapsto (m,n)$ on $\mathbb{Z} \times \mathbb{Z}$ is invertible. Also,

$$\begin{aligned} G_k(-1/z) &= \sum^* \frac{1}{(-m/z+n)^k} \\ &= \sum^* \frac{z^k}{(-m+nz)^k} \\ &= z^k \sum^* \frac{1}{(mz+n)^k} = z^k G_k(z), \end{aligned}$$

where we use that $(n, -m) \mapsto (m, n)$ is invertible. □

Proposition 2.2. $G_k(\infty) = 2\zeta(k)$, *where* ζ *is the Riemann zeta function.*

Proof. As $z \to \infty$ (along the imaginary axis) in (2.1.1), the terms that involve z with $m \neq 0$ go to 0. Thus

$$G_k(\infty) = \sum^*_{n\in\mathbb{Z}} \frac{1}{n^k}.$$

This sum is twice $\zeta(k) = \sum_{n\geq 1} \frac{1}{n^k}$, as claimed. □

2.1.1. The Cusp Form Δ. Suppose $E = \mathbb{C}/\Lambda$ is an elliptic curve over $\mathbb{C}$, viewed as a quotient of $\mathbb{C}$ by a lattice $\Lambda = \mathbb{Z}\omega_1 + \mathbb{Z}\omega_2$, with $\omega_1/\omega_2 \in \mathfrak{h}$ (see [**DS05**, §1.4]). The *Weierstrass $\wp$-function* of the lattice Λ is

$$\wp = \wp_\Lambda(u) = \frac{1}{u^2} + \sum_{k=4,6,8,\ldots} (k-1)G_k(\omega_1/\omega_2)u^{k-2},$$

where the sum is over even integers $k \geq 4$. It satisfies the differential equation

$$(\wp')^2 = 4\wp^3 - 60G_4(\omega_1/\omega_2)\wp - 140G_6(\omega_1/\omega_2).$$

If we set $x = \wp$ and $y = \wp'$, the above is an (affine) equation of the form $y^2 = ax^3 + bx + c$ for an elliptic curve that is complex analytically isomorphic to $\mathbb{C}/\Lambda$ (see [**Ahl78**, pg. 277] for why the cubic has distinct roots).

The discriminant of the cubic

$$4x^3 - 60G_4(\omega_1/\omega_2)x - 140G_6(\omega_1/\omega_2)$$

is $16D(\omega_1/\omega_2)$, where

$$D(z) = (60G_4(z))^3 - 27(140G_6(z))^2.$$

Since $D(z)$ is the difference of two modular forms of weight 12 it has weight 12. Morever,

$$\begin{aligned} D(\infty) &= (60G_4(\infty))^3 - 27\,(140G_6(\infty))^2 \\ &= \left(\frac{60}{3^2\cdot 5}\pi^4\right)^3 - 27\left(\frac{140\cdot 2}{3^3\cdot 5\cdot 7}\pi^6\right)^2 \\ &= 0, \end{aligned}$$

so D is a cusp form of weight 12. Let

$$\Delta = \frac{D}{(2\pi)^{12}}.$$

Lemma 2.3. *If $z \in \mathfrak{h}$, then $\Delta(z) \neq 0$.*

Proof. Let $\omega_1 = z$ and $\omega_2 = 1$. Since $E = \mathbb{C}/(\mathbb{Z}\omega_1 + \mathbb{Z}\omega_2)$ is an elliptic curve, it has nonzero discriminant $\Delta(z) = \Delta(\omega_1/\omega_2) \neq 0$. □

Proposition 2.4. *We have $\Delta = q \cdot \prod_{n=1}^{\infty}(1-q^n)^{24}$.*

Proof. See [**Ser73**, Thm. 6, pg. 95]. □

Remark 2.5. SAGE computes the q-expansion of Δ efficiently to high precision using the command `delta_qexp`:

```
sage: delta_qexp(6)
q - 24*q^2 + 252*q^3 - 1472*q^4 + 4830*q^5 + O(q^6)
```

2.1.2. Fourier Expansions of Eisenstein Series. Recall from (1.2.4) that elements f of $M_k(\mathrm{SL}_2(\mathbb{Z}))$ can be expressed as formal power series in terms of $q(z) = e^{2\pi i z}$ and that this expansion is called the Fourier expansion of f. The following proposition gives the Fourier expansion of the Eisenstein series $G_k(z)$.

Definition 2.6 (Sigma)**.** For any integer $t \geq 0$ and any positive integer n, the *sigma function*

$$\sigma_t(n) = \sum_{1\leq d\mid n} d^t$$

is the sum of the tth powers of the positive divisors of n. Also, let $d(n) = \sigma_0(n)$, which is the number of divisors of n, and let $\sigma(n) = \sigma_1(n)$. For example, if p is prime, then $\sigma_t(p) = 1 + p^t$.

Proposition 2.7. *For every even integer $k \geq 4$, we have*

$$G_k(z) = 2\zeta(k) + 2\cdot\frac{(2\pi i)^k}{(k-1)!}\cdot\sum_{n=1}^{\infty}\sigma_{k-1}(n)q^n.$$

Proof. See [**Ser73**, Section VII.4], which uses clever manipulations of series, starting with the identity

$$\pi \cot(\pi z) = \frac{1}{z} + \sum_{m=1}^{\infty} \left(\frac{1}{z+m} + \frac{1}{z-m} \right).$$

□

From a computational point of view, the q-expansion of Proposition 2.7 is unsatisfactory because it involves transcendental numbers. To understand these numbers, we introduce the *Bernoulli numbers* B_n for $n \geq 0$ *defined* by the following equality of formal power series:

$$\frac{x}{e^x - 1} = \sum_{n=0}^{\infty} B_n \frac{x^n}{n!}. \tag{2.1.2}$$

Expanding the power series, we have

$$\frac{x}{e^x - 1} = 1 - \frac{x}{2} + \frac{x^2}{12} - \frac{x^4}{720} + \frac{x^6}{30240} - \frac{x^8}{1209600} + \cdots .$$

As this expansion suggests, the Bernoulli numbers B_n with $n > 1$ odd are 0 (see Exercise 1.2). Expanding the series further, we obtain the following table:

$$B_0 = 1, \quad B_1 = -\frac{1}{2}, \quad B_2 = \frac{1}{6}, \quad B_4 = -\frac{1}{30}, \quad B_6 = \frac{1}{42}, \quad B_8 = -\frac{1}{30},$$

$$B_{10} = \frac{5}{66}, \quad B_{12} = -\frac{691}{2730}, \quad B_{14} = \frac{7}{6}, \quad B_{16} = -\frac{3617}{510}, \quad B_{18} = \frac{43867}{798},$$

$$B_{20} = -\frac{174611}{330}, \quad B_{22} = \frac{854513}{138}, \quad B_{24} = -\frac{236364091}{2730}, \quad B_{26} = \frac{8553103}{6}.$$

See Section 2.7 for a discussion of fast (analytic) methods for computing Bernoulli numbers.

We compute some Bernoulli numbers in SAGE:

```
sage: bernoulli(12)
-691/2730
sage: bernoulli(50)
495057205241079648212477525/66
sage: len(str(bernoulli(10000)))
27706
```

A key fact is that Bernoulli numbers are rational numbers and they are connected to values of ζ at positive even integers.

Proposition 2.8. *If $k \geq 2$ is an even integer, then*

$$\zeta(k) = -\frac{(2\pi i)^k}{2 \cdot k!} \cdot B_k.$$

Proof. This is proved by manipulating a series expansion of $z\cot(z)$ (see [**Ser73**, Section VII.4]). □

Definition 2.9 (Normalized Eisenstein Series). The *normalized Eisenstein series* of even weight $k \geq 4$ is

$$E_k = \frac{(k-1)!}{2 \cdot (2\pi i)^k} \cdot G_k.$$

Combining Propositions 2.7 and 2.8, we see that

$$E_k = -\frac{B_k}{2k} + q + \sum_{n=2}^{\infty} \sigma_{k-1}(n) q^n. \tag{2.1.3}$$

Warning 2.10. Our series E_k is normalized so that the coefficient of q is 1, but often in the literature E_k is normalized so that the constant coefficient is 1. We use the normalization with the coefficient of q equal to 1, because then the eigenvalue of the nth Hecke operator (see Section 2.4) is the coefficient of q^n. Our normalization is also convenient when considering congruences between cusp forms and Eisenstein series.

2.2. Structure Theorem for Level 1 Modular Forms

In this section we describe a structure theorem for modular forms of level 1. If f is a nonzero meromorphic function on $\mathfrak{h}$ and $w \in \mathfrak{h}$, let $\mathrm{ord}_w(f)$ be the largest integer n such that $f(z)/(w-z)^n$ is holomorphic at w. If $f = \sum_{n=m}^{\infty} a_n q^n$ with $a_m \neq 0$, we set $\mathrm{ord}_\infty(f) = m$. We will use the following theorem to give a presentation for the vector space of modular forms of weight k; this presentation yields an algorithm to compute this space.

Let $M_k = M_k(\mathrm{SL}_2(\mathbb{Z}))$ denote the complex vector space of modular forms of weight k for $\mathrm{SL}_2(\mathbb{Z})$. The *standard fundamental domain* $\mathcal{F}$ for $\mathrm{SL}_2(\mathbb{Z})$ is the set of $z \in \mathfrak{h}$ with $|z| \geq 1$ and $|\mathrm{Re}(z)| \leq 1/2$. Let $\rho = e^{2\pi i/3}$.

Theorem 2.11 (Valence Formula). *Let k be any integer and suppose $f \in M_k(\mathrm{SL}_2(\mathbb{Z}))$ is nonzero. Then*

$$\mathrm{ord}_\infty(f) + \frac{1}{2}\mathrm{ord}_i(f) + \frac{1}{3}\mathrm{ord}_\rho(f) + \sum_{w \in \mathcal{F}}^{*} \mathrm{ord}_w(f) = \frac{k}{12},$$

where $\sum_{w \in \mathcal{F}}^{*}$ is the sum over elements of $\mathcal{F}$ other than i and ρ.

Proof. The proof in [**Ser73**, §VII.3] uses the residue theorem. □

Let $S_k = S_k(\mathrm{SL}_2(\mathbb{Z}))$ denote the subspace of weight k cusp forms for $\mathrm{SL}_2(\mathbb{Z})$. We have an exact sequence

$$0 \to S_k \to M_k \xrightarrow{\iota_\infty} \mathbb{C}$$

that sends $f \in M_k$ to $f(\infty)$. When $k \geq 4$ is even, the space M_k contains the Eisenstein series G_k, and $G_k(\infty) = 2\zeta(k) \neq 0$, so the map $M_k \to \mathbb{C}$ is surjective. This proves the following lemma.

Lemma 2.12. *If $k \geq 4$ is even, then $M_k = S_k \oplus \mathbb{C}G_k$ and the following sequence is exact:*

$$0 \to S_k \to M_k \xrightarrow{\iota_\infty} \mathbb{C} \to 0.$$

Proposition 2.13. *For $k < 0$ and $k = 2$, we have $M_k = 0$.*

Proof. Suppose $f \in M_k$ is nonzero yet $k = 2$ or $k < 0$. By Theorem 2.11,

$$\mathrm{ord}_\infty(f) + \frac{1}{2}\mathrm{ord}_i(f) + \frac{1}{3}\mathrm{ord}_\rho(f) + \sum_{w\in D}^{*} \mathrm{ord}_w(f) = \frac{k}{12} \leq \frac{1}{6}.$$

This is not possible because each quantity on the left is nonnegative so whatever the sum is, it is too big (or 0, in which case $k = 0$). □

Theorem 2.14. *Multiplication by Δ defines an isomorphism $M_{k-12} \to S_k$.*

Proof. By Lemma 2.3, Δ is not identically 0, so because Δ is holomorphic, multiplication by Δ defines an injective map $M_{k-12} \hookrightarrow S_k$. To see that this map is surjective, we show that if $f \in S_k$, then $f/\Delta \in M_{k-12}$. Since Δ has weight 12 and $\mathrm{ord}_\infty(\Delta) \geq 1$, Theorem 2.11 implies that Δ has a simple zero at ∞ and does not vanish on $\mathfrak{h}$. Thus if $f \in S_k$ and if we let $g = f/\Delta$, then g is holomorphic and satisfies the appropriate transformation formula, so $g \in M_{k-12}$. □

Corollary 2.15. *For $k = 0, 4, 6, 8, 10, 14$, the space M_k has dimension 1, with basis 1, G_4, G_6, G_8, G_{10}, and G_{14}, respectively, and $S_k = 0$.*

Proof. Combining Proposition 2.13 with Theorem 2.14, we see that the spaces M_k for $k \leq 10$ cannot have dimension greater than 1, since otherwise $M_{k'} \neq 0$ for some $k' < 0$. Also M_{14} has dimension at most 1, since M_2 has dimension 0. Each of the indicated spaces of weight ≥ 4 contains the indicated Eisenstein series and so has dimension 1, as claimed. □

Corollary 2.16. $\dim M_k = \begin{cases} 0 & \text{if } k \text{ is odd or negative,} \\ \lfloor k/12 \rfloor & \text{if } k \equiv 2 \pmod{12}, \\ \lfloor k/12 \rfloor + 1 & \text{if } k \not\equiv 2 \pmod{12}. \end{cases}$

Here $\lfloor x \rfloor$ is the biggest integer $\leq x$.

Proof. As we have already seen above, the formula is true when $k \leq 12$. By Theorem 2.14, the dimension increases by 1 when k is replaced by $k+12$. □

Theorem 2.17. *The space M_k has as basis the modular forms $G_4^a G_6^b$, where a, b run over all pairs of nonnegative integers such that $4a + 6b = k$.*

Proof. Fix an even integer k. We first prove by induction that the modular forms $G_4^a G_6^b$ generate M_k; the cases $k \leq 10$ and $k = 14$ follow from the above arguments (e.g., when $k = 0$, we have $a = b = 0$ and basis 1). Choose some pair of nonnegative integers a, b such that $4a + 6b = k$. The form $g = G_4^a G_6^b$ is not a cusp form, since it is nonzero at ∞. Now suppose $f \in M_k$ is arbitrary. Since $g(\infty) \neq 0$, there exists $\alpha \in \mathbb{C}$ such that $f - \alpha g \in S_k$. Then by Theorem 2.14, there is $h \in M_{k-12}$ such that $f - \alpha g = \Delta \cdot h$. By induction, h is a polynomial in G_4 and G_6 of the required type, and so is Δ, so f is as well. Thus

$$\{G_4^a G_6^b \mid a \geq 0,\ b \geq 0,\ 4a + 6b = k\}$$

spans M_k.

Suppose there is a nontrivial linear relation between the $G_4^a G_6^b$ for a given k. By multiplying the linear relation by a suitable power of G_4 and G_6, we may assume that we have such a nontrivial relation with $k \equiv 0 \pmod{12}$. Now divide the linear relation by the weight k form $G_6^{k/6}$ to see that G_4^3/G_6^2 satisfies a polynomial with coefficients in $\mathbb{C}$ (see Exercise 2.4). Hence G_4^3/G_6^2 is a root of a polynomial, hence a constant, which is a contradiction since the q-expansion of G_4^3/G_6^2 is not constant. □

Algorithm 2.18 (Basis for M_k). *Given integers n and k, this algorithm computes a basis of q-expansions for the complex vector space M_k mod q^n. The q-expansions output by this algorithm have coefficients in $\mathbb{Q}$.*

(1) [Simple Case] If $k = 0$, output the basis with just 1 in it and terminate; otherwise if $k < 4$ or k is odd, output the empty basis and terminate.
(2) [Power Series] Compute E_4 and E_6 mod q^n using the formula from (2.1.3) and Section 2.7.
(3) [Initialize] Set $b = 0$.
(4) [Enumerate Basis] For each integer b between 0 and $\lfloor k/6 \rfloor$, compute $a = (k - 6b)/4$. If a is an integer, compute and output the basis element $E_4^a E_6^b$ mod q^n. When computing E_4^a, find $E_4^m \pmod{q^n}$ for each $m \leq a$, and save these intermediate powers, so they can be reused later, and likewise for powers of E_6.

Proof. This is simply a translation of Theorem 2.17 into an algorithm, since E_k is a nonzero scalar multiple of G_k. That the q-expansions have coefficients in $\mathbf{Q}$ follows from (2.1.3). □

Example 2.19. We compute a basis for M_{24}, which is the space with smallest weight whose dimension is greater than 1. It has as basis E_4^6, $E_4^3E_6^2$, and E_6^4, whose explicit expansions are

$$\begin{aligned} E_4^6 &= \frac{1}{191102976000000} + \frac{1}{132710400000}q + \frac{203}{44236800000}q^2 + \cdots, \\ E_4^3E_6^2 &= \frac{1}{3511517184000} - \frac{1}{12192768000}q - \frac{377}{4064256000}q^2 + \cdots, \\ E_6^4 &= \frac{1}{64524128256} - \frac{1}{32006016}q + \frac{241}{10668672}q^2 + \cdots. \end{aligned}$$

We compute this basis in SAGE as follows:

```
sage: E4 = eisenstein_series_qexp(4, 3)
sage: E6 = eisenstein_series_qexp(6, 3)
sage: E4^6
1/191102976000000 + 1/132710400000*q
                  + 203/44236800000*q^2 + O(q^3)
sage: E4^3*E6^2
1/3511517184000 - 1/12192768000*q
                - 377/4064256000*q^2 + O(q^3)
sage: E6^4
1/64524128256 - 1/32006016*q + 241/10668672*q^2 + O(q^3)
```

In Section 2.3, we will discuss the reduced echelon form basis for M_k.

2.3. The Miller Basis

Lemma 2.20 (V. Miller). *The space S_k has a basis $f_1, \ldots, f_d$ such that if $a_i(f_j)$ is the ith coefficient of f_j, then $a_i(f_j) = \delta_{i,j}$ for $i = 1, \ldots, d$. Moreover the f_j all lie in $\mathbf{Z}[[q]]$. We call this basis the* Miller basis *for S_k.*

This is a straightforward construction involving E_4, E_6 and Δ. The following proof very closely follows [**Lan95**, Ch. X, Thm. 4.4], which in turn follows the first lemma of V. Miller's thesis.

Proof. Let $d = \dim S_k$. Since $B_4 = -1/30$ and $B_6 = 1/42$, we note that

$$F_4 = -\frac{8}{B_4} \cdot E_4 = 1 + 240q + 2160q^2 + 6720q^3 + 17520q^4 + \cdots$$

and

$$F_6 = -\frac{12}{B_6} \cdot E_6 = 1 - 504q - 16632q^2 - 122976q^3 - 532728q^4 + \cdots$$

have q-expansions in $\mathbb{Z}[[q]]$ with leading coefficient 1. Choose integers $a, b \geq 0$ such that

$$4a + 6b \leq 14 \qquad \text{and} \qquad 4a + 6b \equiv k \pmod{12},$$

with $a = b = 0$ when $k \equiv 0 \pmod{12}$, and let

$$g_j = \Delta^j F_6^{2(d-j)+b} F_4^a = \left(\frac{\Delta}{F_6^2}\right)^j F_6^{2d+b} F_4^a, \qquad \text{for } j = 1, \ldots, d.$$

Then it is elementary to check that g_j has weight k

$$a_j(g_j) = 1 \qquad \text{and} \qquad a_i(g_j) = 0 \qquad \text{when} \qquad i < j.$$

Hence the g_j are linearly independent over $\mathbb{C}$, so form a basis for S_k. Since F_4, F_6, and Δ are all in $\mathbb{Z}[[q]]$, so are the g_j. The f_i may then be constructed from the g_j by Gauss elimination. The coefficients of the resulting power series lie in $\mathbb{Z}$ because each time we clear a column we use the power series g_j whose leading coefficient is 1 (so no denominators are introduced). □

Remark 2.21. The basis coming from Miller's lemma is "canonical", since it is just the reduced row echelon form of any basis. Also the set of all *integral* linear combinations of the elements of the Miller basis are precisely the modular forms of level 1 with integral q-expansion.

We extend the Miller basis to all M_k by taking a multiple of G_k with constant term 1 and subtracting off the f_i from the Miller basis so that the coefficients of $q, q^2, \ldots q^d$ of the resulting expansion are 0. We call the extra basis element f_0.

Example 2.22. If $k = 24$, then $d = 2$. Choose $a = b = 0$, since $k \equiv 0 \pmod{12}$. Then

$$g_1 = \Delta F_6^2 = q - 1032q^2 + 245196q^3 + 10965568q^4 + 60177390q^5 - \cdots$$

and

$$g_2 = \Delta^2 = q^2 - 48q^3 + 1080q^4 - 15040q^5 + \cdots .$$

We let $f_2 = g_2$ and

$$f_1 = g_1 + 1032g_2 = q + 195660q^3 + 12080128q^4 + 44656110q^5 - \cdots .$$

Example 2.23. When $k = 36$, the Miller basis including f_0 is

$$\begin{aligned} f_0 &= 1 + 6218175600q^4 + 15281788354560q^5 + \cdots, \\ f_1 &= q + 57093088q^4 + 37927345230q^5 + \cdots, \\ f_2 &= q^2 + 194184q^4 + 7442432q^5 + \cdots, \\ f_3 &= q^3 - 72q^4 + 2484q^5 + \cdots. \end{aligned}$$

Example 2.24. The SAGE command `victor_miller_basis` computes the Miller basis to any desired precision for a given k.

```
sage: victor_miller_basis(28,5)
[
1 + 15590400*q^3 + 36957286800*q^4 + O(q^5),
q + 151740*q^3 + 61032448*q^4 + O(q^5),
q^2 + 192*q^3 - 8280*q^4 + O(q^5)
]
```

Remark 2.25. To write $f \in M_k$ as a polynomial in E_4 and E_6, it is wasteful to compute the Miller basis. Instead, use the upper triangular (but not echelon!) basis $\Delta^j F_6^{2(d-j)+a} F_4^b$, and match coefficients from q^0 to q^d.

2.4. Hecke Operators

In this section we define Hecke operators on level 1 modular forms and derive their basic properties. We will not give proofs of the analogous properties for Hecke operators on higher level modular forms, since the proofs are clearest in the level 1 case, and the general case is similar (see, e.g., [**Lan95**]).

For any positive integer n, let

$$X_n = \left\{ \begin{pmatrix} a & b \\ 0 & d \end{pmatrix} \in \mathrm{Mat}_2(\mathbb{Z}) \ : \ a \geq 1, \ ad = n, \ \text{and } 0 \leq b < d \right\}.$$

Note that the set X_n is in bijection with the set of subgroups of $\mathbb{Z}^2$ of index n, where $\left(\begin{smallmatrix} a & b \\ c & d \end{smallmatrix}\right)$ corresponds to $L = \mathbb{Z} \cdot (a, b) + \mathbb{Z} \cdot (0, d)$, as one can see using Hermite normal form, which is the analogue over $\mathbb{Z}$ of echelon form (see Exercise 7.5).

Recall from (1.3.1) that if $\gamma = \left(\begin{smallmatrix} a & b \\ c & d \end{smallmatrix}\right) \in \mathrm{GL}_2(\mathbb{Q})$, then

$$f^{[\gamma]_k} = \det(\gamma)^{k-1}(cz + d)^{-k} f(\gamma(z)).$$

Definition 2.26 (Hecke Operator $T_{n,k}$)**.** The nth Hecke operator $T_{n,k}$ of weight k is the operator on the set of functions on $\mathfrak{h}$ defined by

$$T_{n,k}(f) = \sum_{\gamma \in X_n} f^{[\gamma]_k}.$$

Remark 2.27. It would make more sense to write $T_{n,k}$ on the right, e.g., $f|T_{n,k}$, since $T_{n,k}$ is defined using a right group action. However, if n, m are integers, then the action of $T_{n,k}$ and $T_{m,k}$ on weakly modular functions commutes (by Proposition 2.29 below), so it makes no difference whether we view the Hecke operators of given weight k as acting on the right or left.

Proposition 2.28. *If f is a weakly modular function of weight k, then so is $T_{n,k}(f)$; if f is a modular function, then so is $T_{n,k}(f)$.*

Proof. Suppose $\gamma \in \mathrm{SL}_2(\mathbb{Z})$. Since γ induces an automorphism of $\mathbb{Z}^2$,

$$X_n \cdot \gamma = \{\delta\gamma : \delta \in X_n\}$$

is also in bijection with the subgroups of $\mathbb{Z}^2$ of index n. For each element $\delta\gamma \in X_n \cdot \gamma$, there is $\sigma \in \mathrm{SL}_2(\mathbb{Z})$ such that $\sigma\delta\gamma \in X_n$ (the element σ transforms $\delta\gamma$ to Hermite normal form), and the set of elements $\sigma\delta\gamma$ is thus equal to X_n. Thus

$$T_{n,k}(f) = \sum_{\sigma\delta\gamma \in X_n} f^{[\sigma\delta\gamma]_k} = \sum_{\delta \in X_n} f^{[\delta\gamma]_k} = T_{n,k}(f)^{[\gamma]_k}.$$

A finite sum of meromorphic function is meromorphic, so $T_{n,k}(f)$ is weakly modular. If f is holomorphic on $\mathfrak{h}$, then each $f^{[\delta]_k}$ is holomorphic on $\mathfrak{h}$ for $\delta \in X_n$. A finite sum of holomorphic functions is holomorphic, so $T_{n,k}(f)$ is holomorphic.

□

We will frequently drop k from the notation in $T_{n,k}$, since the weight k is implicit in the modular function to which we apply the Hecke operator. Henceforth we make the convention that if we write $T_n(f)$ and if f is modular, then we mean $T_{n,k}(f)$, where k is the weight of f.

Proposition 2.29. *On weight k modular functions we have*

$$T_{mn} = T_m T_n \qquad \text{if } (m,n) = 1, \tag{2.4.1}$$

and

$$T_{p^n} = T_{p^{n-1}} T_p - p^{k-1} T_{p^{n-2}} \qquad \text{if } p \text{ is prime.} \tag{2.4.2}$$

Proof. Let L be a subgroup of index mn. The quotient $\mathbb{Z}^2/L$ is an abelian group of order mn, and $(m,n) = 1$, so $\mathbb{Z}^2/L$ decomposes uniquely as a direct sum of a subgroup of order m with a subgroup of order n. Thus there exists a unique subgroup L' such that $L \subset L' \subset \mathbb{Z}^2$, and L' has index m in $\mathbb{Z}^2$. The subgroup L' corresponds to an element of X_m, and the index n subgroup $L \subset L'$ corresponds to multiplying that element on the right by some uniquely determined element of X_n. We thus have

$$\mathrm{SL}_2(\mathbb{Z}) \cdot X_m \cdot X_n = \mathrm{SL}_2(\mathbb{Z}) \cdot X_{mn},$$

i.e., the set products of elements in X_m with elements of X_n equal the elements of X_{mn}, up to $\mathrm{SL}_2(\mathbb{Z})$-equivalence. Thus for any f, we have $T_{mn}(f) = T_n(T_m(f))$. Applying this formula with m and n swapped yields the equality $T_{mn} = T_m T_n$.

We will show that $T_{p^n} + p^{k-1} T_{p^{n-2}} = T_p T_{p^{n-1}}$. Suppose f is a weight k weakly modular function. Using that $f^{[\left(\begin{smallmatrix} p & 0 \\ 0 & p \end{smallmatrix}\right)]_k} = (p^2)^{k-1} p^{-k} f = p^{k-2} f$, we have

$$\sum_{x \in X_{p^n}} f^{[x]_k} + p^{k-1} \sum_{x \in X_{p^{n-2}}} f^{[x]_k} = \sum_{x \in X_{p^n}} f^{[x]_k} + p \sum_{x \in pX_{p^{n-2}}} f^{[x]_k}.$$

Also

$$T_p T_{p^{n-1}}(f) = \sum_{y \in X_p} \sum_{x \in X_{p^{n-1}}} (f^{[x]_k})^{[y]_k} = \sum_{x \in X_{p^{n-1}} \cdot X_p} f^{[x]_k}.$$

Thus it suffices to show that X_{p^n} disjoint union p copies of $pX_{p^{n-2}}$ is equal to $X_{p^{n-1}} \cdot X_p$, where we consider elements with multiplicities and up to left $\mathrm{SL}_2(\mathbb{Z})$-equivalence (i.e., the left action of $\mathrm{SL}_2(\mathbb{Z})$).

Suppose L is a subgroup of $\mathbb{Z}^2$ of index p^n, so L corresponds to an element of X_{p^n}. First suppose L is not contained in $p\mathbb{Z}^2$. Then the image of L in $\mathbb{Z}^2/p\mathbb{Z}^2 = (\mathbb{Z}/p\mathbb{Z})^2$ is of order p, so if $L' = p\mathbb{Z}^2 + L$, then $[\mathbb{Z}^2 : L'] = p$ and $[L : L'] = p^{n-1}$, and L' is the only subgroup with this property. Second, suppose that $L \subset p\mathbb{Z}^2$ if of index p^n and that $x \in X_{p^n}$ corresponds to L. Then every one of the $p+1$ subgroups $L' \subset \mathbb{Z}^2$ of index p contains L. Thus there are $p+1$ chains $L \subset L' \subset \mathbb{Z}^2$ with $[\mathbb{Z}^2 : L'] = p$.

The chains $L \subset L' \subset \mathbb{Z}^2$ with $[\mathbb{Z}^2 : L'] = p$ and $[\mathbb{Z}^2 : L] = p^{n-1}$ are in bijection with the elements of $X_{p^{n-1}} \cdot X_p$. On the other hand the union of X_{p^n} with p copies of $pX_{p^{n-2}}$ corresponds to the subgroups L of index p^n, but with those that contain $p\mathbb{Z}^2$ counted $p+1$ times. The structure of the set of chains $L \subset L' \subset \mathbb{Z}^2$ that we derived in the previous paragraph gives the result. □

Corollary 2.30. *The Hecke operator T_{p^n}, for prime p, is a polynomial in T_p with integer coefficients, i.e., $T_{p^n} \in \mathbb{Z}[T_p]$. If n, m are any integers, then $T_n T_m = T_m T_n$.*

Proof. The first statement follows from (2.4.2) of Proposition 2.29. It then follows that $T_n T_m = T_m T_n$ when m and n are both powers of a single prime p. Combining this with (2.4.1) gives the second statement in general. □

Proposition 2.31. *Let $f = \sum_{n \in \mathbb{Z}} a_n q^n$ be a modular function of weight k. Then*

$$T_n(f) = \sum_{m \in \mathbb{Z}} \left(\sum_{1 \le d \,|\, \gcd(n,m)} d^{k-1} a_{mn/d^2} \right) q^m.$$

In particular, if $n = p$ is prime, then

$$T_p(f) = \sum_{m\in\mathbb{Z}} \left(a_{mp} + p^{k-1}a_{m/p}\right) q^m,$$

where $a_{m/p} = 0$ if $m/p \notin \mathbb{Z}$.

Proof. This is proved in [**Ser73**, §VII.5.3] by writing out $T_n(f)$ explicitly and using that $\sum_{0\leq b<d} e^{2\pi ibm/d}$ is d if $d \mid m$ and 0 otherwise. □

Corollary 2.32. *The Hecke operators preserve M_k and S_k.*

Remark 2.33. Alternatively, for M_k the above corollary is Proposition 2.28, and for S_k we see from the definitions that if $f(\infty) = 0$, then T_nf also vanishes at ∞.

Example 2.34. Recall from (2.1.3) that

$$E_4 = \frac{1}{240} + q + 9q^2 + 28q^3 + 73q^4 + 126q^5 + 252q^6 + 344q^7 + \cdots .$$

Using the formula of Proposition 2.31, we see that

$$T_2(E_4) = (1/240 + 2^3 \cdot (1/240)) + 9q + (73 + 2^3 \cdot 1)q^2 + \cdots .$$

Since M_4 has dimension 1 and since we have proved that T_2 preserves M_4, we know that T_2 acts as a scalar. Thus we know just from the constant coefficient of $T_2(E_4)$ that

$$T_2(E_4) = 9E_4.$$

More generally, for p prime we see by inspection of the constant coefficient of $T_p(E_4)$ that

$$T_p(E_4) = (1 + p^3)E_4.$$

In fact $T_n(E_k) = \sigma_{k-1}(n)E_k$, for any integer $n \geq 1$ and even weight $k \geq 4$.

Example 2.35. By Corollary 2.32, the Hecke operators T_n also preserve the subspace S_k of M_k. Since S_{12} has dimension 1 (spanned by Δ), we see that Δ is an eigenvector for every T_n. Since the coefficient of q in the q-expansion of Δ is 1, the eigenvalue of T_n on Δ is the nth coefficient of Δ. Since $T_{nm} = T_nT_m$ for $\gcd(n, m) = 1$, we have proved the nonobvious fact that the *Ramanujan function* $\tau(n)$ that gives the nth coefficient of Δ is a multiplicative function, i.e., if $\gcd(n, m) = 1$, then $\tau(nm) = \tau(n)\tau(m)$.

Remark 2.36. The Hecke operators respect the decomposition $M_k = S_k \oplus \mathbb{C}E_k$, i.e., for all k the series E_k are eigenvectors for all T_n.

2.5. Computing Hecke Operators

This section is about how to compute matrices of Hecke operators on M_k.

Algorithm 2.37 (Hecke Operator). *This algorithm computes the matrix of the Hecke operator T_n on the Miller basis for M_k.*

(1) [Dimension] Compute $d = \dim(M_k) - 1$ using Corollary 2.16.
(2) [Basis] Using Lemma 2.20, compute the echelon basis $f_0, \ldots, f_d$ for $M_k \pmod{q^{dn+1}}$.
(3) [Hecke operator] Using Proposition 2.31, compute for each i the image $T_n(f_i) \pmod{q^{d+1}}$.
(4) [Write in terms of basis] The elements $T_n(f_i) \pmod{q^{d+1}}$ determine linear combinations of
$$f_0, f_1, \ldots, f_d \pmod{q^d}.$$
These linear combinations are easy to find once we compute $T_n(f_i)$ $\pmod{q^{d+1}}$, since our basis of f_i is in echelon form. The linear combinations are just the coefficients of the power series $T_n(f_i)$ up to and including q^d.
(5) [Write down matrix] The matrix of T_n acting from the right relative to the basis $f_0, \ldots, f_d$ is the matrix whose rows are the linear combinations found in the previous step, i.e., whose rows are the coefficients of $T_n(f_i)$.

Proof. By Proposition 2.31, the dth coefficient of $T_n(f)$ involves only a_{dn} and smaller-indexed coefficients of f. We need only compute a modular form f modulo q^{dn+1} in order to compute $T_n(f)$ modulo q^{d+1}. Uniqueness in step (4) follows from Lemma 2.20 above. □

Example 2.38. We compute the Hecke operator T_2 on M_{12} using the above algorithm.

(1) [Compute dimension] We have $d = 2 - 1 = 1$.
(2) [Compute basis] Compute up to (but not including) the coefficient of $q^{dn+1} = q^{1\cdot 2+1} = q^3$. As given in the proof of Lemma 2.20, we have
$$F_4 = 1 + 240q + 2160q^2 + \cdots \quad \text{and} \quad F_6 = 1 - 504q - 16632q^2 + \cdots .$$
Thus M_{12} has basis
$$F_4^3 = 1+720q+179280q^2+\cdots \quad \text{and} \quad \Delta = (F_4^3 - F_6^2)/1728 = q - 24q^2 + \cdots .$$
Subtracting 720Δ from F_4^3 yields the echelon basis, which is
$$f_0 = 1 + 196560q^2 + \cdots \quad \text{and} \quad f_1 = q - 24q^2 + \cdots .$$
SAGE does the arithmetic in the above calculation as follows:

```
sage: R.<q> = QQ[['q']]
sage: F4 =  240 * eisenstein_series_qexp(4,3)
sage: F6 = -504 * eisenstein_series_qexp(6,3)
sage: F4^3
1 + 720*q + 179280*q^2 + O(q^3)
sage: Delta = (F4^3 - F6^2)/1728; Delta
q - 24*q^2 + O(q^3)
sage: F4^3 - 720*Delta
1 + 196560*q^2 + O(q^3)
```

(3) [Compute Hecke operator] In each case letting a_n denote the nth coefficient of f_0 or f_1, respectively, we have

$$\begin{aligned} T_2(f_0) &= T_2(1 + 196560q^2 + \cdots) \\ &= (a_0 + 2^{11}a_0)q^0 + (a_2 - 2^{11}a_{1/2})q^1 + \cdots \\ &= 2049 + 196560q + \cdots, \end{aligned}$$

and

$$\begin{aligned} T_2(f_1) &= T_2(q - 24q^2 + \cdots) \\ &= (a_0 + 2^{11}a_0)q^0 + (a_2 + 2^{11}a_{1/2})q^1 + \cdots \\ &= 0 - 24q + \cdots. \end{aligned}$$

(Note that $a_{1/2} = 0$.)

(4) [Write in terms of basis] We read off at once that

$$T_2(f_0) = 2049f_0 + 196560f_1 \quad \text{and} \quad T_2(f_1) = 0f_0 + (-24)f_1.$$

(5) [Write down matrix] Thus the matrix of T_2, acting from the right on the basis f_0, f_1, is

$$T_2 = \begin{pmatrix} 2049 & 196560 \\ 0 & -24 \end{pmatrix}.$$

As a check note that the characteristic polynomial of T_2 is $(x-2049)(x+24)$ and that $2049 = 1 + 2^{11}$ is the sum of the 11th powers of the divisors of 2.

Example 2.39. The Hecke operator T_2 on M_{36} with respect to the echelon basis is

$$\begin{pmatrix} 34359738369 & 0 & 6218175600 & 9026867482214400 \\ 0 & 0 & 34416831456 & 5681332472832 \\ 0 & 1 & 194184 & -197264484 \\ 0 & 0 & -72 & -54528 \end{pmatrix}.$$

It has characteristic polynomial

$$(x - 34359738369) \cdot (x^3 - 139656x^2 - 59208339456x - 1467625047588864),$$

where the cubic factor is irreducible.

The `echelon_form()` command creates the space of modular forms but with basis in echelon form (which is not the default).

```
sage: M = ModularForms(1,36, prec=6).echelon_form()
sage: M.basis()
[
1 + 6218175600*q^4 + 15281788354560*q^5 + O(q^6),
q + 57093088*q^4 + 37927345230*q^5 + O(q^6),
q^2 + 194184*q^4 + 7442432*q^5 + O(q^6),
q^3 - 72*q^4 + 2484*q^5 + O(q^6)
]
```

Next we compute the matrix of the Hecke operator T_2.

```
sage: T2 = M.hecke_matrix(2); T2
[34359738369   0     6218175600 9026867482214400]
[          0   0    34416831456    5681332472832]
[          0   1         194184       -197264484]
[          0   0            -72           -54528]
```

Finally we compute and factor its characteristic polynomial.

```
sage: T2.charpoly().factor()
(x - 34359738369) *
   (x^3 - 139656*x^2 - 59208339456*x - 1467625047588864)
```

The following is a famous open problem about Hecke operators on modular forms of level 1. It generalizes our above observation that the characteristic polynomial of T_2 on M_k, for $k = 12, 36$, factors as a product of a linear factor and an irreducible factor.

Conjecture 2.40 (Maeda). *The characteristic polynomial of T_2 on S_k is irreducible for any k.*

Kevin Buzzard observed that in several specific cases the Galois group of the characteristic polynomial of T_2 is the full symmetric group (see [**Buz96**]). See also [**FJ02**] for more evidence for the following conjecture:

Conjecture 2.41. *For all primes p and all even $k \geq 2$ the characteristic polynomial of $T_{p,k}$ acting on S_k is irreducible.*

2.6. Fast Computation of Fourier Coefficients

How difficult is it to compute prime-indexed coefficients of

$$\Delta = \sum_{n=1}^{\infty} \tau(n) q^n = q - 24q^2 + 252q^3 - 1472q^4 + 4830q^5 - 6048q^6 + \cdots ?$$

Theorem 2.42 (Bosman, Couveignes, Edixhoven, de Jong, Merkl). *Let p be a prime. There is a probabilistic algorithm to compute $\tau(p)$, for prime p, that has expected running time polynomial in* $\log(p)$.

Proof. See [**ECdJ$^+$06**]. □

More generally, if $f = \sum a_n q^n$ is an eigenform in some space $M_k(\Gamma_1(N))$, where $k \geq 2$, then one expects that there is an algorithm to compute a_p in time polynomial in $\log(p)$. Bas Edixhoven, Jean-Marc Couveignes and Robin de Jong have proved that $\tau(p)$ can be computed in polynomial time; their approach involves sophisticated techniques from arithmetic geometry (e.g., étale cohomology, motives, Arakelov theory). The ideas they use are inspired by the ones introduced by Schoof, Elkies and Atkin for quickly counting points on elliptic curves over finite fields (see [**Sch95**]).

Edixhoven describes (in an email to the author) the strategy as follows:

(1) We compute the mod ℓ Galois representation ρ associated to Δ. In particular, we produce a polynomial f such that $\mathbb{Q}[x]/(f)$ is the fixed field of $\ker(\rho)$. This is then used to obtain $\tau(p) \pmod{\ell}$ and to do a Schoof-like algorithm for computing $\tau(p)$.

(2) We compute the field of definition of suitable points of order ℓ on the modular Jacobian $J_1(\ell)$ to do part (1) (see [**DS05**, Ch. 6] for the definition of $J_1(\ell)$).

(3) The method is to approximate the polynomial f in some sense (e.g., over the complex numbers or modulo many small primes r) and to use an estimate from Arakelov theory to determine a precision that will suffice.

2.7. Fast Computation of Bernoulli Numbers

This section, which was written jointly with Kevin McGown, is about computing the Bernoulli numbers B_n, for $n \geq 0$, defined in Section 2.1.2 by

$$\frac{x}{e^x - 1} = \sum_{n=0}^{\infty} B_n \frac{x^n}{n!}. \tag{2.7.1}$$

One way to compute B_n is to multiply both sides of (2.7.1) by $e^x - 1$ and equate coefficients of x^{n+1} to obtain the recurrence

$$B_0 = 1, \qquad B_n = -\frac{1}{n+1} \cdot \sum_{k=0}^{n-1} \binom{n+1}{k} B_k.$$

This recurrence provides a straightforward and easy-to-implement method for calculating B_n if one is interested in computing B_n for all n up to some bound. For example,

$$B_1 = -\frac{1}{2} \cdot \left(\binom{2}{0} B_0 \right) = -\frac{1}{2}$$

and

$$B_2 = -\frac{1}{3} \cdot \left(\binom{3}{0} B_0 + \binom{3}{1} B_1 \right) = -\frac{1}{3} \cdot \left(1 - \frac{3}{2} \right) = \frac{1}{6}.$$

However, computing B_n via the recurrence is slow; it requires summing over many large terms, it requires storing the numbers $B_0, \ldots, B_{n-1}$, and it takes only limited advantage of asymptotically fast arithmetic algorithms. There is also an inductive procedure to compute Bernoulli numbers that resembles Pascal's triangle called the Akiyama-Tanigawa algorithm (see [**Kan00**]).

Another approach to computing B_n is to use Newton iteration and asymptotically fast polynomial arithmetic to approximate $1/(e^x - 1)$. This method yields a very fast algorithm to compute $B_0, B_2, \ldots, B_{p-3}$ modulo p. See [**BCS92**] for an application of this method modulo a prime p to the verification of Fermat's last theorem for irregular primes up to one million.

Example 2.43. David Harvey implemented the algorithm of [**BCS92**] in SAGE as the command `bernoulli_mod_p`:

```
sage: bernoulli_mod_p(23)
[1, 4, 13, 17, 13, 6, 10, 5, 10, 9, 15]
```

A third way to compute B_n uses an algorithm based on Proposition 2.8, which we explain below (Algorithm 2.45). This algorithm appears to have been independently invented by several people: by Bernd C. Kellner (see [**Kel06**]); by Bill Dayl; and by H. Cohen and K. Belabas.

We compute B_n as an exact rational number by approximating $\zeta(n)$ to very high precision using Proposition 2.8, the Euler product

$$\zeta(s) = \sum_{m=1}^{\infty} m^{-s} = \prod_{p \text{ prime}} (1 - p^{-s})^{-1},$$

and the following theorem:

Theorem 2.44 (Clausen, von Staudt). *For even $n \geq 2$,*

$$\operatorname{denom}(B_n) = \prod_{p-1\,|\,n} p.$$

Proof. See [**Lan95**, Ch. X, Thm. 2.1]. □

2.7.1. The Number of Digits of B_n. The following is a new quick way to compute the number of digits of the numerator of B_n. For example, using it we can compute the number of digits of $B_{10^{50}}$ in less than a second.

By Theorem 2.44 we have $d_n = \operatorname{denom}(B_n) = \prod_{p-1|n} p$. The number of digits of the numerator is thus

$$\lceil \log_{10}(d_n \cdot |B_n|) \rceil.$$

But

$$\begin{aligned}\log(|B_n|) &= \log\left(\frac{2 \cdot n!}{(2\pi)^n}\zeta(n)\right)\\ &= \log(2) + \log(n!) - n\log(2) - n\log(\pi) + \log(\zeta(n)),\end{aligned}$$

and $\zeta(n) \sim 1$ so $\log(\zeta(n)) \sim 0$. Finally, Stirling's formula (see [**Ahl78**, pg. 198–206]) gives a fast way to compute $\log(n!) = \log(\Gamma(n+1))$:
(2.7.2)

$$\log(\Gamma(z))\text{``}=\text{''}\frac{1}{\log(2\pi)} + \left(z - \frac{1}{2}\right)\log(z) - z + \sum_{m=1}^{\infty}\frac{B_{2m}}{2m(2m-1)z^{2m-1}}.$$

We put quotes around the equality sign because $\log(\Gamma(z))$ does not converge to its Laurent series. Indeed, note that for any fixed value of z the summands on the right side go to ∞ as $m \to \infty$! Nonetheless, we can use this formula to very efficiently compute $\log(\Gamma(z))$, since if we truncate the sum, then the error is smaller than the next term in the infinite sum.

2.7.2. Computing B_n Exactly. We return to the problem of computing B_n. Let

$$K = \frac{2 \cdot n!}{(2\pi)^n}$$

so that $|B_n| = K\zeta(n)$. Write

$$B_n = \frac{a}{d},$$

with $a, d \in \mathbb{Z}$, $d \geq 1$, and $\gcd(a,d) = 1$. It is elementary to show that $a = (-1)^{n/2+1}\,|a|$ for even $n \geq 2$. Suppose that using the Euler product we approximate $\zeta(n)$ from below by a number z such that

$$0 \leq \zeta(m) - z < \frac{1}{Kd}.$$

Then $0 \le |B_n| - zK < d^{-1}$; hence $0 \le |a| - zKd < 1$. It follows that $|a| = \lceil zKd \rceil$ and hence $a = (-1)^{n/2+1} \lceil zKd \rceil$.

It remains to compute z. Consider the following problem: given $s > 1$ and $\varepsilon > 0$, find $M \in \mathbb{Z}_+$ so that

$$z = \prod_{p \le M} (1 - p^{-s})^{-1}$$

satisfies $0 \le \zeta(s) - z < \varepsilon$. We always have $0 \le \zeta(s) - z$. Also,

$$\sum_{n \le M} n^{-s} \le \prod_{p \le M} (1 - p^{-s})^{-1},$$

so

$$\zeta(s) - z \le \sum_{n=M+1}^{\infty} n^{-s} \le \int_M^{\infty} x^{-s}\, dx = \frac{1}{(s-1)M^{s-1}}.$$

Thus if $M > \varepsilon^{-1/(s-1)}$, then

$$\frac{1}{(s-1)M^{s-1}} \le \frac{1}{M^{s-1}} < \varepsilon,$$

so $\zeta(s) - z < \varepsilon$, as required. For our purposes, we have $s = n$ and $\varepsilon = (Kd)^{-1}$, so it suffices to take $M > (Kd)^{1/(n-1)}$.

Algorithm 2.45 (Bernoulli Number B_n). *Given an integer $n \ge 0$, this algorithm computes the Bernoulli number B_n as an exact rational number.*

(1) [Special cases] If $n = 0$, return 1; if $n = 1$, return $-1/2$; if $n \ge 3$ is odd, return 0.

(2) [Factorial factor] Compute $K = \dfrac{2 \cdot n!}{(2\pi)^n}$ to sufficiently many digits of precision so the ceiling in step (6) is uniquely determined (this precision can be determined using Section 2.7.1).

(3) [Denominator] Compute $d = \prod_{p-1 \mid n} p$.

(4) [Bound] Compute $M = \left\lceil (Kd)^{1/(n-1)} \right\rceil$.

(5) [Approximate $\zeta(n)$] Compute $z = \prod_{p \le M} (1 - p^{-n})^{-1}$.

(6) [Numerator] Compute $a = (-1)^{n/2+1} \lceil dKz \rceil$.

(7) [Output B_n] Return $\dfrac{a}{d}$.

In step (5) use a sieve to compute all primes $p \le M$ efficiently (which is fast, since M is so small). In step (4) we may replace M by any integer greater than the one specified by the formula, so we do not have to compute $(Kd)^{1/(n-1)}$ to very high precision.

In Section 5.2.2 below we will generalize the above algorithm.

Example 2.46. We illustrate Algorithm 2.45 by computing B_{50}. Using 135 binary digits of precision, we compute

$$K = 750086674607695770474736.71552473164563479.$$

The divisors of n are $1, 2, 5, 10, 25, 50$, so

$$d = 2 \cdot 3 \cdot 11 = 66.$$

We find $M = 4$ and compute

$$z = 1.000000000000000088817842109308159029835012.$$

Finally we compute

$$dKz = 495057205241079648212477524.999999994425778,$$

so

$$B_{50} = \frac{495057205241079648212477525}{66}.$$

2.8. Exercises

2.1 Using Proposition 2.8 and the table on page 16, compute $\sum_{n=1}^{\infty} \frac{1}{n^{26}}$ explicitly.

2.2 Prove that if $n > 1$ is odd, then the Bernoulli number B_n is 0.

2.3 Use (2.1.3) to write down the coefficients of 1, q, q^2, and q^3 of the Eisenstein series E_8.

2.4 Suppose k is a positive integer with $k \equiv 0 \pmod{12}$. Suppose $a, b \geq 0$ are integers with $4a + 6b = k$.
   (a) Prove $3 \mid a$.
   (b) Show that $G_4^a \cdot G_6^b \,/\, G_6^{\frac{k}{6}} = \left(G_4^3/G_6^2\right)^{\frac{a}{3}}$.

2.5 Compute the Miller basis for $M_{28}(\mathrm{SL}_2(\mathbb{Z}))$ with precision $O(q^8)$. Your answer will look like Example 2.23.

2.6 Consider the cusp form $f = q^2 + 192q^3 - 8280q^4 + \cdots$ in $S_{28}(\mathrm{SL}_2(\mathbb{Z}))$. Write f as a polynomial in E_4 and E_6 (see Remark 2.25).

2.7 Let G_k be the weight k Eisenstein series from equation (2.1.1). Let c be the complex number so that the constant coefficient of the q-expansion of $g = c \cdot G_k$ is 1. Is it always the case that the q-expansion of g lies in $\mathbb{Z}[[q]]$?

2.8 Compute the matrix of the Hecke operator T_2 on the Miller basis for $M_{32}(\mathrm{SL}_2(\mathbb{Z}))$. Then compute its characteristic polynomial and verify it factors as a product of two irreducible polynomials.

What Next? Much of the rest of this book is about methods for computing subspaces of $M_k(\Gamma_1(N))$ for general N and k. These general methods are more complicated than the methods presented in this chapter, since there are many more modular forms of small weight and it can be difficult to obtain them. Forms of level $N > 1$ have subtle connections with elliptic curves, abelian varieties, and motives. Read on for more!

Chapter 3

Modular Forms of Weight 2

We saw in Chapter 2 (especially Section 2.2) that we can compute each space $M_k(\mathrm{SL}_2(\mathbb{Z}))$ explicitly. This involves computing Eisenstein series E_4 and E_6 to some precision, then forming the basis $\{E_4^a E_6^b : 4a + 6b = k, 0 \leq a, b \in \mathbb{Z}\}$ for $M_k(\mathrm{SL}_2(\mathbb{Z}))$. In this chapter we consider the more general problem of computing $S_2(\Gamma_0(N))$, for any positive integer N. Again we have a decomposition

$$M_2(\Gamma_0(N)) = S_2(\Gamma_0(N)) \oplus E_2(\Gamma_0(N)),$$

where $E_2(\Gamma_0(N))$ is spanned by generalized Eisenstein series and $S_2(\Gamma_0(N))$ is the space of cusp forms, i.e., elements of $M_2(\Gamma_0(N))$ that vanish at *all* cusps.

In Chapter 5 we compute the space $E_2(\Gamma_0(N))$ in a similar way to how we computed $M_k(\mathrm{SL}_2(\mathbb{Z}))$. On the other hand, elements of $S_2(\Gamma_0(N))$ often cannot be written as sums or products of generalized Eisenstein series. In fact, the structure of $M_2(\Gamma_0(N))$ is, in general, much more complicated than that of $M_k(\mathrm{SL}_2(\mathbb{Z}))$. For example, when p is a prime, $E_2(\Gamma_0(p))$ has dimension 1, whereas $S_2(\Gamma_0(p))$ has dimension about $p/12$.

Fortunately an idea of Birch, which he called modular symbols, provides a method for computing $S_2(\Gamma_0(N))$ and indeed for much more that is relevant to understanding special values of L-functions. Modular symbols are also a powerful theoretical tool. In this chapter, we explain how $S_2(\Gamma_0(N))$ is related to modular symbols and how to use this relationship to explicitly compute a basis for $S_2(\Gamma_0(N))$. In Chapter 8 we will introduce more general modular symbols and explain how to use them to compute $S_k(\Gamma_0(N))$, $S_k(\Gamma_1(N))$ and $S_k(N,\varepsilon)$ for any integers $k \geq 2$ and N and character ε.

Section 3.1 contains a very brief summary of basic facts about modular forms of weight 2, modular curves, Hecke operators, and integral homology. Section 3.2 introduces modular symbols and describes how to compute with them. In Section 3.5 we talk about how to cut out the subspace of modular symbols corresponding to cusp forms using the boundary map. Section 3.6 is about a straightforward method to compute a basis for $S_2(\Gamma_0(N))$ using modular symbols, and Section 3.7 outlines a more sophisticated algorithm for computing newforms that uses Atkin-Lehner theory.

Before reading this chapter, you should have read Chapter 1 and Chapter 2. We also assume familiarity with algebraic curves, Riemann surfaces, and homology groups of compact Riemann surfaces.

3.1. Hecke Operators

Recall from Chapter 1 that the group $\Gamma_0(N)$ acts on $\mathfrak{h}^* = \mathfrak{h} \cup \mathbb{P}^1(\mathbb{Q})$ by linear fractional transformations. The quotient $\Gamma_0(N)\backslash\mathfrak{h}^*$ is a Riemann surface, which we denote by $X_0(N)$. See [**DS05**, Ch. 2] for a detailed description of the topology on $X_0(N)$. The Rieman surface $X_0(N)$ also has a canonical structure of algebraic curve over $\mathbb{Q}$, as is explained in [**DS05**, Ch. 7] (see also [**Shi94**, §6.7]).

Recall from Section 1.3 that a cusp form of weight 2 for $\Gamma_0(N)$ is a function f on $\mathfrak{h}$ such that $f(z)dz$ defines a holomorphic differential on $X_0(N)$. Equivalently, a cusp form is a holomorphic function f on $\mathfrak{h}$ such that

(a) the expression $f(z)dz$ is invariant under replacing z by $\gamma(z)$ for each $\gamma \in \Gamma_0(N)$ and

(b) $f(z)$ vanishes at every cusp for $\Gamma_0(N)$.

The space $S_2(\Gamma_0(N))$ of weight 2 cusp forms on $\Gamma_0(N)$ is a finite-dimensional complex vector space, of dimension equal to the genus g of $X_0(N)$. The space $X_0(N)(\mathbb{C})$ is a compact oriented Riemann surface, so it is a 2-dimensional oriented real manifold, i.e., $X_0(N)(\mathbb{C})$ is a g-holed torus (see Figure 3.1.1 on page 38).

Condition (b) in the definition of f means that f has a Fourier expansion about each element of $\mathbb{P}^1(\mathbb{Q})$. Thus, at ∞ we have

$$\begin{aligned} f(z) &= a_1 e^{2\pi i z} + a_2 e^{2\pi i 2z} + a_3 e^{2\pi i 3z} + \cdots \\ &= a_1 q + a_2 q^2 + a_3 q^3 + \cdots, \end{aligned}$$

where, for brevity, we write $q = q(z) = e^{2\pi i z}$.

Example 3.1. Let E be the elliptic curve defined by the equation $y^2+xy = x^3+x^2-4x-5$. Let $a_p = p+1-\#\tilde{E}(\mathbb{F}_p)$, where $\tilde{E}$ is the reduction of E mod p (note that for the primes that divide the conductor of E we have $a_3 = -1$, $a_{13} = 1$). For n composite, define a_n using the relations at the end of Section 3.7. Then the Shimura-Taniyama conjecture asserts that

$$\begin{aligned} f &= q + a_2q^2 + a_3q^3 + a_4q^4 + a_5q^5 + \cdots \\ &= q + q^2 - q^3 - q^4 + 2q^5 + \cdots \end{aligned}$$

is the q-expansion of an element of $S_2(\Gamma_0(39))$. This conjecture, which is now a theorem (see [**BCDT01**]), asserts that any q-expansion constructed as above from an elliptic curve over $\mathbb{Q}$ is a modular form. This conjecture was mostly proved first by Wiles [**Wil95**] as a key step in the proof of Fermat's last theorem.

Just as is the case for level 1 modular forms (see Section 2.4) there are commuting Hecke operators $T_1, T_2, T_3, \ldots$ that act on $S_2(\Gamma_0(N))$. To define them conceptually, we introduce an interpretation of the modular curve $X_0(N)$ as an object whose points *parameterize* elliptic curves with extra structure.

Proposition 3.2. *The complex points of $Y_0(N) = \Gamma_0(N)\backslash\mathfrak{h}$ are in natural bijection with isomorphism classes of pairs (E, C), where E is an elliptic curve over $\mathbb{C}$ and C is a cyclic subgroup of $E(\mathbb{C})$ of order N. The class of the point $\lambda \in \mathfrak{h}$ corresponds to the pair*

$$\left(\mathbb{C}/(\mathbb{Z}+\mathbb{Z}\lambda),\quad \left(\frac{1}{N}\mathbb{Z}+\mathbb{Z}\lambda\right)/(\mathbb{Z}+\mathbb{Z}\lambda)\right).$$

Proof. See Exercise 3.1. □

Suppose n and N are coprime positive integers. There are two natural maps π_1 and π_2 from $Y_0(n\cdot N)$ to $Y_0(N)$; the first, π_1, sends $(E, C) \in Y_0(n\cdot N)(\mathbb{C})$ to (E, C'), where C' is the unique cyclic subgroup of C of order N, and the second, π_2, sends (E, C) to $(E/D, C/D)$, where D is the unique cyclic subgroup of C of order n. These maps extend in a unique way to algebraic maps from $X_0(n\cdot N)$ to $X_0(N)$:

(3.1.1)
$$\begin{array}{ccccc} & & X_0(n\cdot N) & & \\ & \swarrow^{\pi_2} & & \searrow^{\pi_1} & \\ X_0(N) & & & & X_0(N). \end{array}$$

The nth *Hecke operator* T_n is $\pi_{1*}\circ\pi_2^*$, where π_2^* and π_{1*} denote pullback and pushforward of differentials, respectively. (There is a similar definition of T_n when $\gcd(n, N) \neq 1$.) Using our interpretation of $S_2(\Gamma_0(N))$ as differentials on $X_0(N)$, this gives an action of Hecke operators on $S_2(\Gamma_0(N))$. One can show that these induce the maps of Proposition 2.31 on q-expansions.

Example 3.3. There is a basis of $S_2(39)$ so that

$$T_2 = \begin{pmatrix} 1 & 1 & 0 \\ -2 & -3 & -2 \\ 0 & 0 & 1 \end{pmatrix} \quad \text{and} \quad T_5 = \begin{pmatrix} -4 & -2 & -6 \\ 4 & 4 & 4 \\ 0 & 0 & 2 \end{pmatrix}.$$

Notice that these matrices commute. Also, the characteristic polynomial of T_2 is $(x-1)\cdot(x^2+2x-1)$.

3.1.1. Homology. The first homology group $\mathrm{H}_1(X_0(N),\mathbb{Z})$ is the group of closed 1-cycles modulo boundaries of 2-cycles (formal sums of images of 2-simplexes). Topologically $X_0(N)$ is a g-holed torus, where g is the genus of $X_0(N)$. Thus $\mathrm{H}_1(X_0(N),\mathbb{Z})$ is a free abelian group of rank $2g$ (see, e.g., [**GH81**, Ex. 19.30] and [**DS05**, §6.1]), with two generators corresponding to each hole, as illustrated in the case $N = 39$ in Figure 3.1.1.

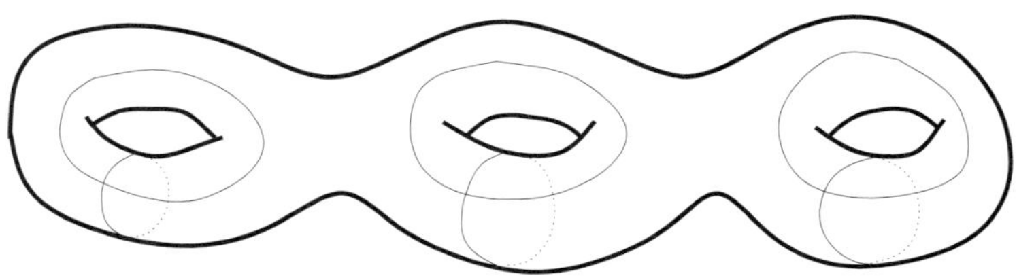

$$\mathrm{H}_1(X_0(39),\mathbb{Z}) \cong \mathbb{Z}\times\mathbb{Z}\times\mathbb{Z}\times\mathbb{Z}\times\mathbb{Z}\times\mathbb{Z}$$

Figure 3.1.1. The homology of $X_0(39)$.

The homology of $X_0(N)$ is closely related to modular forms, since the Hecke operators T_n also act on $\mathrm{H}_1(X_0(N),\mathbb{Z})$. The action is by pullback of homology classes by π_2 followed by taking the image under π_1, where π_1 and π_2 are as in (3.1.1).

Integration defines a pairing

$$\langle\, ,\rangle : S_2(\Gamma_0(N)) \times \mathrm{H}_1(X_0(N),\mathbb{Z}) \to \mathbb{C}. \tag{3.1.2}$$

Explicitly, for a path x,

$$\langle f, x\rangle = 2\pi i \cdot \int_x f(z)dz.$$

Theorem 3.4. *The pairing* (3.1.2) *is nondegenerate and Hecke equivariant in the sense that for every Hecke operator T_n, we have $\langle fT_n, x\rangle = \langle f, T_n x\rangle$. Moreover, it induces a perfect pairing*

$$\langle\, ,\rangle : S_2(\Gamma_0(N)) \times \mathrm{H}_1(X_0(N),\mathbb{R}) \to \mathbb{C}. \tag{3.1.3}$$

This is a special case of the results in Section 8.5.

As we will see, modular symbols allow us to make explicit the action of the Hecke operators on $\mathrm{H}_1(X_0(N), \mathbb{Z})$; the above pairing then translates this into a wealth of information about cusp forms.

We will also consider the relative homology group $\mathrm{H}_1(X_0(N), \mathbb{Z}; \{\text{cusps}\})$ of $X_0(N)$ *relative to the cusps*; it is the same as usual homology, but in addition we allow paths with endpoints in the cusps instead of restricting to closed loops. Modular symbols provide a "combinatorial" presentation of $\mathrm{H}_1(X_0(N), \mathbb{Z})$ in terms of paths between elements of $\mathbb{P}^1(\mathbb{Q})$.

3.2. Modular Symbols

Let $\mathbb{M}_2$ be the free abelian group with basis the set of symbols $\{\alpha, \beta\}$ with $\alpha, \beta \in \mathbb{P}^1(\mathbb{Q})$ modulo the 3-term relations

$$\{\alpha, \beta\} + \{\beta, \gamma\} + \{\gamma, \alpha\} = 0$$

above and modulo any torsion. Since $\mathbb{M}_2$ is torsion-free, we have

$$\{\alpha, \alpha\} = 0 \qquad \text{and} \qquad \{\alpha, \beta\} = -\{\beta, \alpha\}.$$

Remark 3.5 (Warning). The symbols $\{\alpha, \beta\}$ satisfy the relations $\{\alpha, \beta\} = -\{\beta, \alpha\}$, so order matters. The notation $\{\alpha, \beta\}$ looks like the set containing two elements, which strongly (and incorrectly) suggests that the order does not matter. This is the standard notation in the literature.

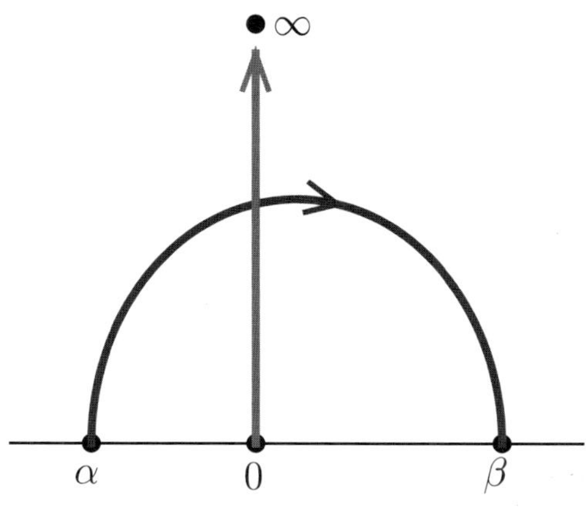

Figure 3.2.1. The modular symbols $\{\alpha, \beta\}$ and $\{0, \infty\}$.

As illustrated in Figure 3.2.1, we "think of" this modular symbol as the homology class, relative to the cusps, of a path from α to β in $\mathfrak{h}^*$.

Define a *left action of* $\mathrm{GL}_2(\mathbb{Q})$ on $\mathbb{M}_2$ by letting $g \in \mathrm{GL}_2(\mathbb{Q})$ act by

$$g\{\alpha, \beta\} = \{g(\alpha), g(\beta)\},$$

and g acts on α and β via the corresponding linear fractional transformation. The space $\mathbb{M}_2(\Gamma_0(N))$ of *modular symbols for* $\Gamma_0(N)$ is the quotient of $\mathbb{M}_2$ by the submodule generated by the infinitely many elements of the form $x - g(x)$, for x in $\mathbb{M}_2$ and g in $\Gamma_0(N)$, and modulo any torsion. A modular symbol for $\Gamma_0(N)$ is an element of this space. We frequently denote the equivalence class of a modular symbol by giving a representative element.

Example 3.6. Some modular symbols are 0 no matter what the level N is! For example, since $\gamma = \left(\begin{smallmatrix} 1 & 1 \\ 0 & 1 \end{smallmatrix}\right) \in \Gamma_0(N)$, we have

$$\{\infty, 0\} = \{\gamma(\infty), \gamma(0)\} = \{\infty, 1\},$$

so

$$0 = \{\infty, 1\} - \{\infty, 0\} = \{\infty, 1\} + \{0, \infty\} = \{0, \infty\} + \{\infty, 1\} = \{0, 1\}.$$

See Exercise 3.2 for a generalization of this observation.

There is a natural homomorphism

$$\varphi : \mathbb{M}_2(\Gamma_0(N)) \to \mathrm{H}_1(X_0(N), \{\text{cusps}\}, \mathbb{Z}) \tag{3.2.1}$$

that sends a formal linear combination of geodesic paths in the upper half plane to their image as paths on $X_0(N)$. In [**Man72**] Manin proved that (3.2.1) is an isomorphism (this is a fairly involved topological argument).

Manin identified the subspace of $\mathbb{M}_2(\Gamma_0(N))$ that is sent isomorphically onto $\mathrm{H}_1(X_0(N), \mathbb{Z})$. Let $\mathbb{B}_2(\Gamma_0(N))$ denote the free abelian group whose basis is the finite set $C(\Gamma_0(N)) = \Gamma_0(N)\backslash\mathbb{P}^1(\mathbb{Q})$ of cusps for $\Gamma_0(N)$. The *boundary map*

$$\delta : \mathbb{M}_2(\Gamma_0(N)) \to \mathbb{B}_2(\Gamma_0(N))$$

sends $\{\alpha, \beta\}$ to $\{\beta\} - \{\alpha\}$, where $\{\beta\}$ denotes the basis element of $\mathbb{B}_2(\Gamma_0(N))$ corresponding to $\beta \in \mathbb{P}^1(\mathbb{Q})$. The kernel $\mathbb{S}_2(\Gamma_0(N))$ of δ is the subspace of *cuspidal modular symbols*. Thus an element of $\mathbb{S}_2(\Gamma_0(N))$ can be thought of as a linear combination of paths in $\mathfrak{h}^*$ whose endpoints are cusps and whose images in $X_0(N)$ are homologous to a $\mathbb{Z}$-linear combination of closed paths.

Theorem 3.7 (Manin). *The map φ above induces a canonical isomorphism*

$$\mathbb{S}_2(\Gamma_0(N)) \cong \mathrm{H}_1(X_0(N), \mathbb{Z}).$$

Proof. This is [**Man72**, Thm. 1.9]. □

For any (commutative) ring R let

$$\mathbb{M}_2(\Gamma_0(N), R) = \mathbb{M}_2(\Gamma_0(N)) \otimes_{\mathbb{Z}} R$$

and

$$\mathbb{S}_2(\Gamma_0(N), R) = \mathbb{S}_2(\Gamma_0(N)) \otimes_{\mathbb{Z}} R.$$

Proposition 3.8. *We have*

$$\dim_{\mathbb{C}} \mathbb{S}_2(\Gamma_0(N), \mathbb{C}) = 2 \dim_{\mathbb{C}} S_2(\Gamma_0(N)).$$

Proof. We have

$$\dim_{\mathbb{C}} \mathbb{S}_2(\Gamma_0(N), \mathbb{C}) = \operatorname{rank}_{\mathbb{Z}} \mathbb{S}_2(\Gamma_0(N)) = \operatorname{rank}_{\mathbb{Z}} \mathrm{H}_1(X_0(N), \mathbb{Z}) = 2g.$$

□

Example 3.9. We illustrate modular symbols in the case when $N = 11$. Using SAGE (below), which implements the algorithm that we describe below over $\mathbb{Q}$, we find that $\mathbb{M}_2(\Gamma_0(11); \mathbb{Q})$ has basis $\{\infty, 0\}$, $\{-1/8, 0\}$, $\{-1/9, 0\}$. A basis for the integral homology $\mathrm{H}_1(X_0(11), \mathbb{Z})$ is the subgroup generated by $\{-1/8, 0\}$ and $\{-1/9, 0\}$.

```
sage: set_modsym_print_mode ('modular')
sage: M = ModularSymbols(11, 2)
sage: M.basis()
({Infinity,0}, {-1/8,0}, {-1/9,0})
sage: S = M.cuspidal_submodule()
sage: S.integral_basis()     # basis over ZZ.
({-1/8,0}, {-1/9,0})
sage: set_modsym_print_mode ('manin')     # set it back
```

3.3. Computing with Modular Symbols

3.3.1. Manin's Trick. In this section, we describe a trick of Manin that we will use to prove that spaces of modular symbols are computable.

By Exercise 1.6 the group $\Gamma_0(N)$ has finite index in $\mathrm{SL}_2(\mathbb{Z})$. Fix right coset representatives $r_0, r_1, \ldots, r_m$ for $\Gamma_0(N)$ in $\mathrm{SL}_2(\mathbb{Z})$, so that

$$\mathrm{SL}_2(\mathbb{Z}) = \Gamma_0(N) r_0 \cup \Gamma_0(N) r_1 \cup \cdots \cup \Gamma_0(N) r_m,$$

where the union is disjoint. For example, when N is prime, a list of coset representatives is

$$\begin{pmatrix} 1 & 0 \\ 0 & 1 \end{pmatrix}, \begin{pmatrix} 1 & 0 \\ 1 & 1 \end{pmatrix}, \begin{pmatrix} 1 & 0 \\ 2 & 1 \end{pmatrix}, \begin{pmatrix} 1 & 0 \\ 3 & 1 \end{pmatrix}, \ldots, \begin{pmatrix} 1 & 0 \\ N-1 & 1 \end{pmatrix}, \begin{pmatrix} 0 & -1 \\ 1 & 0 \end{pmatrix}.$$

Let

$$\mathbb{P}^1(\mathbb{Z}/N\mathbb{Z}) = \{(a : b) \,:\, a, b \in \mathbb{Z}/N\mathbb{Z},\ \gcd(a, b, N) = 1\}/\sim \tag{3.3.1}$$

where $(a : b) \sim (a' : b')$ if there is $u \in (\mathbb{Z}/N\mathbb{Z})^*$ such that $a = ua', b = ub'$.

Proposition 3.10. *There is a bijection between $\mathbb{P}^1(\mathbb{Z}/N\mathbb{Z})$ and the right cosets of $\Gamma_0(N)$ in $\mathrm{SL}_2(\mathbb{Z})$, which sends a coset representative $\left(\begin{smallmatrix} a & b \\ c & d \end{smallmatrix}\right)$ to the class of $(c : d)$ in $\mathbb{P}^1(\mathbb{Z}/N\mathbb{Z})$.*

Proof. See Exercise 3.3. □

See Proposition 8.6 for the analogous statement for $\Gamma_1(N)$.

We now describe an observation of Manin (see [**Man72**, §1.5]) that is crucial to making $\mathbb{M}_2(\Gamma_0(N))$ computable. It allows us to write any modular symbol $\{\alpha, \beta\}$ as a $\mathbb{Z}$-linear combination of symbols of the form $r_i\{0, \infty\}$, where the $r_i \in \mathrm{SL}_2(\mathbb{Z})$ are coset representatives as above. In particular, the finitely many symbols $r_0\{0, \infty\}, \ldots, r_m\{0, \infty\}$ generate $\mathbb{M}_2(\Gamma_0(N))$.

Proposition 3.11 (Manin)**.** *Let N be a positive integer and $r_0, \ldots, r_m$ a set of right coset representatives for $\Gamma_0(N)$ in $\mathrm{SL}_2(\mathbb{Z})$. Every $\{\alpha, \beta\} \in \mathbb{M}_2(\Gamma_0(N))$ is a $\mathbb{Z}$-linear combination of $r_0\{0, \infty\}, \ldots, r_m\{0, \infty\}$.*

We give two proofs of the proposition. The first is useful for computation (see [**Cre97a**, §2.1.6]); the second (see [**MTT86**, §2]) is easier to understand conceptually since it does not require any knowledge of continued fractions.

Continued Fractions Proof of Proposition 3.11. Since

$$\{\alpha, \beta\} = \{0, \beta\} - \{0, \alpha\},$$

it suffices to consider modular symbols of the form $\{0, b/a\}$, where the rational number b/a is in lowest terms. Expand b/a as a continued fraction and consider the successive convergents in lowest terms:

$$\frac{b_{-2}}{a_{-2}} = \frac{0}{1}, \quad \frac{b_{-1}}{a_{-1}} = \frac{1}{0}, \quad \frac{b_0}{a_0} = \frac{b_0}{1}, \quad \ldots, \quad \frac{b_{n-1}}{a_{n-1}}, \quad \frac{b_n}{a_n} = \frac{b}{a}$$

where the first two are included formally. Then

$$b_k a_{k-1} - b_{k-1} a_k = (-1)^{k-1},$$

so that

$$g_k = \begin{pmatrix} b_k & (-1)^{k-1} b_{k-1} \\ a_k & (-1)^{k-1} a_{k-1} \end{pmatrix} \in \mathrm{SL}_2(\mathbb{Z}).$$

Hence

$$\left\{ \frac{b_{k-1}}{a_{k-1}}, \frac{b_k}{a_k} \right\} = g_k\{0, \infty\} = r_i\{0, \infty\},$$

for some i, is of the required special form. Since

$$\{0, b/a\} = \{0, \infty\} + \{\infty, b_0\} + \left\{ \frac{b_0}{1}, \frac{b_1}{a_1} \right\} + \cdots + \left\{ \frac{b_{n-1}}{a_{n-1}}, \frac{b_n}{a_n} \right\},$$

this completes the proof. □

Inductive Proof of Proposition 3.11. As in the first proof it suffices to prove the proposition for any symbol $\{0, b/a\}$, where b/a is in lowest terms. We will induct on $a \in \mathbb{Z}_{\geq 0}$. If $a = 0$, then the symbol is $\{0, \infty\}$, which corresponds to the identity coset, so assume that $a > 0$. Find $a' \in \mathbb{Z}$ such that

$$ba' \equiv 1 \pmod{a};$$

then $b' = (ba' - 1)/a \in \mathbb{Z}$ so the matrix

$$\delta = \begin{pmatrix} b & b' \\ a & a' \end{pmatrix}$$

is an element of $\mathrm{SL}_2(\mathbb{Z})$. Thus $\delta = \gamma \cdot r_j$ for some right coset representative r_j and $\gamma \in \Gamma_0(N)$. Then

$$\{0, b/a\} - \{0, b'/a'\} = \{b'/a', b/a\} = \begin{pmatrix} b & b' \\ a & a' \end{pmatrix} \cdot \{0, \infty\} = r_j\{0, \infty\},$$

as elements of $\mathbb{M}_2(\Gamma_0(N))$. By induction, $\{0, b'/a'\}$ is a linear combination of symbols of the form $r_k\{0, \infty\}$, which completes the proof. □

Example 3.12. Let $N = 11$, and consider the modular symbol $\{0, 4/7\}$. We have

$$\frac{4}{7} = 0 + \frac{1}{1 + \frac{1}{1+\frac{1}{3}}},$$

so the partial convergents are

$$\frac{b_{-2}}{a_{-2}} = \frac{0}{1}, \quad \frac{b_{-1}}{a_{-1}} = \frac{1}{0}, \quad \frac{b_0}{a_0} = \frac{0}{1}, \quad \frac{b_1}{a_1} = \frac{1}{1}, \quad \frac{b_2}{a_2} = \frac{1}{2}, \quad \frac{b_3}{a_3} = \frac{4}{7}.$$

Thus, noting as in Example 3.6 that $\{0, 1\} = 0$, we have

$$\begin{aligned}
\{0, 4/7\} &= \{0, \infty\} + \{\infty, 0\} + \{0, 1\} + \{1, 1/2\} + \{1/2, 4/7\} \\
&= \begin{pmatrix} 1 & -1 \\ 2 & -1 \end{pmatrix} \{0, \infty\} + \begin{pmatrix} 4 & 1 \\ 7 & 2 \end{pmatrix} \{0, \infty\} \\
&= \begin{pmatrix} 1 & 0 \\ 9 & 1 \end{pmatrix} \{0, \infty\} + \begin{pmatrix} 1 & 0 \\ 9 & 1 \end{pmatrix} \{0, \infty\} \\
&= 2 \cdot \left[\begin{pmatrix} 1 & 0 \\ 9 & 1 \end{pmatrix} \{0, \infty\} \right].
\end{aligned}$$

We compute the convergents of 4/7 in SAGE as follows (note that 0 and ∞ are excluded):

```
sage: convergents(4/7)
[0, 1, 1/2, 4/7]
```

3.3.2. Manin Symbols. As above, fix coset representatives $r_0, \dots, r_m$ for $\Gamma_0(N)$ in $\mathrm{SL}_2(\mathbb{Z})$. Consider formal symbols $[r_i]'$ for $i = 0, \dots, m$. Let $[r_i]$ be the modular symbol $r_i\{0, \infty\} = \{r_i(0), r_i(\infty)\}$. We equip the symbols $[r_0]', \dots, [r_m]'$ with a *right action of* $\mathrm{SL}_2(\mathbb{Z})$, which is given by $[r_i]'.g = [r_j]'$, where $\Gamma_0(N)r_j = \Gamma_0(N)r_i g$. We extend the notation by writing $[\gamma]' = [\Gamma_0(N)\gamma]' = [r_i]'$, where $\gamma \in \Gamma_0(N)r_i$. Then the right action of $\Gamma_0(N)$ is simply $[\gamma]'.g = [\gamma g]'$.

Theorem 1.2 implies that $\mathrm{SL}_2(\mathbb{Z})$ is generated by the two matrices $\sigma = \left(\begin{smallmatrix} 0 & -1 \\ 1 & 0 \end{smallmatrix}\right)$ and $\tau = \left(\begin{smallmatrix} 1 & -1 \\ 1 & 0 \end{smallmatrix}\right)$. Note that $\sigma = S$ from Theorem 1.2 and $\tau = TS$, so $T = \tau\sigma \in \langle \sigma, \tau \rangle$.

The following theorem provides us with a finite presentation for the space $\mathbb{M}_2(\Gamma_0(N))$ of modular symbols.

Theorem 3.13 (Manin)**.** *Consider the quotient M of the free abelian group on Manin symbols $[r_0]', \dots, [r_m]'$ by the subgroup generated by the elements (for all i):*

$$[r_i]' + [r_i]'\sigma \qquad \textit{and} \qquad [r_i]' + [r_i]'\tau + [r_i]'\tau^2,$$

and modulo any torsion. Then there is an isomorphism

$$\Psi : M \xrightarrow{\sim} \mathbb{M}_2(\Gamma_0(N))$$

given by $[r_i]' \mapsto [r_i] = r_i\{0, \infty\}$.

Proof. We will only prove that Ψ is surjective; the proof that Ψ is injective requires much more work and will be omitted from this book (see [**Man72**, §1.7] for a complete proof).

Proposition 3.11 implies that Ψ is surjective, assuming that Ψ is well defined. We next verify that Ψ is well defined, i.e., that the listed 2-term and 3-term relations hold in the image. To see that the first relation holds, note that

$$\begin{aligned} [r_i] + [r_i]\sigma &= \{r_i(0), r_i(\infty)\} + \{r_i\sigma(0), r_i\sigma(\infty)\} \\ &= \{r_i(0), r_i(\infty)\} + \{r_i(\infty), r_i(0)\} \\ &= 0. \end{aligned}$$

For the second relation we have

$$\begin{aligned} [r_i] + [r_i]\tau + [r_i]\tau^2 &= \{r_i(0), r_i(\infty)\} + \{r_i\tau(0), r_i\tau(\infty)\} + \{r_i\tau^2(0), r_i\tau^2(\infty)\} \\ &= \{r_i(0), r_i(\infty)\} + \{r_i(\infty), r_i(1)\} + \{r_i(1), r_i(0)\} \\ &= 0. \end{aligned}$$

□

Example 3.14. By default SAGE computes modular symbols spaces over $\mathbb{Q}$, i.e., $\mathbb{M}_2(\Gamma_0(N);\mathbb{Q}) \cong \mathbb{M}_2(\Gamma_0(N)) \otimes \mathbb{Q}$. SAGE represents (weight 2) Manin symbols as pairs (c, d). Here c, d are integers that satisfy $0 \leq c, d < N$; they define a point $(c : d) \in \mathbb{P}^1(\mathbb{Z}/N\mathbb{Z})$, hence a right coset of $\Gamma_0(N)$ in $\mathrm{SL}_2(\mathbb{Z})$ (see Proposition 3.10).

Create $\mathbb{M}_2(\Gamma_0(N);\mathbb{Q})$ in SAGE by typing `ModularSymbols(N, 2)`. We then use the SAGE command `manin_generators` to enumerate a list of generators $[r_0], \ldots, [r_n]$ as in Theorem 3.13 for several spaces of modular symbols.

```
sage: M = ModularSymbols(2,2)
sage: M
Modular Symbols space of dimension 1 for Gamma_0(2)
of weight 2 with sign 0 over Rational Field
sage: M.manin_generators()
[(0,1), (1,0), (1,1)]

sage: M = ModularSymbols(3,2)
sage: M.manin_generators()
[(0,1), (1,0), (1,1), (1,2)]

sage: M = ModularSymbols(6,2)
sage: M.manin_generators()
[(0,1), (1,0), (1,1), (1,2), (1,3), (1,4), (1,5), (2,1),
 (2,3), (2,5), (3,1), (3,2)]
```

Given `x=(c,d)`, the command `x.lift_to_sl2z(N)` computes an element of $\mathrm{SL}_2(\mathbb{Z})$ whose lower two entries are congruent to (c, d) modulo N.

```
sage: M = ModularSymbols(2,2)
sage: [x.lift_to_sl2z(2) for x in M.manin_generators()]
[[1, 0, 0, 1], [0, -1, 1, 0], [0, -1, 1, 1]]
sage: M = ModularSymbols(6,2)
sage: x = M.manin_generators()[9]
sage: x
(2,5)
sage: x.lift_to_sl2z(6)
[1, 2, 2, 5]
```

The `manin_basis` command returns a list of indices into the Manin generator list such that the corresponding symbols form a basis for the quotient of the $\mathbb{Q}$-vector space spanned by Manin symbols modulo the 2-term and 3-term relations of Theorem 3.13.

```
sage: M = ModularSymbols(2,2)
sage: M.manin_basis()
[1]
sage: [M.manin_generators()[i] for i in M.manin_basis()]
[(1,0)]
sage: M = ModularSymbols(6,2)
sage: M.manin_basis()
[1, 10, 11]
sage: [M.manin_generators()[i] for i in M.manin_basis()]
[(1,0), (3,1), (3,2)]
```

Thus, e.g., every element of $\mathbb{M}_2(\Gamma_0(6))$ is a $\mathbb{Q}$-linear combination of the three symbols $[(1,0)]$, $[(3,1)]$, and $[(3,2)]$. We can write each of these as a modular symbol using the `modular_symbol_rep` function.

```
sage: M.basis()
((1,0), (3,1), (3,2))
sage: [x.modular_symbol_rep() for x in M.basis()]
[{Infinity,0}, {0,1/3}, {-1/2,-1/3}]
```

The `manin_gens_to_basis` function returns a matrix whose rows express each Manin symbol generator in terms of the subset of Manin symbols that forms a basis (as returned by `manin_basis`).

```
sage: M = ModularSymbols(2,2)
sage: M.manin_gens_to_basis()
[-1]
[ 1]
[ 0]
```

Since the basis is $(1,0)$, this means that in $\mathbb{M}_2(\Gamma_0(2);\mathbb{Q})$, we have $[(0,1)] = -[(1,0)]$ and $[(1,1)] = 0$. (Since no denominators are involved, we have in fact computed a presentation of $\mathbb{M}_2(\Gamma_0(2);\mathbb{Z})$.)

To convert a Manin symbol $x = (c,d)$ to an element of a modular symbols space M, use `M(x)`:

```
sage: M = ModularSymbols(2,2)
sage: x = (1,0); M(x)
(1,0)
```

Next consider $\mathbb{M}_2(\Gamma_0(6);\mathbb{Q})$:

```
sage: M = ModularSymbols(6,2)
sage: M.manin_gens_to_basis()
[-1  0  0]
[ 1  0  0]
[ 0  0  0]
[ 0 -1  1]
[ 0 -1  0]
[ 0 -1  1]
[ 0  0  0]
[ 0  1 -1]
[ 0  0 -1]
[ 0  1 -1]
[ 0  1  0]
[ 0  0  1]
```

Recall that our choice of basis for $\mathbb{M}_2(\Gamma_0(6);\mathbb{Q})$ is $[(1,0)], [(3,1)], [(3,2)]$. Thus, e.g., the first row of this matrix says that $[(0,1)] = -[(1,0)]$, and the fourth row asserts that $[(1,2)] = -[(3,1)] + [(3,2)]$.

```
sage: M = ModularSymbols(6,2)
sage: M((0,1))
-(1,0)
sage: M((1,2))
-(3,1) + (3,2)
```

3.4. Hecke Operators

3.4.1. Hecke Operators on Modular Symbols. When p is a prime not dividing N, define

$$T_p(\{\alpha,\beta\}) = \begin{pmatrix} p & 0 \\ 0 & 1 \end{pmatrix}\{\alpha,\beta\} + \sum_{r \bmod p} \begin{pmatrix} 1 & r \\ 0 & p \end{pmatrix}\{\alpha,\beta\}.$$

The Hecke operators are compatible with the integration pairing $\langle\, ,\rangle$ of Section 3.1, in the sense that $\langle fT_p, x\rangle = \langle f, T_p x\rangle$. When $p \mid N$, the definition is the same, except that the matrix $\left(\begin{smallmatrix} p & 0 \\ 0 & 1 \end{smallmatrix}\right)$ is not included in the sum (see Theorem 8.23). There is a similar definition of T_n for n composite (see Section 8.3.1).

Example 3.15. For example, when $N = 11$, we have

$$\begin{aligned} T_2\{0, 1/5\} &= \{0, 2/5\} + \{0, 1/10\} + \{1/2, 3/5\} \\ &= -2\{0, 1/5\}. \end{aligned}$$

See Figure 3.4.1.

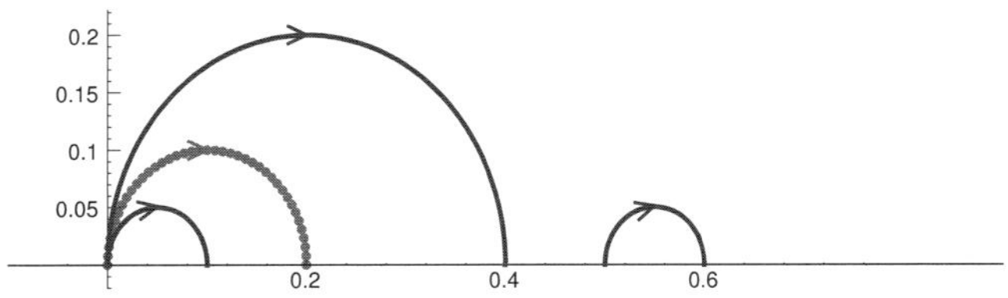

Figure 3.4.1. Image of $\{0, 1/5\}$ under T_2

3.4.2. Hecke Operators on Manin Symbols.

In [**Mer94**], L. Merel gives a description of the action of T_p directly on Manin symbols $[r_i]$ (see Section 8.3.2 for details). For example, when $p = 2$ and N is odd, we have

$$T_2([r_i]) = [r_i]\begin{pmatrix} 1 & 0 \\ 0 & 2 \end{pmatrix} + [r_i]\begin{pmatrix} 2 & 0 \\ 0 & 1 \end{pmatrix} + [r_i]\begin{pmatrix} 2 & 1 \\ 0 & 1 \end{pmatrix} + [r_i]\begin{pmatrix} 1 & 0 \\ 1 & 2 \end{pmatrix}. \tag{3.4.1}$$

For any prime, let C_p be the set of matrices constructed using the following algorithm (see [**Cre97a**, §2.4]):

Algorithm 3.16 (Cremona's Heilbronn Matrices)**.** *Given a prime p, this algorithm outputs a list of 2×2 matrices of determinant p that can be used to compute the Hecke operator T_p.*

(1) Output $\begin{pmatrix} 1 & 0 \\ 0 & p \end{pmatrix}$.

(2) For $r = \left\lceil -\frac{p}{2} \right\rceil, \ldots, \left\lfloor \frac{p}{2} \right\rfloor$:

   (a) Let $x_1 = p$, $x_2 = -r$, $y_1 = 0$, $y_2 = 1$, $a = -p$, $b = r$.

   (b) Output $\begin{pmatrix} x_1 & x_2 \\ y_1 & y_2 \end{pmatrix}$.

   (c) As long as $b \neq 0$, do the following:

      (i) Let q be the integer closest to a/b (if a/b is a half integer, round away from 0).

      (ii) Let $c = a - bq$, $a = -b$, $b = c$.

(iii) Set $x_3 = qx_2 - x_1$, $x_1 = x_2$, $x_2 = x_3$, and
$y_3 = qy_2 - y_1$, $y_1 = y_2$, $y_2 = y_3$.

(iv) Output $\begin{pmatrix} x_1 & x_2 \\ y_1 & y_2 \end{pmatrix}$.

Proposition 3.17 (Cremona, Merel). *Let C_p be as above. Then for $p \nmid N$ and $[x] \in \mathbb{M}_2(\Gamma_0(N))$ a Manin symbol, we have*

$$T_p([x]) = \sum_{g \in C_p} [xg].$$

Proof. See Proposition 2.4.1 of [**Cre97a**]. □

There are other lists of matrices, due to Merel, that work even when $p \mid N$ (see Section 8.3.2).

The command `HeilbronnCremonaList(p)`, for p prime, outputs the list of matrices from Algorithm 3.16.

```
sage: HeilbronnCremonaList(2)
[[1, 0, 0, 2], [2, 0, 0, 1], [2, 1, 0, 1], [1, 0, 1, 2]]
sage: HeilbronnCremonaList(3)
[[1, 0, 0, 3], [3, 1, 0, 1], [1, 0, 1, 3], [3, 0, 0, 1],
 [3, -1, 0, 1], [-1, 0, 1, -3]]
sage: HeilbronnCremonaList(5)
[[1, 0, 0, 5], [5, 2, 0, 1], [2, 1, 1, 3], [1, 0, 3, 5],
 [5, 1, 0, 1], [1, 0, 1, 5], [5, 0, 0, 1], [5, -1, 0, 1],
 [-1, 0, 1, -5], [5, -2, 0, 1], [-2, 1, 1, -3],
 [1, 0, -3, 5]]
sage: len(HeilbronnCremonaList(37))
128
sage: len(HeilbronnCremonaList(389))
1892
sage: len(HeilbronnCremonaList(2003))
11662
```

Example 3.18. We compute the matrix of T_2 on $\mathbb{M}_2(\Gamma_0(2))$:

```
sage: M = ModularSymbols(2,2)
sage: M.T(2).matrix()
[1]
```

Example 3.19. We compute some Hecke operators on $\mathbb{M}_2(\Gamma_0(6))$:

```
sage: M = ModularSymbols(6, 2)
sage: M.T(2).matrix()
[ 2  1 -1]
[-1  0  1]
[-1 -1  2]
sage: M.T(3).matrix()
[3 2 0]
[0 1 0]
[2 2 1]
sage: M.T(3).fcp()  # factored characteristic polynomial
(x - 3) * (x - 1)^2
```

For $p \geq 5$ we have $T_p = p + 1$, since $M_2(\Gamma_0(6))$ is spanned by generalized Eisenstein series (see Chapter 5).

Example 3.20. We compute the Hecke operators on $\mathbb{M}_2(\Gamma_0(39))$:

```
sage: M = ModularSymbols(39, 2)
sage: T2 = M.T(2)
sage: T2.matrix()
[ 3  0 -1  0  0  1  1 -1  0]
[ 0  0  2  0 -1  1  0  1 -1]
[ 0  1  0 -1  1  1  0  1 -1]
[ 0  0  1  0  0  1  0  1 -1]
[ 0 -1  2  0  0  1  0  1 -1]
[ 0  0  1  1  0  1  1 -1  0]
[ 0  0  0 -1  0  1  1  2  0]
[ 0  0  0  1  0  0  2  0  1]
[ 0  0 -1  0  0  0  1  0  2]
sage: T2.fcp()     # factored characteristic polynomial
(x - 3)^3 * (x - 1)^2 * (x^2 + 2*x - 1)^2
```

The Hecke operators commute, so their eigenspace structures are related.

```
sage: T2 = M.T(2).matrix()
sage: T5 = M.T(5).matrix()
sage: T2*T5 - T5*T2 == 0
True
sage: T5.charpoly().factor()
(x^2 - 8)^2 * (x - 6)^3 * (x - 2)^2
```

The decomposition of T_2 is a list of the kernels of $(f^e)(T_2)$, where f runs through the irreducible factors of the characteristic polynomial of T_2 and f^e exactly divides this characteristic polynomial. Using SAGE, we find them:

```
sage: M = ModularSymbols(39, 2)
sage: M.T(2).decomposition()
[
Modular Symbols subspace of dimension 3 of Modular
Symbols space of dimension 9 for Gamma_0(39) of weight
2 with sign 0 over Rational Field,
Modular Symbols subspace of dimension 2 of Modular
Symbols space of dimension 9 for Gamma_0(39) of weight
2 with sign 0 over Rational Field,
Modular Symbols subspace of dimension 4 of Modular
Symbols space of dimension 9 for Gamma_0(39) of weight
2 with sign 0 over Rational Field
]
```

3.5. Computing the Boundary Map

In Section 3.2 we defined a map $\delta : \mathbb{M}_2(\Gamma_0(N)) \to \mathbb{B}_2(\Gamma_0(N))$. The kernel of this map is the space $\mathbb{S}_2(\Gamma_0(N))$ of cuspidal modular symbols. This kernel will be important in computing cusp forms in Section 3.7 below.

To compute the boundary map on $[\gamma]$, note that $[\gamma] = \{\gamma(0), \gamma(\infty)\}$, so if $\gamma = \left(\begin{smallmatrix} a & b \\ c & d \end{smallmatrix}\right)$, then

$$\delta([\gamma]) = \{\gamma(\infty)\} - \{\gamma(0)\} = \{a/c\} - \{b/d\}.$$

Computing this boundary map would appear to first require an algorithm to compute the set $C(\Gamma_0(N)) = \Gamma_0(N)\backslash\mathbb{P}^1(\mathbb{Q})$ of cusps for $\Gamma_0(N)$. In fact, there is a trick that computes the set of cusps in the course of running the algorithm. First, give an algorithm for deciding whether or not two elements of $\mathbb{P}^1(\mathbb{Q})$ are equivalent modulo the action of $\Gamma_0(N)$. Then simply construct $C(\Gamma_0(N))$ in the course of computing the boundary map, i.e., keep a list of cusps found so far, and whenever a new cusp class is discovered, add it to the list. The following proposition, which is proved in [**Cre97a**, Prop. 2.2.3], explains how to determine whether two cusps are equivalent.

Proposition 3.21 (Cremona). *Let (c_i, d_i), $i = 1, 2$, be pairs of integers with $\gcd(c_i, d_i) = 1$ and possibly $d_i = 0$. There is $g \in \Gamma_0(N)$ such that $g(c_1/d_1) = c_2/d_2$ in $\mathbb{P}^1(\mathbb{Q})$ if and only if*

$$s_1 d_2 \equiv s_2 d_1 \pmod{\gcd(d_1 d_2, N)}$$

where s_j *satisfies* $c_j s_j \equiv 1 \pmod{d_j}$.

In SAGE the command `boundary_map()` computes the boundary map from $\mathbb{M}_2(\Gamma_0(N))$ to $\mathbb{B}_2(\Gamma_0(N))$, and the `cuspidal_submodule()` command computes its kernel. For example, for level 2 the boundary map is given by the matrix $[1 \ \ -1]$, and its kernel is the 0 space:

```
sage: M = ModularSymbols(2, 2)
sage: M.boundary_map()
Hecke module morphism boundary map defined by the matrix
[ 1 -1]
Domain: Modular Symbols space of dimension 1 for
Gamma_0(2) of weight ...
Codomain: Space of Boundary Modular Symbols for
Congruence Subgroup Gamma0(2) ...
sage: M.cuspidal_submodule()
Modular Symbols subspace of dimension 0 of Modular
Symbols space of dimension 1 for Gamma_0(2) of weight
2 with sign 0 over Rational Field
```

The smallest level for which the boundary map has nontrivial kernel, i.e., for which $\mathbb{S}_2(\Gamma_0(N)) \neq 0$, is $N = 11$.

```
sage: M = ModularSymbols(11, 2)
sage: M.boundary_map().matrix()
[ 1 -1]
[ 0  0]
[ 0  0]
sage: M.cuspidal_submodule()
Modular Symbols subspace of dimension 2 of Modular
Symbols space of dimension 3 for Gamma_0(11) of weight
2 with sign 0 over Rational Field
sage: S = M.cuspidal_submodule(); S
Modular Symbols subspace of dimension 2 of Modular
Symbols space of dimension 3 for Gamma_0(11) of weight
2 with sign 0 over Rational Field
sage: S.basis()
((1,8), (1,9))
```

The following illustrates that the Hecke operators preserve $\mathbb{S}_2(\Gamma_0(N))$:

```
sage: S.T(2).matrix()
[-2  0]
[ 0 -2]
sage: S.T(3).matrix()
[-1  0]
[ 0 -1]
sage: S.T(5).matrix()
[1 0]
[0 1]
```

A nontrivial fact is that for p prime the eigenvalue of each of these matrices is $p+1-\#E(\mathbb{F}_p)$, where E is the elliptic curve $X_0(11)$ defined by the (affine) equation $y^2+y=x^3-x^2-10x-20$. For example, we have

```
sage: E = EllipticCurve([0,-1,1,-10,-20])
sage: 2 + 1 - E.Np(2)
-2
sage: 3 + 1 - E.Np(3)
-1
sage: 5 + 1 - E.Np(5)
1
sage: 7 + 1 - E.Np(7)
-2
```

The same numbers appear as the eigenvalues of Hecke operators:

```
sage: [S.T(p).matrix()[0,0] for p in [2,3,5,7]]
[-2, -1, 1, -2]
```

In fact, something similar happens for every elliptic curve over $\mathbb{Q}$. The book [**DS05**] (especially Chapter 8) is about this striking numerical relationship between the number of points on elliptic curves modulo p and coefficients of modular forms.

3.6. Computing a Basis for $S_2(\Gamma_0(N))$

This section is about a method for using modular symbols to compute a basis for $S_2(\Gamma_0(N))$. It is not the most efficient for certain applications, but it is easy to explain and understand. See Section 3.7 for a method that takes advantage of additional structure of $S_2(\Gamma_0(N))$.

Let $\mathbb{M}_2(\Gamma_0(N);\mathbb{Q})$ and $\mathbb{S}_2(\Gamma_0(N);\mathbb{Q})$ be the spaces of modular symbols and cuspidal modular symbols over $\mathbb{Q}$. Before we begin, we describe a simple but crucial fact about the relation between cusp forms and Hecke operators.

If $f = \sum b_n q^n \in \mathbb{C}[[q]]$ is a power series, let $a_n(f) = b_n$ be the n coefficient of f. Notice that a_n is a $\mathbb{C}$-linear map $\mathbb{C}[[q]] \to \mathbb{C}$.

As explained in [**DS05**, Prop. 5.3.1] and [**Lan95**, §VII.3] (recall also Proposition 2.31), the Hecke operators T_n act on elements of $M_2(\Gamma_0(N))$ as follows (where $k = 2$ below):

$$T_n\left(\sum_{m=0}^{\infty} a_m q^m\right) = \sum_{m=0}^{\infty}\left(\sum_{1\le d\,|\,\gcd(n,m)} \varepsilon(d)\cdot d^{k-1}\cdot a_{mn/d^2}\right) q^m, \tag{3.6.1}$$

where $\varepsilon(d) = 1$ if $\gcd(d, N) = 1$ and $\varepsilon(d) = 0$ if $\gcd(d, N) \neq 1$. (Note: More generally, if $f \in M_k(\Gamma_1(N))$ is a modular form with Dirichlet character ε, then the above formula holds; above we are considering this formula in the special case when ε is the trivial character and $k = 2$.)

Lemma 3.22. *Suppose $f \in \mathbb{C}[[q]]$ and n is a positive integer. Let T_n be the operator on q-expansions (formal power series) defined by* (3.6.1). *Then*

$$a_1(T_n(f)) = a_n(f).$$

Proof. The coefficient of q in (3.6.1) is $\varepsilon(1)\cdot 1\cdot a_{1\cdot n/1^2} = a_n$. □

The *Hecke algebra* $\mathbb{T}$ is the ring generated by all Hecke operators T_n acting on $M_k(\Gamma_1(N))$. Let $\mathbb{T}'$ denote the image of the Hecke algebra in $\mathrm{End}(S_2(\Gamma_0(N)))$, and let $\mathbb{T}'_{\mathbb{C}} = \mathbb{T}'\otimes_{\mathbb{Z}}\mathbb{C}$ be the $\mathbb{C}$-span of the Hecke operators. Let $\tilde{\mathbb{T}}_{\mathbb{C}}$ denote the subring of $\mathrm{End}(\mathbb{C}[[q]])$ generated over $\mathbb{C}$ by all Hecke operators acting on formal power series via definition (3.6.1).

Proposition 3.23. *There is a bilinear pairing of complex vector spaces*

$$\mathbb{C}[[q]] \times \tilde{\mathbb{T}}_{\mathbb{C}} \to \mathbb{C}$$

given by

$$\langle f, t\rangle = a_1(t(f)).$$

If f is such that $\langle f, t\rangle = 0$ for all $t \in \tilde{\mathbb{T}}_{\mathbb{C}}$, then $f = 0$.

Proof. The pairing is bilinear since both t and a_1 are linear.

Suppose $f \in \mathbb{C}[[q]]$ is such that $\langle f, t\rangle = 0$ for all $t \in \tilde{\mathbb{T}}_{\mathbb{C}}$. Then $\langle f, T_n\rangle = 0$ for each positive integer n. But by Lemma 3.22 we have

$$a_n(f) = a_1(T_n(f)) = 0$$

for all n; thus $f = 0$. □

Proposition 3.24. *There is a perfect bilinear pairing of complex vector spaces*

$$S_2(\Gamma_0(N)) \times \mathbb{T}'_{\mathbb{C}} \to \mathbb{C}$$

given by

$$\langle f, t\rangle = a_1(t(f)).$$

Proof. The pairing has 0 kernel on the left by Proposition 3.23. Suppose that $t \in \mathbb{T}'_{\mathbb{C}}$ is such that $\langle f, t\rangle = 0$ for all $f \in S_2(\Gamma_0(N))$. Then $a_1(t(f)) = 0$ for all f. For any n, the image $T_n(f)$ is also a cusp form, so $a_1(t(T_n(f))) = 0$ for all n and f. Finally the fact that $\mathbb{T}'$ is commutative and Lemma 3.22 together imply that for all n and f,

$$0 = a_1(t(T_n(f))) = a_1(T_n(t(f))) = a_n(t(f)),$$

so $t(f) = 0$ for all f. Thus t is the 0 operator.

Since $S_2(\Gamma_0(N))$ has finite dimension and the kernel on each side of the pairing is 0, it follows that the pairing is perfect, i.e., defines an *isomorphism*

$$\mathbb{T}'_{\mathbb{C}} \cong \mathrm{Hom}_{\mathbb{C}}(S_2(\Gamma_0(N)); \mathbb{C}).$$

□

By Proposition 3.24 there is an isomorphism of vector spaces

$$\Psi : S_2(\Gamma_0(N)) \xrightarrow{\cong} \mathrm{Hom}(\mathbb{T}'_{\mathbb{C}}, \mathbb{C}) \tag{3.6.2}$$

that sends $f \in S_2(\Gamma_0(N))$ to the homomorphism

$$t \mapsto a_1(t(f)).$$

For any $\mathbb{C}$-linear map $\varphi : \mathbb{T}'_{\mathbb{C}} \to \mathbb{C}$, let

$$f_\varphi = \sum_{n=1}^{\infty} \varphi(T_n) q^n \in \mathbb{C}[[q]].$$

Lemma 3.25. *The series f_φ is the q-expansion of $\Psi^{-1}(\varphi) \in S_2(\Gamma_0(N))$.*

Proof. Note that it is not even *a priori* obvious that f_φ is the q-expansion of a modular form. Let $g = \Psi^{-1}(\varphi)$, which is by definition the unique element of $S_2(\Gamma_0(N))$ such that $\langle g, T_n\rangle = \varphi(T_n)$ for all n. By Lemma 3.22, we have

$$\langle f_\varphi, T_n\rangle = a_1(T_n(f_\varphi)) = a_n(f_\varphi) = \varphi(T_n),$$

so $\langle f_\varphi - g, T_n\rangle = 0$ for all n. Proposition 3.23 implies that $f_\varphi - g = 0$, so $f_\varphi = g = \Psi^{-1}(\varphi)$, as claimed. □

Conclusion: The cusp forms f_φ, as φ varies through a basis of $\mathrm{Hom}(\mathbb{T}'_{\mathbb{C}}, \mathbb{C})$, form a basis for $S_2(\Gamma_0(N))$. In particular, we can compute $S_2(\Gamma_0(N))$ by computing $\mathrm{Hom}(\mathbb{T}'_{\mathbb{C}}, \mathbb{C})$, where we compute $\mathbb{T}'$ in any way we want, e.g., using a space that contains an isomorphic copy of $S_2(\Gamma_0(N))$.

Algorithm 3.26 (Basis of Cusp Forms). *Given positive integers N and B, this algorithm computes a basis for $S_2(\Gamma_0(N))$ to precision $O(q^B)$.*

(1) Compute $\mathbb{M}_2(\Gamma_0(N);\mathbb{Q})$ via the presentation of Section 3.3.2.
(2) Compute the subspace $\mathbb{S}_2(\Gamma_0(N);\mathbb{Q})$ of cuspidal modular symbols as in Section 3.5.
(3) Let $d = \frac{1}{2}\cdot\dim \mathbb{S}_2(\Gamma_0(N);\mathbb{Q})$. By Proposition 3.8, d is the dimension of $S_2(\Gamma_0(N))$.
(4) Let $[T_n]$ denote the matrix of T_n acting on a basis of $\mathbb{S}_2(\Gamma_0(N);\mathbb{Q})$. For a matrix A, let $a_{ij}(A)$ denote the ijth entry of A. For various integers i, j with $0 \leq i, j \leq d-1$, compute formal q-expansions

$$f_{ij}(q) = \sum_{n=1}^{B-1} a_{ij}([T_n])q^n + O(q^B) \in \mathbb{Q}[[q]]$$

until we find enough to span a space of dimension d (or exhaust all of them). These f_{ij} are a basis for $S_2(\Gamma_0(N))$ to precision $O(q^B)$.

3.6.1. Examples.

We use SAGE to demonstrate Algorithm 3.26.

Example 3.27. The smallest N with $S_2(\Gamma_0(N)) \neq 0$ is $N = 11$.

```
sage: M = ModularSymbols(11); M.basis()
((1,0), (1,8), (1,9))
sage: S = M.cuspidal_submodule(); S
Modular Symbols subspace of dimension 2 of Modular
Symbols space of dimension 3 for Gamma_0(11) of weight
2 with sign 0 over Rational Field
```

We compute a few Hecke operators, and then read off a nonzero cusp form, which forms a basis for $S_2(\Gamma_0(11))$:

```
sage: S.T(2).matrix()
[-2  0]
[ 0 -2]
sage: S.T(3).matrix()
[-1  0]
[ 0 -1]
```

Thus

$$f_{0,0} = q - 2q^2 - q^3 + \cdots \in S_2(\Gamma_0(11))$$

forms a basis for $S_2(\Gamma_0(11))$.

Example 3.28. We compute a basis for $S_2(\Gamma_0(33))$ to precision $O(q^6)$.

```
sage: M = ModularSymbols(33)
sage: S = M.cuspidal_submodule(); S
Modular Symbols subspace of dimension 6 of Modular
Symbols space of dimension 9 for Gamma_0(33) of weight
2 with sign 0 over Rational Field
```

Thus $\dim S_2(\Gamma_0(33)) = 3$.

```
sage: R.<q> = PowerSeriesRing(QQ)
sage: v = [S.T(n).matrix()[0,0] for n in range(1,6)]
sage: f00 = sum(v[n-1]*q^n for n in range(1,6)) + O(q^6)
sage: f00
q - q^2 - q^3 + q^4 + O(q^6)
```

This gives us one basis element of $S_2(\Gamma_0(33))$. It remains to find two others. We find

```
sage: v = [S.T(n).matrix()[0,1] for n in range(1,6)]
sage: f01 = sum(v[n-1]*q^n for n in range(1,6)) + O(q^6)
sage: f01
-2*q^3 + O(q^6)
```

and

```
sage: v = [S.T(n).matrix()[1,0] for n in range(1,6)]
sage: f10 = sum(v[n-1]*q^n for n in range(1,6)) + O(q^6)
sage: f10
q^3 + O(q^6)
```

This third one is (to our precision) a scalar multiple of the second, so we look further.

```
sage: v = [S.T(n).matrix()[1,1] for n in range(1,6)]
sage: f11 = sum(v[n-1]*q^n for n in range(1,6)) + O(q^6)
sage: f11
q - 2*q^2 + 2*q^4 + q^5 + O(q^6)
```

This latter form is clearly not in the span of the first two. Thus we have the following basis for $S_2(\Gamma_0(33))$ (to precision $O(q^6)$):

$$
\begin{aligned}
f_{00} &= q - q^2 - q^3 + q^4 + \cdots, \\
f_{11} &= q - 2q^2 + 2q^4 + q^5 + \cdots, \\
f_{10} &= q^3 + \cdots.
\end{aligned}
$$

Example 3.29. Next consider $N = 23$, where we have

$$d = \dim S_2(\Gamma_0(23)) = 2.$$

The command `q_expansion_cuspforms` computes matrices T_n and returns a function f such that $f(i,j)$ is the q-expansion of $f_{i,j}$ to some precision. (For efficiency reasons, $f(i,j)$ in SAGE actually computes matrices of T_n acting on a basis for the linear dual of $\mathbb{S}_2(\Gamma_0(N))$.)

```
sage: M = ModularSymbols(23)
sage: S = M.cuspidal_submodule()
sage: S
Modular Symbols subspace of dimension 4 of Modular
Symbols space of dimension 5 for Gamma_0(23) of weight
2 with sign 0 over Rational Field
sage: f = S.q_expansion_cuspforms(6)
sage: f(0,0)
q - 2/3*q^2 + 1/3*q^3 - 1/3*q^4 - 4/3*q^5 + O(q^6)
sage: f(0,1)
O(q^6)
sage: f(1,0)
-1/3*q^2 + 2/3*q^3 + 1/3*q^4 - 2/3*q^5 + O(q^6)
```

Thus a basis for $S_2(\Gamma_0(23))$ is

$$f_{0,0} = q - \frac{2}{3}q^2 + \frac{1}{3}q^3 - \frac{1}{3}q^4 - \frac{4}{3}q^5 + \cdots,$$
$$f_{1,0} = -\frac{1}{3}q^2 + \frac{2}{3}q^3 + \frac{1}{3}q^4 - \frac{2}{3}q^5 + \cdots.$$

Or, in echelon form,

$$q - q^3 - q^4 + \cdots$$
$$q^2 - 2q^3 - q^4 + 2q^5 + \cdots$$

which we computed using

```
sage: S.q_expansion_basis(6)
[
q - q^3 - q^4 + O(q^6),
q^2 - 2*q^3 - q^4 + 2*q^5 + O(q^6)
]
```

3.7. Computing $S_2(\Gamma_0(N))$ Using Eigenvectors

In this section we describe how to use modular symbols to construct a basis of $S_2(\Gamma_0(N))$ consisting of modular forms that are eigenvectors for every element of the ring $\mathbb{T}^{(N)}$ generated by the Hecke operator T_p, with $p \nmid N$. Such eigenvectors are called *eigenforms*.

Suppose M is a positive integer that divides N. As explained in [**Lan95**, VIII.1–2], for each divisor d of N/M there is a natural *degeneracy map* $\alpha_{M,d}$: $S_2(\Gamma_0(M)) \to S_2(\Gamma_0(N))$ given by $\alpha_{M,d}(f(q)) = f(q^d)$. The *new subspace* of $S_2(\Gamma_0(N))$, denoted $S_2(\Gamma_0(N))_{\text{new}}$, is the complementary $\mathbb{T}$-submodule of the $\mathbb{T}$-module generated by the images of all maps $\alpha_{M,d}$, with M and d as above. It is a nontrivial fact that this complement is well defined; one possible proof uses the Petersson inner product (see [**Lan95**, §VII.5]).

The theory of Atkin and Lehner [**AL70**] (see Theorem 9.4 below) asserts that, as a $\mathbb{T}^{(N)}$-module, $S_2(\Gamma_0(N))$ decomposes as follows:

$$S_2(\Gamma_0(N)) \quad = \bigoplus_{M|N,\ d|N/M} \beta_{M,d}(S_2(\Gamma_0(M))_{\text{new}}).$$

To compute $S_2(\Gamma_0(N))$ it suffices to compute $S_2(\Gamma_0(M))_{\text{new}}$ for each $M \mid N$.

We now turn to the problem of computing $S_2(\Gamma_0(N))_{\text{new}}$. Atkin and Lehner [**AL70**] proved that $S_2(\Gamma_0(N))_{\text{new}}$ is spanned by eigenforms for all T_p with $p \nmid N$ and that the common eigenspaces of all the T_p with $p \nmid N$ each have dimension 1. Moreover, if $f \in S_2(\Gamma_0(N))_{\text{new}}$ is an eigenform then the coefficient of q in the q-expansion of f is nonzero, , o it is possible to normalize f so the coefficient of q is 1 (such a *normalized* eigenform in the new subspace is called a *newform*). With f so normalized, if $T_p(f) = a_p f$, then the pth Fourier coefficient of f is a_p. If $f = \sum_{n=1}^{\infty} a_n q^n$ is a normalized eigenvector for all T_p, then the a_n, with n composite, are determined by the a_p, with p prime, by the following formulas: $a_{nm} = a_n a_m$ when n and m are relatively prime and $a_{p^r} = a_{p^{r-1}} a_p - p a_{p^{r-2}}$ for $p \nmid N$ prime. When $p \mid N$, $a_{p^r} = a_p^r$. We conclude that in order to compute $S_2(\Gamma_0(N))_{\text{new}}$, it suffices to compute all systems of eigenvalues $\{a_2, a_3, a_5, \ldots\}$ of the prime-indexed Hecke operators $T_2, T_3, T_5, \ldots$ acting on $S_2(\Gamma_0(N))_{\text{new}}$. Given a system of eigenvalues, the corresponding eigenform is $f = \sum_{n=1}^{\infty} a_n q^n$, where the a_n, for n composite, are determined by the recurrence given above.

In light of the pairing $\langle\, ,\, \rangle$ introduced in Section 3.1, computing the above systems of eigenvalues $\{a_2, a_3, a_5, \ldots\}$ amounts to computing the systems of eigenvalues of the Hecke operators T_p on the subspace V of $\mathbb{S}_2(\Gamma_0(N))$ that corresponds to the new subspace of $S_2(\Gamma_0(N))$. For each proper divisor M of N and each divisor d of N/M, let $\phi_{M,d} : \mathbb{S}_2(\Gamma_0(N)) \to \mathbb{S}_2(\Gamma_0(M))$ be the map sending x to $\left(\begin{smallmatrix} d & 0 \\ 0 & 1 \end{smallmatrix}\right) x$. Then V is the intersection of the kernels of all maps $\phi_{M,d}$.

Computing the systems of eigenvalues of a collection of commuting diagonalizable endomorphisms is a problem in linear algebra (see Chapter 7).

Example 3.30. All forms in $S_2(\Gamma_0(39))$ are new. Up to Galois conjugacy, the eigenvalues of the Hecke operators T_2, T_3, T_5, and T_7 on $\mathbb{S}_2(\Gamma_0(39))$ are $\{1,-1,2,-4\}$ and $\{a,1,-2a-2,2a+2\}$, where $a^2+2a-1=0$. Each of these eigenvalues occur in $\mathbb{S}_2(\Gamma_0(39))$ with multiplicity two; for example, the characteristic polynomial of T_2 on $\mathbb{S}_2(\Gamma_0(39))$ is $(x-1)^2\cdot(x^2+2x-1)^2$. Thus $S_2(\Gamma_0(39))$ is spanned by

$$f_1 = q+q^2-q^3-q^4+2q^5-q^6-4q^7+\cdots,$$
$$f_2 = q+aq^2+q^3+(-2a-1)q^4+(-2a-2)q^5+aq^6+(2a+2)q^7+\cdots,$$
$$f_3 = q+\sigma(a)q^2+q^3+(-2\sigma(a)-1)q^4+(-2\sigma(a)-2)q^5+\sigma(a)q^6+\cdots,$$

where $\sigma(a)$ is the other $\mathrm{Gal}(\overline{\mathbb{Q}}/\mathbb{Q})$-conjugate of a.

3.7.1. Summary. To compute the q-expansion of a basis for $S_2(\Gamma_0(N))$, we use the degeneracy maps so that we only have to solve the problem for $S_2(\Gamma_0(M))_{\mathrm{new}}$, for all integers $M \mid N$. Using modular symbols, we compute all systems of eigenvalues $\{a_2,a_3,a_5,\ldots\}$, and then write down the corresponding eigenforms $\sum a_n q^n$.

3.8. Exercises

3.1 Suppose that $\lambda,\lambda'\in\mathfrak{h}$ are in the same orbit for the action of $\Gamma_0(N)$, i.e., that there exists $g\in\Gamma_0(N)$ such that $g(\lambda)=\lambda'$. Let $\Lambda=\mathbb{Z}+\mathbb{Z}\lambda$ and $\Lambda'=\mathbb{Z}+\mathbb{Z}\lambda'$. Prove that the pairs $(\mathbb{C}/\Lambda,(\frac{1}{N}\mathbb{Z}+\Lambda)/\Lambda)$ and $(\mathbb{C}/\Lambda',(\frac{1}{N}\mathbb{Z}+\Lambda')/\Lambda')$ are isomorphic. (By an isomorphism $(E,C)\to(F,D)$ of pairs, we mean an isomorphism $\phi: E\to F$ of elliptic curves that sends C to D. You may use the fact that an isomorphism of elliptic curves over $\mathbb{C}$ is a $\mathbb{C}$-linear map $\mathbb{C}\to\mathbb{C}$ that sends the lattice corresponding to one curve onto the lattice corresponding to the other.)

3.2 Let n,m be integers and N a positive integer. Prove that the modular symbol $\{n,m\}$ is 0 as an element of $\mathbb{M}_2(\Gamma_0(N))$. [Hint: See Example 3.6.]

3.3 Let p be a prime.
   (a) List representative elements of $\mathbb{P}^1(\mathbb{Z}/p\mathbb{Z})$.
   (b) What is the cardinality of $\mathbb{P}^1(\mathbb{Z}/p\mathbb{Z})$ as a function of p?
   (c) Prove that there is a bijection between the right cosets of $\Gamma_0(p)$ in $\mathrm{SL}_2(\mathbb{Z})$ and the elements of $\mathbb{P}^1(\mathbb{Z}/p\mathbb{Z})$ that sends $\left(\begin{smallmatrix} a & b \\ c & d\end{smallmatrix}\right)$ to $(c:d)$. (As mentioned in this chapter, the analogous statement is also true when the level is composite; see [**Cre97a**, §2.2] for complete details.)

3.4 Use the inductive proof of Proposition 3.11 to write $\{0,4/7\}$ in terms of Manin symbols for $\Gamma_0(7)$.

3.5 Show that the Hecke operator T_2 acts as multiplication by 3 on the space $\mathbb{M}_2(\Gamma_0(3))$ as follows:

(a) Write down right coset representatives for $\Gamma_0(3)$ in $\mathrm{SL}_2(\mathbb{Z})$.

(b) List all eight relations coming from Theorem 3.13.

(c) Find a single Manin symbols $[r_i]$ so that the three other Manin symbols are a nonzero multiple of $[r_i]$ modulo the relations found in the previous step.

(d) Use formula (3.4.1) to compute $T_2([r_i])$. You will obtain a sum of four symbols. Using the relations above, write this sum as a multiple of $[r_i]$. (The multiple must be 3 or you made a mistake.)

Chapter 4

Dirichlet Characters

In this chapter we develop a theory for computing with Dirichlet characters, which are extremely important to computations with modular forms for (at least) two reasons:

(1) To compute the Eisenstein subspace $E_k(\Gamma_1(N))$ of $M_k(\Gamma_1(N))$, we write down Eisenstein series attached to pairs of Dirichlet characters (the space $E_k(\Gamma_1(N))$ will be defined in Chapter 5).

(2) To compute $S_k(\Gamma_1(N))$, we instead compute a decomposition

$$M_k(\Gamma_1(N)) = \bigoplus M_k(\Gamma_1(N), \varepsilon)$$

and then compute each factor (see Section 9.1). Here the sum is over all Dirichlet characters ε of modulus N.

Dirichlet characters appear frequently in many other areas of number theory. For example, by the Kronecker-Weber theorem, Dirichlet characters correspond to the 1-dimensional representations of $\mathrm{Gal}(\overline{\mathbb{Q}}/\mathbb{Q})$.

After defining Dirichlet characters in Section 4.1, in Section 4.2 we describe a good way to represent Dirichlet characters using a computer. Section 4.3 is about how to evaluate Dirichlet characters and leads naturally to a discussion of the baby-step giant-step algorithm for solving the discrete log problem and methods for efficiently computing the Kronecker symbol. In Section 4.4 we explain how to factor Dirichlet characters into their prime power constituents and apply this to the computations of conductors. We describe how to carry out a number of standard operations with Dirichlet characters in Section 4.6 and discuss alternative ways to represent them in Section 4.7. Finally, in Section 4.8 we give a very short tutorial about how to compute with Dirichlet characters using SAGE.

4.1. The Definition

Fix an integral domain R and a root ζ of unity in R.

Definition 4.1 (Dirichlet Character). A *Dirichlet character* of modulus N over R is a map $\varepsilon : \mathbb{Z} \to R$ such that there is a homomorphism $f : (\mathbb{Z}/N\mathbb{Z})^* \to \langle\zeta\rangle$ for which

$$\varepsilon(a) = \begin{cases} 0 & \text{if } \gcd(a, N) > 1, \\ f\,(a \bmod N) & \text{if } \gcd(a, N) = 1. \end{cases}$$

We denote the group of such Dirichlet characters by $D(N, R)$. Note that elements of $D(N, R)$ are in bijection with homomorphisms $(\mathbb{Z}/N\mathbb{Z})^* \to \langle\zeta\rangle$.

A familiar Dirichlet character is the Legendre symbol $\left(\frac{a}{p}\right)$, with p an odd prime, that appears in quadratic reciprocity theory. It is a Dirichlet character of modulus p that takes the value 1 on integers that are congruent to a nonzero square modulo p, the value -1 on integers that are congruent to a nonzero nonsquare modulo p, and 0 on integers divisible by p.

4.2. Representing Dirichlet Characters

Lemma 4.2. *The groups $(\mathbb{Z}/N\mathbb{Z})^*$ and $D(N, \mathbb{C})$ are isomorphic.*

Proof. We prove the more general fact that for any finite abelian group G, we have that $G \approx \mathrm{Hom}(G, \mathbb{C}^*)$. To deduce this latter isomorphism, first reduce to the case when G is cyclic by writing G as a product of cyclic groups. The cyclic case follows because if G is cyclic of order n, then $\mathbb{C}^*$ contains an nth root of unity, so $\mathrm{Hom}(G, \mathbb{C}^*)$ is also cyclic of order n. Any two cyclic groups of the same order are isomorphic, so G and $\mathrm{Hom}(G, \mathbb{C}^*)$ are isomorphic. □

Corollary 4.3. *We have $\#D(N, R) \mid \varphi(N)$, with equality if and only if the order of our choice of $\zeta \in R$ is a multiple of the exponent of the group $(\mathbb{Z}/N\mathbb{Z})^*$.*

Proof. This is because $\#(\mathbb{Z}/N\mathbb{Z})^* = \varphi(N)$. □

Fix a positive integer N. To find the set of "canonical" generators for the group $(Z/NZ)^*$, write $N = \prod_{i=0}^{n} p_i^{e_i}$ where $p_0 < p_1 < \cdots < p_n$ are the prime divisors of N. By Exercise 4.2, each factor $(\mathbb{Z}/p_i^{e_i}\mathbb{Z})^*$ is a cyclic group $C_i = \langle g_i \rangle$, except if $p_0 = 2$ and $e_0 \geq 3$, in which case $(\mathbb{Z}/p_0^{e_0}\mathbb{Z})^*$ is a product of the cyclic subgroup $C_0 = \langle -1 \rangle$ of order 2 with the cyclic subgroup $C_1 = \langle 5 \rangle$. In all cases we have

$$(\mathbb{Z}/N\mathbb{Z})^* \cong \prod_{0 \leq i \leq n} C_i = \prod_{0 \leq i \leq n} \langle g_i \rangle.$$

For i such that $p_i > 2$, choose the generator g_i of C_i to be the element of $\{2, 3, \ldots, p_i^{e_i} - 1\}$ that is smallest and generates. Finally, use the Chinese Remainder Theorem (see [**Coh93**, §1.3.3]) to lift each g_i to an element in $(\mathbb{Z}/N\mathbb{Z})^*$, also denoted g_i, that is 1 modulo each $p_j^{e_j}$ for $j \neq i$.

Algorithm 4.4 (Minimal Generator for $(\mathbb{Z}/p^r\mathbb{Z})^*$). *Given a prime power p^r with p odd, this algorithm computes the minimal generator of $(\mathbb{Z}/p^r\mathbb{Z})^*$.*

(1) [Factor Group Order] Factor $n = \phi(p^r) = p^{r-1} \cdot 2 \cdot ((p-1)/2)$ as a product $\prod p_i^{e_i}$ of primes. This is equivalent in difficulty to factoring $(p-1)/2$. (See, e.g., [**Coh93**, Ch.8, Ch. 10] for an excellent discussion of factorization algorithms, though of course much progress has been made since then.)
(2) [Initialize] Set $g = 2$.
(3) [Generator?] Using the binary powering algorithm (see [**Coh93**, §1.2]), compute $g^{n/p_i} \pmod{p^r}$, for each prime divisor p_i of n. If any of these powers are 1, then g is not a generator, so set $g = g+1$ and go to step (2). If no powers are 1, output g and terminate.

See Exercise 4.3 for a proof that this algorithm is correct.

Example 4.5. A minimal generator for $(\mathbb{Z}/49\mathbb{Z})^*$ is 3. We have $n = \varphi(49) = 42 = 2 \cdot 3 \cdot 7$ and

$$2^{n/2} \equiv 1, \qquad 2^{n/3} \equiv 18, \qquad 2^{n/7} \equiv 15 \pmod{49},$$

so 2 is not a generator for $(\mathbb{Z}/49\mathbb{Z})^*$. (We see this just from $2^{n/2} \equiv 1 \pmod{49}$.) However 3 is a generator since

$$3^{n/2} \equiv 48, \qquad 3^{n/3} \equiv 30, \qquad 3^{n/7} \equiv 43 \pmod{49}.$$

Example 4.6. In this example we compute minimal generators for $N = 25$, 100, and 200:

(1) The minimal generator for $(\mathbb{Z}/25\mathbb{Z})^*$ is 2.
(2) The minimal generators for $(\mathbb{Z}/100\mathbb{Z})^*$, lifted to numbers modulo 100, are $g_0 = 51$ and $g_1 = 77$. Notice that $g_0 \equiv -1 \pmod 4$ and $g_0 \equiv 1 \pmod{25}$ and that $g_1 \equiv 2 \pmod{25}$ is the minimal generator modulo 25.
(3) The minimal generators for $(\mathbb{Z}/200\mathbb{Z})^*$, lifted to numbers modulo 200, are $g_0 = 151$, $g_1 = 101$, and $g_2 = 177$. Note that $g_0 \equiv -1 \pmod 4$, that $g_1 \equiv 5 \pmod 8$ and $g_2 \equiv 2 \pmod{25}$.

In SAGE, the command `Integers(N)` creates $\mathbb{Z}/N\mathbb{Z}$.

```
sage: R = Integers(49)
sage: R
Ring of integers modulo 49
```

The `unit_gens` command computes the minimal generators for $(\mathbb{Z}/N\mathbb{Z})^*$, as defined above.

```
sage: R.unit_gens()
[3]
sage: Integers(25).unit_gens()
[2]
sage: Integers(100).unit_gens()
[51, 77]
sage: Integers(200).unit_gens()
[151, 101, 177]
sage: Integers(2005).unit_gens()
[402, 1206]
sage: Integers(200000000).unit_gens()
[174218751, 51562501, 187109377]
```

Fix an element ζ of finite multiplicative order in a ring R, and let $D(N, R)$ denote the group of Dirichlet characters of modulus N over R, with image in $\langle\zeta\rangle \cup \{0\}$. In most of this chapter, we specify an element $\varepsilon \in D(N, R)$ by giving the list

$$[\varepsilon(g_0), \varepsilon(g_1), \ldots, \varepsilon(g_n)] \tag{4.2.1}$$

of images of the generators of $(\mathbb{Z}/N\mathbb{Z})^*$. (Note that if N is even, the number of elements of the list (4.2.1) *does* depend on whether or not $8 \mid N$—there are two factors corresponding to 2 if $8 \mid N$, but only one if $8 \nmid N$.) This representation completely determines ε and is convenient for arithmetic operations. It is analogous to representing a linear transformation by a matrix.

Remark 4.7. In any actual implementation (e.g., the one in SAGE), it is better to represent the $\varepsilon(g_i)$ by recording an integer j such that $\varepsilon(g_i) = \zeta^j$, where $\zeta \in R$ is a fixed root of unity. Then (4.2.1) is internally represented as an element of $(\mathbb{Z}/m\mathbb{Z})^{n+1}$, where m is the multiplicative order of ζ. When the representation of (4.2.1) is needed for an algorithm, it can be quickly computed on the fly using a table of the powers of ζ. See Section 4.7 for further discussion about ways to represent characters.

Example 4.8. The group $D(5, \mathbb{C})$ has elements $\{[1], [i], [-1], [-i]\}$, so it is cyclic of order $\varphi(5) = 4$. In contrast, the group $D(5, \mathbb{Q})$ has only the two elements [1] and [−1] and order 2. The command `DirichletGroup(N)` with no second argument creates the group of Dirichlet characters with values in the cyclotomic field $\mathbb{Q}(\zeta_n)$, where n is the exponent of the group $(\mathbb{Z}/N\mathbb{Z})^*$. Every element in $D(N, \mathbb{C})$ takes values in $\mathbb{Q}(\zeta_n)$, so $D(N, \mathbb{Q}(\zeta_n)) \approx D(N, \mathbb{C})$.

```
sage: list(DirichletGroup(5))
[[1], [zeta4], [-1], [-zeta4]]
sage: list(DirichletGroup(5, QQ))
[[1], [-1]]
```

4.3. Evaluation of Dirichlet Characters

This section is about how to compute $\varepsilon(n)$, where ε is a Dirichlet character and n is an integer. We begin with an example.

Example 4.9. If $N = 200$, then $g_0 = 151$, $g_1 = 101$ and $g_2 = 177$, as we saw in Example 4.6. The exponent of $(\mathbb{Z}/200\mathbb{Z})^*$ is 20, since that is the least common multiple of the exponents of $4 = \#(\mathbb{Z}/8\mathbb{Z})^*$ and $20 = \#(\mathbb{Z}/25\mathbb{Z})^*$. The orders of g_0, g_1, and g_2 are 2, 2, and 20. Let $\zeta = \zeta_{20}$ be a primitive 20th root of unity in $\mathbb{C}$. Then the following are generators for $D(200, \mathbb{C})$:

$$\varepsilon_0 = [-1, 1, 1], \qquad \varepsilon_1 = [1, -1, 1], \qquad \varepsilon_2 = [1, 1, \zeta],$$

and $\varepsilon = [1, -1, \zeta^5]$ is an example element of order 4. To evaluate $\varepsilon(3)$, we write 3 in terms of g_0, g_1, and g_2. First, reducing 3 modulo 8, we see that $3 \equiv g_0 \cdot g_1 \pmod 8$. Next reducing 3 modulo 25 and trying powers of $g_2 = 2$, we find that $e \equiv g_2^7 \pmod{25}$. Thus

$$\begin{aligned}
\varepsilon(3) &= \varepsilon(g_0 \cdot g_1 \cdot g_2^7) \\
&= \varepsilon(g_0)\varepsilon(g_1)\varepsilon(g_2)^7 \\
&= 1 \cdot (-1) \cdot (\zeta^5)^7 \\
&= -\zeta^{35} = -\zeta^{15}.
\end{aligned}$$

We next illustrate the above computation of $\varepsilon(3)$ in SAGE. First we make the group $D(200, \mathbb{Q}(\zeta_8))$ and list its generators.

```
sage: G = DirichletGroup(200)
sage: G
Group of Dirichlet characters of modulus 200 over
Cyclotomic Field of order 20 and degree 8
sage: G.exponent()
20
sage: G.gens()
([-1, 1, 1], [1, -1, 1], [1, 1, zeta20])
```

We construct ε.

```
sage: K = G.base_ring()
sage: zeta = K.0
sage: eps = G([1,-1,zeta^5])
sage: eps
[1, -1, zeta20^5]
```

Finally, we evaluate ε at 3.

```
sage: eps(3)
zeta20^5
sage: -zeta^15
zeta20^5
```

Example 4.9 illustrates that if ε is represented using a list as described above, evaluation of ε is inefficient without extra information; it requires solving the discrete log problem in $(\mathbb{Z}/N\mathbb{Z})^*$.

Remark 4.10. For a general character ε, is calculation of ε at least as hard as finding discrete logarithms? Quadratic characters are easier—see Algorithm 4.23.

Algorithm 4.11 (Evaluate ε). *Given a Dirichlet character ε of modulus N, represented by a list $[\varepsilon(g_0), \varepsilon(g_1), \ldots, \varepsilon(g_n)]$, and an integer a, this algorithm computes $\varepsilon(a)$.*

(1) [GCD] Compute $g = \gcd(a, N)$. If $g > 1$, output 0 and terminate.
(2) [Discrete Log] For each i, write $a \pmod{p_i^{e_i}}$ as a power m_i of g_i using some algorithm for solving the discrete log problem (see below). If $p_i = 2$, write $a \pmod{p_i^{e_i}}$ as $(-1)^{m_0} \cdot 5^{m_1}$. (This step is analogous to writing a vector in terms of a basis.)

(3) [Multiply] Output $\prod \varepsilon(g_i)^{m_i}$ as an element of R, and terminate. (This is analogous to multiplying a matrix times a vector.)

4.3.1. The Discrete Log Problem. Exercise 4.4 gives an isomorphism of groups

$$(1 + p^{n-1}(\mathbb{Z}/p^n\mathbb{Z}), \times) \cong (\mathbb{Z}/p\mathbb{Z}, +),$$

so one sees by induction that step (2) is "about as difficult" as finding a discrete log in $(\mathbb{Z}/p\mathbb{Z})^*$. There is an algorithm called "baby-step giant-step", which solves the discrete log problem in $(\mathbb{Z}/p\mathbb{Z})^*$ in time $O(\sqrt{\ell})$, where ℓ is the largest prime factor of $p - 1 = \#(\mathbb{Z}/p\mathbb{Z})^*$ (note that the discrete log problem in $(\mathbb{Z}/p\mathbb{Z})^*$ reduces to a series of discrete log problems in each prime-order cyclic factor). This is unfortunately still exponential in the number of digits of ℓ; it also uses $O(\sqrt{\ell})$ memory. We now describe this algorithm without any specific optimizations.

Algorithm 4.12 (Baby-step Giant-step Discrete Log). *Given a prime p, a generator g of $(\mathbb{Z}/p\mathbb{Z})^*$, and an element $a \in (\mathbb{Z}/p\mathbb{Z})^*$, this algorithm finds an n such that $g^n = a$. (Note that this algorithm works in any cyclic group, not just $(\mathbb{Z}/p\mathbb{Z})^*$.)*

(1) [Make Lists] Let $m = \lceil\sqrt{p}\rceil$ be the ceiling of $\sqrt{p}$, and construct two lists

$$1,\ g^m,\ \ldots,\ g^{(m-1)m} \qquad \text{(giant steps)}$$

and

$$a, ag,\ ag^2,\ \ldots,\ ag^{m-1} \qquad \text{(baby steps)}.$$

(2) [Find Match] Sort the two lists and find a match $g^{im} = ag^j$. Then $a = g^{im-j}$.

Proof. We prove that there will always be a match. Since we know that $a = g^k$ for some k with $0 \le k \le p - 1$ and any such k can be written in the form $im - j$ for $0 \le i, j \le m - 1$, we will find such a match. □

Algorithm 4.12 uses nothing special about $(\mathbb{Z}/p\mathbb{Z})^*$, so it works in a generic group. It is a theorem that there is no faster algorithm to find discrete logs in a "generic group" (see [**Sho97, Nec94**]). There are much better subexponential algorithms for solving the discrete log problem in $(\mathbb{Z}/p\mathbb{Z})^*$, which use the special structure of this group. They use the number field sieve (see, e.g., [**Gor93**]), which is also the best-known algorithm for factoring integers. This class of algorithms has been very well studied by cryptographers; though sub-exponential, solving discrete log problems when p is large is still extremely difficult. For a more in-depth survey see [**Gor04**]. For computing Dirichlet characters in our context, p is not too large, so Algorithm 4.12 works well.

4.3.2. Enumeration of All Values. For many applications of Dirichlet characters to computing modular forms, N is fairly small, e.g., $N < 10^6$, and we evaluate ε on a *huge* number of random elements, inside inner loops of algorithms. Thus for such purposes it will often be better to make a table of all values of ε, so that evaluation of ε is extremely fast. The following algorithm computes a table of all values of ε, and it does not require computing any discrete logs since we are computing *all* values.

Algorithm 4.13 (Values of ε). *Given a Dirichlet character ε represented by the list of values of ε on the minimal generators g_i of $(\mathbb{Z}/N\mathbb{Z})^*$, this algorithm creates a list of all the values of ε.*

(1) [Initialize] For each minimal generator g_i, set $a_i = 0$. Let $n = \prod g_i^{a_i}$, and set $z = 1$. Create a list v of N values, all initially set equal to 0. When this algorithm terminates, the list v will have the property that
$$v\,[x \pmod N] = \varepsilon(x).$$
Notice that we index v starting at 0.

(2) [Add Value to Table] Set $v[n] = z$.

(3) [Finished?] If each a_i is one less than the order of g_i, output v and terminate.

(4) [Increment] Set $a_0 = a_0 + 1$, $n = n \cdot g_0 \pmod N$, and $z = z \cdot \varepsilon(g_0)$. If $a_0 \geq \mathrm{ord}(g_0)$, set $a_0 \to 0$, and then set $a_1 = a_1 + 1$, $n = n \cdot g_1 \pmod N$, and $z = z \cdot \varepsilon(g_1)$. If $a_1 \geq \mathrm{ord}(g_1)$, do what you just did with a_0 but with all subscripts replaced by 1. Etc. (Imagine a car odometer.) Go to step (2).

4.4. Conductors of Dirichlet Characters

The following algorithm for computing the order of ε reduces the problem to computing the orders of powers of ζ in R.

Algorithm 4.14 (Order of Character). *This algorithm computes the order of a Dirichlet character $\varepsilon \in D(N,R)$.*

(1) Compute the order r_i of each $\varepsilon(g_i)$, for each minimal generator g_i of $(\mathbb{Z}/N\mathbb{Z})^*$. The order of $\varepsilon(g_i)$ is a divisor of $n = \#(\mathbb{Z}/p_i^{e_i}\mathbb{Z})^*$ so we can compute its order by considering the divisors of n.

(2) Compute and output the least common multiple of the integers r_i.

Remark 4.15. Computing the order of $\varepsilon(g_i) \in R$ is potentially difficult. Simultaneously using a different representation of Dirichlet characters avoids having to compute the order of elements of R (see Section 4.7).

The next algorithm factors ε as a product of "local" characters, one for each prime divisor of N. It is useful for other algorithms, e.g., for explicit computations with trace formulas (see [**Hij74**]). This factorization is easy to compute because of how we represent ε.

Algorithm 4.16 (Factorization of Character). *Given a Dirichlet character $\varepsilon \in D(N, R)$, with $N = \prod p_i^{e_i}$, this algorithm finds Dirichlet characters ε_i modulo $p_i^{e_i}$, such that for all $a \in (\mathbb{Z}/N\mathbb{Z})^*$, we have $\varepsilon(a) = \prod \varepsilon_i(a \pmod{p_i^{e_i}})$. If $2 \mid N$, the steps are as follows:*

(1) Let g_i be the minimal generators of $(\mathbb{Z}/N\mathbb{Z})^*$, so ε is given by a list
$$[\varepsilon(g_0), \ldots, \varepsilon(g_r)].$$
(2) For $i = 2, \ldots, n$, let ε_i be the element of $D(p_i^{e_i}, R)$ defined by the singleton list $[\varepsilon(g_i)]$.
(3) Let ε_1 be the element of $D(2^{e_1}, R)$ defined by the list $[\varepsilon(g_0), \varepsilon(g_1)]$ of length 2. Output the ε_i and terminate.

If $2 \nmid N$, then omit step (3), and include all i in step (2).

The factorization of Algorithm 4.16 is unique since each ε_i is determined by the image of the canonical map $(\mathbb{Z}/p_i^{e_i}\mathbb{Z})^*$ in $(\mathbb{Z}/N\mathbb{Z})^*$, which sends $a \pmod{p_i^{e_i}}$ to the element of $(\mathbb{Z}/N\mathbb{Z})^*$ that is $a \pmod{p_i^{e_i}}$ and $1 \pmod{p_j^{e_j}}$ for $j \neq i$.

Example 4.17. If $\varepsilon = [1, -1, \zeta^5] \in D(200, \mathbb{C})$, then $\varepsilon_1 = [1, -1] \in D(8, \mathbb{C})$ and $\varepsilon_2 = [\zeta^5] \in D(25, \mathbb{C})$.

Definition 4.18 (Conductor). The *conductor* of a Dirichlet character $\varepsilon \in D(N, R)$ is the smallest positive divisor $c \mid N$ such that there is a character $\varepsilon' \in D(c, R)$ for which $\varepsilon(a) = \varepsilon'(a)$ for all $a \in \mathbb{Z}$ with $(a, N) = 1$. A Dirichlet character is *primitive* if its modulus equals its conductor. The character ε' associated to ε with modulus equal to the conductor of ε is called the *primitive character associated to ε*.

We will be interested in conductors later, when computing new subspaces of spaces of modular forms with character. Also certain formulas for special values of L functions are only valid for primitive characters.

Algorithm 4.19 (Conductor). *This algorithm computes the conductor of a Dirichlet character $\varepsilon \in D(N, R)$.*

(1) [Factor Character] Using Algorithm 4.16, find characters ε_i whose product is ε.
(2) [Compute Orders] Using Algorithm 4.14, compute the orders r_i of each ε_i.
(3) [Conductors of Factors] For each i, either set $c_i \to 1$ if ε_i is the trivial character (i.e., of order 1) or set $c_i = p_i^{\mathrm{ord}_{p_i}(r_i)+1}$, where $\mathrm{ord}_p(n)$ is the largest power of p that divides n.

(4) [Adjust at 2?] If $p_1 = 2$ and $\varepsilon_1(5) \neq 1$, set $c_1 = 2c_1$.
(5) [Finished] Output $c = \prod c_i$ and terminate.

Proof. Let ε_i be the local factors of ε, as in step (1). We first show that the product of the conductors f_i of the ε_i is the conductor f of ε. Since ε_i factors through $(\mathbb{Z}/f_i\mathbb{Z})^*$, the product ε of the ε_i factors through $(\mathbb{Z}/\prod f_i\mathbb{Z})^*$, so the conductor of ε divides $\prod f_i$. Conversely, if $\mathrm{ord}_{p_i}(f) < \mathrm{ord}_{p_i}(f_i)$ for some i, then we could factor ε as a product of local (prime power) characters differently, which contradicts that this factorization is unique.

It remains to prove that if ε is a nontrivial character of modulus p^n, where p is a prime, and if r is the order of ε, then the conductor of ε is $p^{\mathrm{ord}_p(r)+1}$, except possibly if $8 \mid p^n$. Since the order and conductor of ε and of the associated primitive character ε' are the same, we may assume ε is primitive, i.e., that p^n is the conductor of ε; note that $n > 0$, since ε is nontrivial.

First suppose p is odd. Then the abelian group $D(p^n, R)$ splits as a direct sum $D(p, R) \oplus D(p^n, R)'$, where $D(p^n, R)'$ is the p-power torsion subgroup of $D(p^n, R)$. Also ε has order $u \cdot p^m$, where u, which is coprime to p, is the order of the image of ε in $D(p, R)$ and p^m is the order of the image in $D(p^n, R)'$. If $m = 0$, then the order of ε is coprime to p, so ε is in $D(p, R)$, which means that $n = 1$, so $n = m+1$, as required. If $m > 0$, then $\zeta \in R$ must have order divisible by p, so R has characteristic not equal to p. The conductor of ε does not change if we adjoin roots of unity to R, so in light of Lemma 4.2 we may assume that $D(N, R) \approx (\mathbb{Z}/N\mathbb{Z})^*$. It follows that for each $n' \leq n$, the p-power subgroup $D(p^{n'}, R)'$ of $D(p^{n'}, R)$ is the $p^{n'-1}$-torsion subgroup of $D(p^n, R)'$. Thus $m = n - 1$, since $D(p^n, R)'$ is by assumption the smallest such group that contains the projection of ε. This proves the formula of step (3). We leave the argument when $p = 2$ as an exercise (see Exercise 4.5). □

Example 4.20. If $\varepsilon = [1, -1, \zeta^5] \in D(200, \mathbb{C})$, then as in Example 4.17, ε is the product of $\varepsilon_1 = [1, -1]$ and $\varepsilon_2 = [\zeta^5]$. Because $\varepsilon_1(5) = -1$, the conductor of ε_1 is 8. The order of ε_2 is 4 (since ζ is a 20th root of unity), so the conductor of ε_2 is 5. Thus the conductor of ε is $40 = 8 \cdot 5$.

4.5. The Kronecker Symbol

In this section all characters have values in $\mathbb{C}$.

Frequently quadratic characters are described in terms of the Kronecker symbol $\left(\frac{a}{n}\right)$, which we define for any integer a and positive integer n as follows. First, if $n = p$ is an odd prime, then for any integer a,

$$\left(\frac{a}{p}\right) = \begin{cases} 0 & \text{if } \gcd(a,p) \neq 1, \\ 1 & \text{if } a \text{ is a square mod } p, \\ -1 & \text{if } a \text{ is not a square mod } p. \end{cases}$$

If $p = 2$, then

$$\left(\frac{a}{2}\right) = \begin{cases} 0 & \text{if } a \text{ is even}, \\ 1 & \text{if } a \equiv \pm 1 \pmod 8, \\ -1 & \text{if } a \equiv \pm 3 \pmod 8). \end{cases}$$

More generally, if $n = \prod p_i^{e_i}$ with the p_i prime, then

$$\left(\frac{a}{n}\right) = \prod \left(\frac{a}{p_i}\right)^{e_i}.$$

Remark 4.21. One can also extend $\left(\frac{a}{n}\right)$ to $n < 0$, but we will not need this. The extension is to set $\left(\frac{a}{-1}\right) = -1$ and $\left(\frac{a}{1}\right) = 1$, for $a \neq 0$, and to extend multiplicatively (in the denominator). Note that the map $\left(\frac{\bullet}{-1}\right)$ is not a Dirichlet character (see Exercise 4.1).

Let M be the product of the primes p such that $\operatorname{ord}_p(n)$ is odd. If M is odd, let $N = M$; otherwise, let $N = 8M$.

Lemma 4.22. *The function*

$$\varepsilon(a) = \begin{cases} \left(\frac{a}{n}\right) & \text{if } \gcd(a,N) = 1, \\ 0 & \text{otherwise} \end{cases}$$

is a Dirichlet character of modulus N. The function

$$\varepsilon(a) = \begin{cases} \left(\frac{-1}{a}\right) & \text{if } a \text{ is odd}, \\ 0 & \text{if } a \text{ is even} \end{cases}$$

is a Dirichlet character of modulus N.

Proof. When restricted to $(\mathbb{Z}/N\mathbb{Z})^*$, each map $\left(\frac{\bullet}{p}\right)$, for p prime, is a homomorphism, so ε a product of homomorphisms. The second statement follows from the definition and the fact that -1 is a square modulo an odd prime p if and only if $p \equiv 1 \pmod 4$. □

This section is about going between representing quadratic characters as row matrices and via Kronecker symbols. This is valuable because the algorithms in [**Coh93**, §1.1.4] for computing Kronecker symbols run in time quadratic in the number of digits of the input. They do not require computing discrete logarithms; instead, they use, e.g., that $\left(\frac{a}{p}\right) \equiv a^{(p-1)/2} \pmod{p}$, when p is an odd prime.

Algorithm 4.23 (Kronecker Symbol as Dirichlet Character). *Given $n > 0$, this algorithm computes a representation of the Kronecker symbol $\left(\frac{\bullet}{n}\right)$ as a Dirichlet character.*

(1) [Modulus] Compute N as in Lemma 4.22.
(2) [Minimal Generators] Compute minimal generators g_i of $(\mathbb{Z}/N\mathbb{Z})^*$ using Algorithm 4.4.
(3) [Images] Compute $\left(\frac{g_i}{N}\right)$ for each g_i using one of the algorithms of [**Coh93**, §1.1.4].

Example 4.24. We compute the Dirichlet character associated to $\left(\frac{\bullet}{200}\right)$. Using SAGE, we compute the $\left(\frac{g_i}{200}\right)$, for $i = 0, 1, 2$, where the g_i are as in Example 4.9:

```
sage: kronecker(151,200)
1
sage: kronecker(101,200)
-1
sage: kronecker(177,200)
1
```

Thus the corresponding character is defined by $[1, -1, 1]$.

Example 4.25. We compute the character associated to $\left(\frac{\bullet}{420}\right)$. We have $420 = 4 \cdot 3 \cdot 5 \cdot 7$, and minimal generators are

$$g_0 = 211, \quad g_1 = 1, \quad g_2 = 281, \quad g_3 = 337, \quad g_4 = 241.$$

We have $g_0 \equiv -1 \pmod{4}$, $g_2 \equiv 2 \pmod{3}$, $g_3 \equiv 2 \pmod{5}$ and $g_4 \equiv 3 \pmod{7}$. We find $\left(\frac{g_0}{420}\right) = \left(\frac{g_1}{420}\right) = 1$ and $\left(\frac{g_2}{420}\right) = \left(\frac{g_3}{420}\right) = \left(\frac{g_4}{420}\right) = -1$. The corresponding character is $[1, 1, -1, -1, -1]$.

Using the following algorithm, we can go in the other direction, i.e., write any quadratic Dirichlet character as a Kronecker symbol.

Algorithm 4.26 (Dirichlet Character as Kronecker Symbol). *Given ε of order 2 with modulus N, this algorithm writes ε as a Kronecker symbol.*

(1) [Conductor] Use Algorithm 4.19 to compute the conductor f of ε.
(2) [Odd] If f is odd, output $\left(\frac{\bullet}{f}\right)$.
(3) [Even] If $\varepsilon(-1) = 1$, output $\left(\frac{\bullet}{f}\right)$; if $\varepsilon(-1) = -1$, output $\left(\frac{\bullet}{f}\right) \cdot \left(\frac{-1}{\bullet}\right)$.

Proof. Since f is the conductor of a quadratic Dirichlet character, it is a square-free product g of odd primes times either 4 or 8, so the group $(\mathbb{Z}/f\mathbb{Z})^*$ does not inject into $(\mathbb{Z}/g\mathbb{Z})^*$ for any proper divisor g of f (see this by reducing to the prime power case). Since g is odd and square-free, the character $\left(\frac{\bullet}{g}\right)$ has conductor g. For each odd prime p, by step (3) of Algorithm 4.19 the factor at p of both ε and $\left(\frac{\bullet}{g}\right)$ is a quadratic character with modulus p. By Exercise 4.2 and Lemma 4.2 the group $D(p, \mathbb{C})$ is cyclic, so it has a unique element of order 2, so the factors of ε and $\left(\frac{\bullet}{g}\right)$ at p are equal.

The quadratic characters with conductor a power of 2 are $[-1]$, $[1, -1]$, and $[-1, -1]$. The character $[1, -1]$ is $\left(\frac{\bullet}{2}\right)$ and the character $[-1]$ is $\left(\frac{-1}{\bullet}\right)$. □

Example 4.27. Consider $\varepsilon = [-1, -1, -1, -1, -1]$ with modulus $840 = 8 \cdot 3 \cdot 5 \cdot 7$. It has conductor 840, and $\varepsilon(-1) = -1$, so for all a with $\gcd(a, 840) = 1$, we have $\varepsilon(a) = \left(\frac{a}{840}\right) \cdot \left(\frac{-1}{a}\right)$.

4.6. Restriction, Extension, and Galois Orbits

The following two algorithms restrict and extend characters to a compatible modulus. Using them, it is easy to define multiplication of two characters $\varepsilon \in D(N, R)$ and $\varepsilon' \in D(N', R')$, as long as R and R' are subrings of a common ring. To carry out the multiplication, extend both characters to a common base ring, and then extend them to characters modulo $\mathrm{lcm}(N, N')$ and multiply.

Algorithm 4.28 (Restriction of Character). *Given a Dirichlet character $\varepsilon \in D(N, R)$ and a divisor N' of N that is a multiple of the conductor of ε, this algorithm finds a characters $\varepsilon' \in D(N', R)$, such that $\varepsilon'(a) = \varepsilon(a)$, for all $a \in \mathbb{Z}$ with $(a, N) = 1$.*

(1) [Conductor] Compute the conductor of ε using Algorithm 4.19, and verify that N' is divisible by the conductor and divides N.
(2) [Minimal Generators] Compute minimal generators g_i for $(\mathbb{Z}/N'\mathbb{Z})^*$.
(3) [Values of Restriction] For each i, compute $\varepsilon'(g_i)$ as follows. Find a multiple aN' of N' such that $(g_i + aN', N) = 1$; then $\varepsilon'(g_i) = \varepsilon(g_i + aN')$.
(4) [Output Character] Output the Dirichlet character of modulus N' defined by $[\varepsilon'(g_0), \ldots, \varepsilon'(g_n)]$.

Proof. The only part that is not clear is that in step (3) there is an a such that $(g_i + aN', N) = 1$. If we write $N = N_1 \cdot N_2$, with $(N_1, N_2) = 1$ and N_1 divisible by all primes that divide N', then $(g_i, N_1) = 1$ since $(g_i, N') = 1$. By the Chinese Remainder Theorem, there is an $x \in \mathbb{Z}$ such that $x \equiv g_i$

(mod N_1) and $x \equiv 1 \pmod{N_2}$. Then $x = g_i + bN_1 = g_i + (bN_1/N') \cdot N'$ and $(x, N) = 1$, which completes the proof. $\square$

Algorithm 4.29 (Extension of Character). *Given a Dirichlet character $\varepsilon \in D(N, R)$ and a multiple N' of N, this algorithm finds a character $\varepsilon' \in D(N', R)$, such that $\varepsilon'(a) = \varepsilon(a)$, for all $a \in \mathbb{Z}$ with $(a, N') = 1$.*

(1) [Minimal Generators] Compute minimal generators g_i for $(\mathbb{Z}/N'\mathbb{Z})^*$.
(2) [Evaluate] Compute $\varepsilon(g_i)$ for each i. Since $(g_i, N') = 1$, we also have $(g_i, N) = 1$.
(3) [Output Character] Output the character $[\varepsilon(g_0), \ldots, \varepsilon(g_n)]$.

Let F be the prime subfield of R, and assume that $R \subset \overline{F}$, where $\overline{F}$ is a separable closure of F. If $\sigma \in \mathrm{Gal}(\overline{F}/F)$ and $\varepsilon \in D(N, R)$, let $(\sigma\varepsilon)(n) = \sigma(\varepsilon(n))$; this defines an action of $\mathrm{Gal}(\overline{F}/F)$ on $D(N, R)$. Our next algorithm computes the orbits for the action of $\mathrm{Gal}(\overline{F}/F)$ on $D(N, R)$. This algorithm can provide huge savings for modular forms computations because the spaces $M_k(N, \varepsilon)$ and $M_k(N, \varepsilon')$ are canonically isomorphic if ε and ε' are conjugate.

Algorithm 4.30 (Galois Orbit). *Given a Dirichlet character $\varepsilon \in D(N, R)$, this algorithm computes the orbit of ε under the action of $G = \mathrm{Gal}(\overline{F}/F)$, where F is the prime subfield of* $\mathrm{Frac}(R)$, *so $F = \mathbb{F}_p$ or $\mathbb{Q}$.*

(1) [Order of ζ] Let n be the order of the chosen root $\zeta \in R$.
(2) [Nontrivial Automorphisms] If $\mathrm{char}(R) = 0$, let
$$A = \{a : 2 \le a < n \text{ and } (a, n) = 1\}.$$
If $\mathrm{char}(R) = p > 0$, compute the multiplicative order r of $p \pmod{n}$, and let
$$A = \{p^m : 1 \le m < r\}.$$
(3) [Compute Orbit] Compute and output the *set* of unique elements ε^a for each $a \in A$ (there could be repeats, so we output unique elements only).

Proof. We prove that the nontrivial automorphisms of $\langle\zeta\rangle$ in characteristic p are as in step (2). It is well known that every automorphism in characteristic p on $\zeta \in \overline{\mathbb{F}}_p$ is of the form $x \mapsto x^{p^s}$, for some s. The images of ζ under such automorphisms are
$$\zeta, \zeta^p, \zeta^{p^2}, \ldots.$$
Suppose $r > 0$ is minimal such that $\zeta = \zeta^{p^r}$. Then the orbit of ζ is $\zeta, \ldots, \zeta^{p^{r-1}}$. Also $p^r \equiv 1 \pmod{n}$, where n is the multiplicative order of ζ, so r is the multiplicative order of p modulo n, which completes the proof. $\square$

Example 4.31. The Galois orbits of characters in $D(20, \mathbb{C}^*)$ are as follows:

$$\begin{aligned} G_0 &= \{[1, 1, 1]\}, \\ G_1 &= \{[-1, 1, 1]\}, \\ G_2 &= \{[1, 1, \zeta_4],\ [1, 1, -\zeta_4]\} \\ G_3 &= \{[-1, 1, \zeta_4],\ [-1, 1, -\zeta_4]\} \\ G_4 &= \{[1, 1, -1]\}, \\ G_5 &= \{[-1, 1, -1]\}. \end{aligned}$$

The conductors of the characters in orbit G_0 are 1, in orbit G_1 they are 4, in orbit G_2 they are 5, in G_3 they are 20, in G_4 the conductor is 5, and in G_5 the conductor is 20. (You should verify this.)

SAGE computes Galois orbits as follows:

```
sage: G = DirichletGroup(20)
sage: G.galois_orbits()
[
[[1, 1]],
[[1, zeta4], [1, -zeta4]],
[[1, -1]],
[[-1, 1]],
[[-1, zeta4], [-1, -zeta4]],
[[-1, -1]]
]
```

4.7. Alternative Representations of Characters

Let N be a positive integer and R an integral domain, with fixed root of unity ζ of order n, and let $D(N, R) = D(N, R, \zeta)$. As in the rest of this chapter, write $N = \prod p_i^{e_i}$, and let $C_i = \langle g_i \rangle$ be the corresponding cyclic factors of $(\mathbb{Z}/N\mathbb{Z})^*$. In this section we discuss other ways to represent elements $\varepsilon \in D(N, R)$. Each representation has advantages and disadvantages, and no single representation is best. It is easy to convert between them, and some algorithms are much easier using one representation than when using another. In this section we present two other representations, each having advantages and disadvantages. There is no reason to restrict to only one representation; for example, SAGE internally uses both.

We could represent ε by giving a list $[b_0, \ldots, b_r]$, where each $b_i \in \mathbb{Z}/n\mathbb{Z}$ and $\varepsilon(g_i) = \zeta^{b_i}$. Then arithmetic in $D(N, R)$ is arithmetic in $(\mathbb{Z}/n\mathbb{Z})^{r+1}$, which is very efficient. A drawback to this approach (in practice) is that it is easy to accidently consider sequences that do not actually correspond to elements of $D(N, R)$. Also the choice of ζ is less clear, which can cause confusion. Finally, the orders of the local factors is more opaque, e.g., compare $[-1, \zeta_{40}]$ with $[20, 1]$. Overall this representation is not too bad and is more like representing a linear transformation by a matrix. It has the *advantage* over the representation discussed earlier in this chapter that arithmetic in $D(N, R)$ is very efficient and does not require any operations in the ring R.

Another way to represent ε would be to give a list $[b_0, \ldots, b_r]$ of integers, but this time with $b_i \in \mathbb{Z}/\gcd(s_i, n)\mathbb{Z}$, where s_i is the order of g_i. Then

$$\varepsilon(g_i) = \zeta^{b_i \cdot n/(\gcd(s_i,n))},$$

which is already pretty complicated. With this representation we set up an identification

$$D(N, R) \cong \bigoplus_i \mathbb{Z}/\gcd(s_i, n)\mathbb{Z},$$

and arithmetic is efficient. This approach is seductive because every sequence of integers determines a character, and the sizes of the integers in the sequence nicely indicate the local orders of the character. However, giving analogues of many of the algorithms discussed in this chapter that operate on characters represented this way is tricky. For example, the representation depends very much on the order of ζ, so it is difficult to correctly compute natural maps $D(N, R) \to D(N, S)$, for $R \subset S$ rings.

4.8. Dirichlet Characters in SAGE

To create a Dirichlet character in SAGE, first create the group $D(N, R)$ of Dirichlet characters then construct elements of that group. First we make $D(11, \mathbb{Q})$:

```
sage: G = DirichletGroup(11, QQ); G
Group of Dirichlet characters of modulus 11 over
Rational Field
```

A Dirichlet character prints as a matrix that gives the values of the character on canonical generators of $(\mathbb{Z}/N\mathbb{Z})^*$ (as discussed below).

```
sage: list(G)
[[1], [-1]]
sage: eps = G.0        # 0th generator for Dirichlet group
sage: eps
[-1]
```

The character ε takes the value -1 on the unit generator.

```
sage: G.unit_gens()
[2]
sage: eps(2)
-1
sage: eps(3)
1
```

It is 0 on any integer not coprime to 11:

```
sage: [eps(11*n) for n in range(10)]
[0, 0, 0, 0, 0, 0, 0, 0, 0, 0]
```

We can also create groups of Dirichlet characters taking values in other rings or fields. For example, we create the cyclotomic field $\mathbb{Q}(\zeta_4)$.

```
sage: R = CyclotomicField(4)
sage: CyclotomicField(4)
Cyclotomic Field of order 4 and degree 2
```

Then we define $G = D(15, \mathbb{Q}(\zeta_4))$.

```
sage: G = DirichletGroup(15, R)
sage: G
Group of Dirichlet characters of modulus 15 over
Cyclotomic Field of order 4 and degree 2
```

Next we list each of its elements.

```
sage: list(G)
[[1, 1], [-1, 1], [1, zeta4], [-1, zeta4], [1, -1],
[-1, -1], [1, -zeta4], [-1, -zeta4]]
```

Now we evaluate the second generator of G on various integers:

```
sage: e = G.1
sage: e(4)
-1
sage: e(-1)
-1
sage: e(5)
0
```

Finally we list all the values of e.

```
sage: [e(n) for n in range(15)]
[0, 1, zeta4, 0, -1, 0, 0, zeta4, -zeta4,
    0, 0, 1, 0, -zeta4, -1]
```

We can also compute with groups of Dirichlet characters with values in a finite field.

```
sage: G = DirichletGroup(15, GF(5)); G
Group of Dirichlet characters of modulus 15
      over Finite Field of size 5
```

We list all the elements of G, again represented by lists that give the images of each unit generator, as an element of $\mathbb{F}_5$.

```
sage: list(G)
  [[1, 1], [4, 1], [1, 2], [4, 2], [1, 4], [4, 4],
   [1, 3], [4, 3]]
```

We evaluate the second generator of G on several integers.

```
sage: e = G.1
sage: e(-1)
4
sage: e(2)
2
sage: e(5)
0
sage: print [e(n) for n in range(15)]
[0, 1, 2, 0, 4, 0, 0, 2, 3, 0, 0, 1, 0, 3, 4]
```

4.9. Exercises

4.1 Let $f : \mathbb{Z} \to \mathbb{C}$ be the map given by

$$f(a) = \begin{cases} 0 & \text{if } a = 0, \\ -1 & \text{if } a < 0, \\ 1 & \text{if } a > 0. \end{cases}$$

Prove that f is not a Dirichlet character of any modulus N.

4.2 This exercise is about the structure of the units of $\mathbb{Z}/p^n\mathbb{Z}$.

(a) If p is odd and n is a positive integer, prove that $(\mathbb{Z}/p^n\mathbb{Z})^*$ is cyclic.

(b) For $n \geq 3$, prove that $(\mathbb{Z}/2^n\mathbb{Z})^*$ is a direct sum of the cyclic subgroups $\langle -1 \rangle$ and $\langle 5 \rangle$, of orders 2 and 2^{n-2}, respectively.

4.3 Prove that Algorithm 4.4 works, i.e., that if $g \in (\mathbb{Z}/p^r\mathbb{Z})^*$ and $g^{n/p_i} \neq 1$ for all $p_i \mid n = \varphi(p^r)$, then g is a generator of $(\mathbb{Z}/p^r\mathbb{Z})^*$.

4.4 (a) Let p be an odd prime and $n \geq 2$ an integer, and prove that

$$\big((1 + p^{n-1}\mathbb{Z}/p^n\mathbb{Z}),\ \times\big) \cong (\mathbb{Z}/p\mathbb{Z},\ +).$$

(b) Use the first part to show that solving the discrete log problem in $(\mathbb{Z}/p^n\mathbb{Z})^*$ is "not much harder" than solving the discrete log problem in $(\mathbb{Z}/p\mathbb{Z})^*$.

4.5 Suppose ε is a nontrivial Dirichlet character of modulus 2^n of order r over the complex numbers $\mathbb{C}$. Prove that the conductor of ε is

$$c = \begin{cases} 2^{\mathrm{ord}_2(r)+1} & \text{if } \varepsilon(5) = 1, \\ 2^{\mathrm{ord}_2(r)+2} & \text{if } \varepsilon(5) \neq 1. \end{cases}$$

4.6 (a) Find an irreducible quadratic polynomial f over $\mathbb{F}_5$.

(b) Then $\mathbb{F}_{25} = \mathbb{F}_5[x]/(f)$. Find an element with multiplicative order 4 in $\mathbb{F}_{25}$.

(c) Make a list of all Dirichlet characters in $D(25, \mathbb{F}_{25}, \zeta)$.

(d) Divide these characters into orbits for the action of $\mathrm{Gal}(\overline{\mathbb{F}}_5/\mathbb{F}_5)$.

Chapter 5

Eisenstein Series and Bernoulli Numbers

We introduce generalized Bernoulli numbers attached to Dirichlet characters and give an algorithm to enumerate the Eisenstein series in $M_k(N, \varepsilon)$.

5.1. The Eisenstein Subspace

Let $M_k(\Gamma_1(N))$ be the space of modular forms of weight k for $\Gamma_1(N)$, and let $\mathbb{T}$ be the *Hecke algebra* acting on $M_k(\Gamma_1(N))$, which is the subring of $\mathrm{End}(M_k(\Gamma_1(N)))$ generated by all Hecke operators. Then there is a $\mathbb{T}$-module decomposition

$$M_k(\Gamma_1(N)) = E_k(\Gamma_1(N)) \oplus S_k(\Gamma_1(N)),$$

where $S_k(\Gamma_1(N))$ is the subspace of modular forms that vanish at all cusps and $E_k(\Gamma_1(N))$ is the *Eisenstein subspace*, which is uniquely determined by this decomposition. The above decomposition induces a decomposition of $M_k(\Gamma_0(N))$ and of $M_k(N, \varepsilon)$, for any Dirichlet character ε of modulus N.

5.2. Generalized Bernoulli Numbers

Suppose ε is a Dirichlet character of modulus N over $\mathbb{C}$. Leopoldt [**Leo58**] defined generalized Bernoulli numbers attached to ε.

Definition 5.1 (Generalized Bernoulli Number). We define the *generalized Bernoulli numbers* $B_{k,\varepsilon}$ attached to ε by the following identity of infinite series:

$$\sum_{a=1}^{N} \frac{\varepsilon(a) \cdot x \cdot e^{ax}}{e^{Nx} - 1} = \sum_{k=0}^{\infty} B_{k,\varepsilon} \cdot \frac{x^k}{k!}.$$

If ε is the trivial character of modulus 1 and B_k are as in Section 2.1, then $B_{k,\varepsilon} = B_k$, except when $k = 1$, in which case $B_{1,\varepsilon} = -B_1 = 1/2$ (see Exercise 5.2).

5.2.1. Algebraically Computing Generalized Bernoulli Numbers. Let $\mathbb{Q}(\varepsilon)$ denote the field generated by the image of the character ε; thus $\mathbb{Q}(\varepsilon)$ is the cyclotomic extension $\mathbb{Q}(\zeta_n)$, where n is the order of ε.

Algorithm 5.2 (Generalized Bernoulli Numbers). *Given an integer $k \geq 0$ and any Dirichlet character ε with modulus N, this algorithm computes the generalized Bernoulli numbers $B_{j,\varepsilon}$, for $j \leq k$.*

(1) Compute $g = x/(e^{Nx} - 1) \in \mathbb{Q}[[x]]$ to precision $O(x^{k+1})$ by computing $e^{Nx} - 1 = \sum_{n \geq 1} N^n x^n/n!$ to precision $O(x^{k+2})$ and computing the inverse $1/(e^{Nx} - 1)$, then multiplying by x.
(2) For each $a = 1, \ldots, N$, compute $f_a = g \cdot e^{ax} \in \mathbb{Q}[[x]]$, to precision $O(x^{k+1})$. This requires computing $e^{ax} = \sum_{n \geq 0} a^n x^n/n!$ to precision $O(x^{k+1})$. (Omit computation of e^{Nx} if $N > 1$ since then $\varepsilon(N) = 0$.)
(3) Then for $j \leq k$, we have

$$B_{j,\varepsilon} = j! \cdot \sum_{a=1}^{N} \varepsilon(a) \cdot c_j(f_a),$$

where $c_j(f_a)$ is the coefficient of x^j in f_a.

Note that in steps (1) and (2) we compute the power series doing arithmetic only in $\mathbb{Q}[[x]]$, not in $\mathbb{Q}(\varepsilon)[[x]]$, which could be much less efficient if ε has large order. In step (1) if k is huge, we could compute the inverse $1/(e^{Nx} - 1)$ using asymptotically fast arithmetic and Newton iteration.

Example 5.3. The nontrivial character ε with modulus 4 has order 2 and takes values in $\mathbb{Q}$. The Bernoulli numbers $B_{k,\varepsilon}$ for k even are all 0 and for

k odd they are

$$\begin{aligned} B_{1,\varepsilon} &= -1/2, \\ B_{3,\varepsilon} &= 3/2, \\ B_{5,\varepsilon} &= -25/2, \\ B_{7,\varepsilon} &= 427/2, \\ B_{9,\varepsilon} &= -12465/2, \\ B_{11,\varepsilon} &= 555731/2, \\ B_{13,\varepsilon} &= -35135945/2, \\ B_{15,\varepsilon} &= 2990414715/2, \\ B_{17,\varepsilon} &= -329655706465/2, \\ B_{19,\varepsilon} &= 45692713833379/2. \end{aligned}$$

Example 5.4. The generalized Bernoulli numbers need not be in $\mathbb{Q}$. Suppose ε is the mod 5 character such that $\varepsilon(2) = i = \sqrt{-1}$. Then $B_{k,\varepsilon} = 0$ for k even and

$$\begin{aligned} B_{1,\varepsilon} &= \frac{-i-3}{5}, \\ B_{3,\varepsilon} &= \frac{6i+12}{5}, \\ B_{5,\varepsilon} &= \frac{-86i-148}{5}, \\ B_{7,\varepsilon} &= \frac{2366i+3892}{5}, \\ B_{9,\varepsilon} &= \frac{-108846i-176868}{5}, \\ B_{11,\varepsilon} &= \frac{7599526i+12309572}{5}, \\ B_{13,\varepsilon} &= \frac{-751182406i-1215768788}{5}, \\ B_{15,\varepsilon} &= \frac{99909993486i+161668772052}{5}, \\ B_{17,\varepsilon} &= \frac{-17209733596766i-27846408467908}{5}. \end{aligned}$$

Example 5.5. We use SAGE to compute some of the above generalized Bernoulli numbers. First we define the character and verify that $\varepsilon(2) = i$ (note that in SAGE `zeta4` is $\sqrt{-1}$).

```
sage: G = DirichletGroup(5)
sage: e = G.0
sage: e(2)
zeta4
```

We compute the Bernoulli number $B_{1,\varepsilon}$.

```
sage: e.bernoulli(1)
-1/5*zeta4 - 3/5
```

We compute $B_{9,\varepsilon}$.

```
sage: e.bernoulli(9)
-108846/5*zeta4 - 176868/5
```

Proposition 5.6. *If $\varepsilon(-1) \neq (-1)^k$ and $k \geq 2$, then $B_{k,\varepsilon} = 0$.*

Proof. See Exercise 5.3. □

5.2.2. Computing Generalized Bernoulli Numbers Analytically.

This section, which was written jointly with Kevin McGown, is about a way to compute generalized Bernoulli numbers, which is similar to the algorithm in Section 2.7.

Let χ be a primitive Dirichlet character modulo its conductor f. Note from the definition of Bernoulli numbers that if $\sigma \in \mathrm{Gal}(\overline{\mathbb{Q}}/\mathbb{Q})$, then

$$\sigma(B_{n,\chi}) = B_{n,\sigma(\chi)}. \tag{5.2.1}$$

For any character χ, we define the Gauss sum $\tau(\chi)$ as

$$\tau(\chi) = \sum_{r=1}^{f-1} \chi(r)\,\zeta^r\,,$$

where $\zeta = \exp(2\pi i/f)$ is the principal fth root of unity. The Dirichlet L-function for χ for $\mathrm{Re}(s) > 1$ is

$$L(s,\chi) = \sum_{n=1}^{\infty} \chi(n)\,n^{-s}\,.$$

In the right half plane $\{s \in \mathbb{C} \mid \mathrm{Re}(s) > 1\}$ this function is analytic, and because χ is multiplicative, we have the Euler product representation

$$L(s,\chi) = \prod_{p \text{ prime}} \left(1 - \chi(p)p^{-s}\right)^{-1}. \tag{5.2.2}$$

We note (but will not use) that through analytic continuation $L(s,\chi)$ can be extended to a meromorphic function on the entire complex plane.

If χ is a nonprincipal primitive Dirichlet character of conductor f such that $\chi(-1) = (-1)^n$, then (see, e.g., [**Wan82**])

$$L(n, \chi) = (-1)^{n-1} \frac{\tau(\chi)}{2} \left(\frac{2\pi i}{f}\right)^n \frac{B_{n,\overline{\chi}}}{n!}.$$

Solving for the Bernoulli number yields

$$B_{n,\chi} = (-1)^{n-1} \frac{2n!}{\tau(\overline{\chi})} \left(\frac{f}{2\pi i}\right)^n L(n, \overline{\chi}).$$

This allows us to give decimal approximations for $B_{n,\chi}$. It remains to compute $B_{n,\chi}$ exactly (i.e., as an algebraic integer). To simplify the above expression, we define

$$K_{n,\chi} = (-1)^{n-1}\, 2n! \left(\frac{f}{2i}\right)^n$$

and write

$$B_{n,\chi} = \frac{K_{n,\chi}}{\pi^n\, \tau(\overline{\chi})}\, L(n, \overline{\chi}). \tag{5.2.3}$$

Note that we can compute $K_{n,\chi}$ exactly in the field $\mathbb{Q}(i)$.

The following result identifies the denominator of $B_{n,\chi}$.

Theorem 5.7. *Let n and χ be as above, and define an integer d as follows:*

$$d = \begin{cases} 1 & \text{if } f \text{ is divisible by two distinct primes,} \\ 2 & \text{if } f = 4, \\ 1 & \text{if } f = 2^\mu,\ \mu > 2, \\ np & \text{if } f = p,\ p > 2, \\ (1 - \chi(1+p)) & \text{if } f = p^\mu,\ p > 2,\ \mu > 1. \end{cases}$$

Then $dn^{-1} B_{n,\chi}$ is integral.

Proof. See [**Car59a**] for the proof and [**Car59b**] for further details. □

To compute the algebraic integer $dn^{-1}B_{n,\chi}$, and we compute $L(n, \overline{\chi})$ to very high precision using the Euler product (5.2.2) and the formula (5.2.3). We carry out the same computation for each of the $\mathrm{Gal}(\overline{\mathbb{Q}}/\mathbb{Q})$ conjugates of χ, which by (5.2.1) yields the conjugates of $dn^{-1}B_{n,\chi}$. We can then write down the characteristic polynomial of $dn^{-1}B_{n,\chi}$ to very high precision and recognize the coefficients as rational integers. Finally, we determine which of the roots of the characteristic polynomial is $dn^{-1}B_{n,\chi}$ by approximating them all numerically to high precision and seeing which is closest to our numerical approximation to $dn^{-1}B_{n,\chi}$. The details are similar to what is explained in Section 2.7.

5.3. Explicit Basis for the Eisenstein Subspace

Suppose χ and ψ are primitive Dirichlet characters with conductors L and R, respectively. Let

$$(5.3.1)\quad E_{k,\chi,\psi}(q) = c_0 + \sum_{m\geq 1}\left(\sum_{n\mid m}\psi(n)\cdot\chi(m/n)\cdot n^{k-1}\right)q^m \in \mathbb{Q}(\chi,\psi)[[q]],$$

where

$$c_0 = \begin{cases} 0 & \text{if } L>1,\\ -\dfrac{B_{k,\psi}}{2k} & \text{if } L=1.\end{cases}$$

Note that when $\chi = \psi = 1$ and $k \geq 4$, then $E_{k,\chi,\psi} = E_k$, where E_k is from Chapter 1.

Miyake proves statements that imply the following in [**Miy89**, Ch. 7].

Theorem 5.8. *Suppose t is a positive integer and χ, ψ are as above and that k is a positive integer such that $\chi(-1)\psi(-1) = (-1)^k$. Except when $k = 2$ and $\chi = \psi = 1$, the power series $E_{k,\chi,\psi}(q^t)$ defines an element of $M_k(RLt, \chi/\psi)$. If $\chi = \psi = 1$, $k = 2$, $t > 1$, and $E_2(q) = E_{k,\chi,\psi}(q)$, then $E_2(q) - tE_2(q^t)$ is a modular form in $M_2(\Gamma_0(t))$.*

Theorem 5.9. *The Eisenstein series in $M_k(N,\varepsilon)$ coming from Theorem 5.8 with $RLt \mid N$ and $\chi/\psi = \varepsilon$ form a basis for the Eisenstein subspace $E_k(N,\varepsilon)$.*

Theorem 5.10. *The Eisenstein series $E_{k,\chi,\psi}(q) \in M_k(RL)$ defined above are eigenforms (i.e., eigenvectors for all Hecke operators T_n). Also $E_2(q) - tE_2(q^t)$, for $t > 1$, is an eigenform.*

Since $E_{k,\chi,\psi}(q)$ is normalized so the coefficient of q is 1, the eigenvalue of T_m is the coefficient

$$\sum_{n\mid m}\psi(n)\cdot\chi(m/n)\cdot n^{k-1}$$

of q^m (see Proposition 9.10). Also for $f = E_2(q) - tE_2(q^t)$ with $t > 1$ prime, the coefficient of q is 1, $T_m(f) = \sigma_1(m)\cdot f$ for $(m,t) = 1$, and $T_t(f) = ((t+1)-t)f = f$.

Algorithm 5.11 (Enumerating Eisenstein Series)**.** *Given a weight k and a Dirichlet character ε of modulus N, this algorithm computes a basis for the Eisenstein subspace $E_k(N,\varepsilon)$ of $M_k(N,\varepsilon)$ to precision $O(q^r)$.*

(1) [Weight 2 Trivial Character?] If $k = 2$ and $\varepsilon = 1$, output the Eisenstein series $E_2(q) - tE_2(q^t)$, for each divisor $t \mid N$ with $t \neq 1$, and then terminate.

(2) [Empty Space?] If $\varepsilon(-1) \neq (-1)^k$, output the empty list.
(3) [Compute Dirichlet Group] Let $G = D(N, \mathbb{Q}(\zeta_n))$ be the group of Dirichlet characters with values in $\mathbb{Q}(\zeta_n)$, where n is the exponent of $(\mathbb{Z}/N\mathbb{Z})^*$.
(4) [Compute Conductors] Compute the conductor of every element of G using Algorithm 4.19.
(5) [List Characters χ] Form a list V of all Dirichlet characters $\chi \in G$ such that $\mathrm{cond}(\chi) \cdot \mathrm{cond}(\chi/\varepsilon)$ divides N.
(6) [Compute Eisenstein Series] For each character χ in V, let $\psi = \chi/\varepsilon$ and compute $E_{k,\chi,\psi}(q^t) \pmod{q^r}$ for each divisor t of $N/(\mathrm{cond}(\chi)\cdot \mathrm{cond}(\psi))$. Here we compute $E_{k,\chi,\psi}(q^t) \pmod{q^r}$ using (5.3.1) and Algorithm 5.2.

Remark 5.12. Algorithm 5.11 is what is currently used in SAGE. It might be better to first reduce to the prime power case by writing all characters as a product of local characters and combine steps (4) and (5) into a single step that involves orders. However, this might make things more obscure.

Example 5.13. The following is a basis of Eisenstein series for $E_2(\Gamma_1(13))$.

$$\begin{aligned}
f_1 &= \frac{1}{2} + q + 3q^2 + 4q^3 + \cdots,\\
f_2 &= -\frac{7}{13}\zeta_{12}^2 - \frac{11}{13} + q + \left(2\zeta_{12}^2 + 1\right) q^2 + \left(-3\zeta_{12}^2 + 1\right) q^3 + \cdots,\\
f_3 &= q + \left(\zeta_{12}^2 + 2\right) q^2 + \left(-\zeta_{12}^2 + 3\right) q^3 + \cdots,\\
f_4 &= -\zeta_{12}^2 + q + \left(2\zeta_{12}^2 - 1\right) q^2 + \left(3\zeta_{12}^2 - 2\right) q^3 + \cdots,\\
f_5 &= q + \left(\zeta_{12}^2 + 1\right) q^2 + \left(\zeta_{12}^2 + 2\right) q^3 + \cdots,\\
f_6 &= -1 + q + -q^2 + 4q^3 + \cdots,\\
f_7 &= q + q^2 + 4q^3 + \cdots,\\
f_8 &= \zeta_{12}^2 - 1 + q + \left(-2\zeta_{12}^2 + 1\right) q^2 + \left(-3\zeta_{12}^2 + 1\right) q^3 + \cdots,\\
f_9 &= q + \left(-\zeta_{12}^2 + 2\right) q^2 + \left(-\zeta_{12}^2 + 3\right) q^3 + \cdots,\\
f_{10} &= \frac{7}{13}\zeta_{12}^2 - \frac{18}{13} + q + \left(-2\zeta_{12}^2 + 3\right) q^2 + \left(3\zeta_{12}^2 - 2\right) q^3 + \cdots,\\
f_{11} &= q + \left(-\zeta_{12}^2 + 3\right) q^2 + \left(\zeta_{12}^2 + 2\right) q^3 + \cdots.
\end{aligned}$$

We computed it as follows:

```
sage: E = EisensteinForms(Gamma1(13),2)
sage: E.eisenstein_series()
```

We can also compute the parameters χ, ψ, t that define each series:

```
sage: e = E.eisenstein_series()
sage: for e in E.eisenstein_series():
...       print e.parameters()
...
([1], [1], 13)
([1], [zeta6], 1)
([zeta6], [1], 1)
([1], [zeta6 - 1], 1)
([zeta6 - 1], [1], 1)
([1], [-1], 1)
([-1], [1], 1)
([1], [-zeta6], 1)
([-zeta6], [1], 1)
([1], [-zeta6 + 1], 1)
([-zeta6 + 1], [1], 1)
```

5.4. Exercises

5.1 Suppose A and B are diagonalizable linear transformations of a finite-dimensional vector space V over an algebraically closed field K and that $AB = BA$. Prove there is a basis for V so that the matrices of A and B with respect to that basis are both simultaneously diagonal.

5.2 If ε is the trivial character of modulus 1 and B_k are as in Section 2.1, then $B_{k,\varepsilon} = B_k$, except when $k = 1$, in which case $B_{1,\varepsilon} = -B_1 = 1/2$.

5.3 Prove that for $k \geq 2$ if $\varepsilon(-1) \neq (-1)^k$, then $B_{k,\varepsilon} = 0$.

5.4 Show that the dimension of the Eisenstein subspace $E_3(\Gamma_1(13))$ is 12 by finding a basis of series $E_{k,\chi,\psi}$. You do not have to write down the q-expansions of the series, but you do have to figure out which χ, ψ to use.

Chapter 6

Dimension Formulas

When computing with spaces of modular forms, it is helpful to have easy-to-compute formulas for dimensions of these spaces. Such formulas provide a check on the output of the algorithms from Chapter 8 that compute explicit bases for spaces of modular forms. We can also use dimension formulas to improve the efficiency of some of the algorithms in Chapter 8, since we can use them to determine the ranks of certain matrices without having to explicitly compute those matrices. Dimension formulas can also be used in generating bases of q-expansions; if we know the dimension of $M_k(N, \varepsilon)$ and if we have a process for computing q-expansions of elements of $M_k(N, \varepsilon)$, e.g., multiplying together q-expansions of certain forms of smaller weight, then we can tell when we are done generating $M_k(N, \varepsilon)$.

This chapter contains formulas for dimensions of spaces of modular forms, along with some remarks about how to evaluate these formulas. In some cases we give dimension formulas for spaces that we will define in later chapters. We also give many examples, some of which were computed using the modular symbols algorithms from Chapter 8.

Many of the dimension formulas and algorithms we give below grew out of Shimura's book [**Shi94**] and a program that Bruce Kaskel wrote (around 1996) in PARI, which Kevin Buzzard extended. That program codified dimension formulas that Buzzard and Kaskel found or extracted from the literature (mainly [**Shi94**, §2.6]). The algorithms for dimensions of spaces with nontrivial character are from [**CO77**], with some refinements suggested by Kevin Buzzard.

For the rest of this chapter, N denotes a positive integer and $k \geq 2$ is an integer. We will give *no simple formulas* for dimensions of spaces of weight 1 modular forms; in fact, it might not be possible to give such formulas since the methods used to derive the formulas below do not apply in the case $k = 1$. If $k = 0$, the only modular forms are the constants, and for $k < 0$ the dimension of $M_k(N, \varepsilon)$ is 0.

For a nonzero integer N and a prime p, let $v_p(N)$ be the largest integer e such that $p^e \mid N$. In the formulas in this chapter, p always denotes a prime number. Let $M_k(N, \varepsilon)$ be the space of modular forms of level N weight k and character ε, and let $S_k(N, \varepsilon)$ and $E_k(N, \varepsilon)$ be the cuspidal and Eisenstein subspaces, respectively.

The dimension formulas below for $S_k(\Gamma_0(N))$, $S_k(\Gamma_1(N))$, $E_k(\Gamma_0(N))$ and $E_k(\Gamma_1(N))$ can be found in [**DS05**, Ch. 3], [**Shi94**, §2.6][1] and [**Miy89**, §2.5]. They are derived using the Riemann-Roch Theorem applied to the covering $X_0(N) \to X_0(1)$ or $X_1(N) \to X_1(1)$ and appropriately chosen divisors. It would be natural to give a sample argument along these lines at this point, but we will not since it easy to find such arguments in other books and survey papers (see, e.g., [**DI95**]). So you will not learn much about how to derive dimension formulas from this chapter. What you will learn is precisely what the dimension formulas are, which is something that is often hard to extract from obscure references.

In addition to reading this chapter, the reader may wish to consult [**Mar05**] for proofs of similar dimension formulas, asymptotic results, and a nonrecursive formula for dimensions of certain new subspaces.

6.1. Modular Forms for $\Gamma_0(N)$

For any prime p and any positive integer N, let $v_p(N)$ be the power of p that divides N. Also, let

$$\mu_0(N) = \prod_{p \mid N} \left(p^{v_p(N)} + p^{v_p(N)-1}\right),$$

$$\mu_{0,2}(N) = \begin{cases} 0 & \text{if } 4 \mid N, \\ \prod_{p \mid N} \left(1 + \left(\frac{-4}{p}\right)\right) & \text{otherwise,} \end{cases}$$

$$\mu_{0,3}(N) = \begin{cases} 0 & \text{if } 2 \mid N \text{ or } 9 \mid N, \\ \prod_{p \mid N} \left(1 + \left(\frac{-3}{p}\right)\right) & \text{otherwise,} \end{cases}$$

$$c_0(N) = \sum_{d \mid N} \varphi(\gcd(d, N/d)),$$

$$g_0(N) = 1 + \frac{\mu_0(N)}{12} - \frac{\mu_{0,2}(N)}{4} - \frac{\mu_{0,3}(N)}{3} - \frac{c_0(N)}{2}.$$

[1] The formulas in [**Shi94**, §2.6] contain some minor mistakes.

Note that $\mu_0(N)$ is the index of $\Gamma_0(N)$ in $\mathrm{SL}_2(\mathbb{Z})$ (see Exercise 6.1).

Proposition 6.1. *We have* $\dim S_2(\Gamma_0(N)) = g_0(N)$*, and for* $k \geq 4$ *even,*

$$\dim S_k(\Gamma_0(N)) = (k-1)\cdot(g_0(N)-1) + \left(\frac{k}{2}-1\right)\cdot c_0(N) + \mu_{0,2}(N)\cdot\left\lfloor\frac{k}{4}\right\rfloor + \mu_{0,3}(N)\cdot\left\lfloor\frac{k}{3}\right\rfloor.$$

The dimension of the Eisenstein subspace is

$$\dim E_k(\Gamma_0(N)) = \begin{cases} c_0(N) & \text{if } k \neq 2, \\ c_0(N)-1 & \text{if } k = 2. \end{cases}$$

The following is a table of $\dim S_k(\Gamma_0(N))$ for some values of N and k:

N	$S_2(\Gamma_0(N))$	$S_4(\Gamma_0(N))$	$S_6(\Gamma_0(N))$	$S_{24}(\Gamma_0(N))$
1	0	0	0	2
10	0	3	5	33
11	1	2	4	22
100	7	36	66	336
389	32	97	161	747
1000	131	430	730	3430
2007	221	806	1346	6206
100000	14801	44800	74800	344800

Example 6.2. Use the commands `dimension_cusp_forms`, `dimension_eis`, and `dimension_modular_forms` to compute the dimensions of the three spaces $S_k(\Gamma_0(N))$, $E_k(\Gamma_0(N))$ and $M_k(\Gamma_0(N))$, respectively. For example,

```
sage: dimension_cusp_forms(Gamma0(2007),2)
221
sage: dimension_eis(Gamma0(2007),2)
7
sage: dimension_modular_forms(Gamma0(2007),2)
228
```

Remark 6.3. Csirik, Wetherell, and Zieve prove in [**CWZ01**] that a random positive integer has probability 0 of being a value of

$$g_0(N) = \dim S_2(\Gamma_0(N)),$$

and they give bounds on the size of the set of values of $g_0(N)$ below some given x. For example, they show that $150, 180, 210, 286, 304, 312, \ldots$ are the first few integers that are not of the form $g_0(N)$ for any N. See Figure 6.1.1 for a plot of the very erratic function $g_0(N)$. In contrast, the function $k \mapsto \dim S_{2k}(\Gamma_0(12))$ is very well behaved (see Figure 6.1.2).

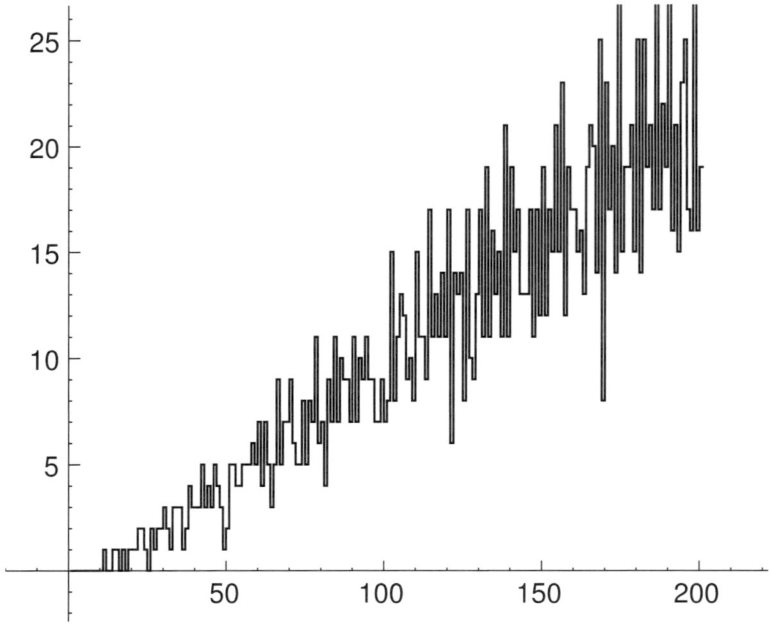

Figure 6.1.1. Dimension of $S_2(\Gamma_0(N))$ as a function of N.

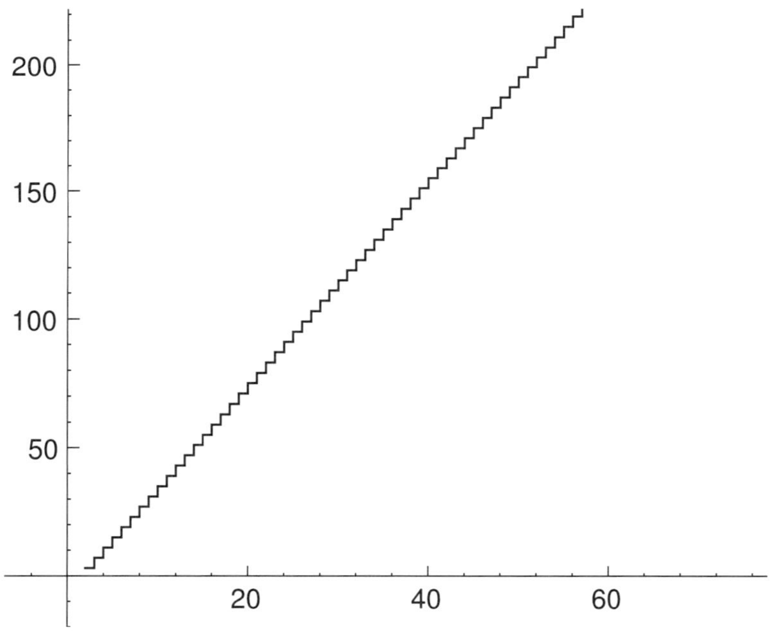

Figure 6.1.2. Dimension of $S_{2k}(\Gamma_0(12))$ as a function of k.

6.1.1. New and Old Subspaces. In this section we assume the reader is either familiar with newforms or has read Section 9.2.

For any integer R, let

$$\overline{\mu}(R) = \begin{cases} 0 & \text{if } p^3 \mid R \text{ for some } p, \\ \prod_{p \| R} -2 & \text{otherwise,} \end{cases}$$

where the product is over primes that exactly divide R. Note that $\overline{\mu}$ is *not* the Moebius function, but it has a similar flavor.

Proposition 6.4. *The dimension of the new subspace is*

$$\dim S_k(\Gamma_0(N))_{\text{new}} = \sum_{M \mid N} \overline{\mu}(N/M) \cdot \dim S_k(\Gamma_0(M)),$$

where the sum is over the positive divisors M of N. As a consequence of Theorem 9.4, we also have

$$\dim S_k(\Gamma_0(N)) = \sum_{M \mid N} \sigma_0(N/M) \dim S_k(\Gamma_0(M))_{\text{new}},$$

where $\sigma_0(N/M)$ is the number of divisors of N/M.

Example 6.5. We compute the dimension of the new subspace of $S_k(\Gamma_0(N))$ using the SAGE command `dimension_new_cusp_forms` as follows:

```
sage: dimension_new_cusp_forms(Gamma0(11),12)
8
sage: dimension_cusp_forms(Gamma0(11),12)
10
sage: dimension_new_cusp_forms(Gamma0(2007),12)
1017
sage: dimension_cusp_forms(Gamma0(2007),12)
2460
```

6.2. Modular Forms for $\Gamma_1(N)$

This section follows Section 6.1 closely, but with suitable modifications with $\Gamma_0(N)$ replaced by $\Gamma_1(N)$.

Define functions of a positive integer N by the following formulas:

$$\mu_1(N) = \begin{cases} \mu_0(N) & \text{if } N = 1,2, \\ \dfrac{\phi(N)\cdot\mu_0(N)}{2} & \text{otherwise,} \end{cases}$$

$$\mu_{1,2}(N) = \begin{cases} 0 & \text{if } N \geq 4, \\ \mu_{0,2}(N) & \text{otherwise,} \end{cases}$$

$$\mu_{1,3}(N) = \begin{cases} 0 & \text{if } N \geq 4, \\ \mu_{0,3}(N) & \text{otherwise,} \end{cases}$$

$$c_1(N) = \begin{cases} c_0(N) & \text{if } N = 1,2, \\ 3 & \text{if } N = 4, \\ \displaystyle\sum_{d|N} \frac{\phi(d)\phi(N/d)}{2} & \text{otherwise,} \end{cases}$$

$$g_1(N) = 1 + \frac{\mu_1(N)}{12} - \frac{\mu_{1,2}(N)}{4} - \frac{\mu_{1,3}(N)}{3} - \frac{c_1(N)}{2}.$$

Note that $g_1(N)$ is the genus of the modular curve $X_1(N)$ (associated to $\Gamma_1(N)$) and $c_1(N)$ is the number of cusps of $X_1(N)$.

Proposition 6.6. *We have* $\dim S_2(\Gamma_1(N)) = g_1(N)$. *If* $N \leq 2$, *then* $\Gamma_0(N) = \Gamma_1(N)$ *so*

$$\dim S_k(\Gamma_1(N)) = \dim S_k(\Gamma_0(N)),$$

where $\dim S_k(\Gamma_0(N))$ *is given by the formula of Proposition 6.1. If* $k \geq 3$, *let*

$$a(N,k) = (k-1)(g_1(N)-1) + \left(\frac{k}{2} - 1\right)\cdot c_1(N).$$

Then for $N \geq 3$,

$$\dim S_k(\Gamma_1(N)) = \begin{cases} a + 1/2 & \textit{if } N = 4 \textit{ and } 2 \nmid k, \\ a + \lfloor k/3 \rfloor & \textit{if } N = 3, \\ a & \textit{otherwise.} \end{cases}$$

The dimension of the Eisenstein subspace is as follows:

$$\dim E_k(\Gamma_1(N)) = \begin{cases} c_1(N) & \textit{if } k \neq 2, \\ c_1(N) - 1 & \textit{if } k = 2. \end{cases}$$

The dimension of the new subspace of $M_k(\Gamma_1(N))$ *is*

$$\dim S_k(\Gamma_1(N))_{\text{new}} = \sum_{M|N} \overline{\mu}(N/M)\cdot \dim S_k(\Gamma_1(M)),$$

where $\overline{\mu}$ *is as in the statement of Proposition 6.4.*

Remark 6.7. Since $M_k = S_k \oplus E_k$, the formulas above for $\dim S_k$ and $\dim E_k$ also yield a formula for the dimension of M_k.

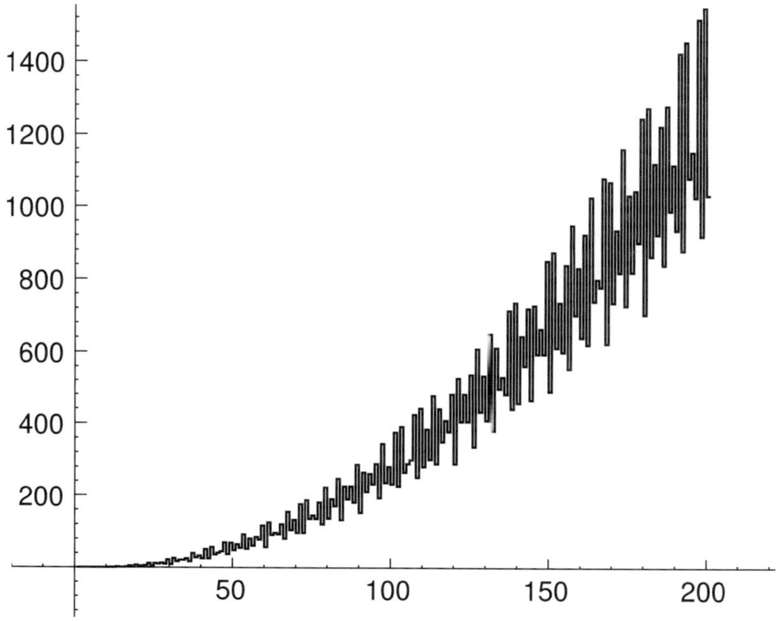

Figure 6.2.1. Dimension of $S_2(\Gamma_1(N))$ as a function of N.

The following table contains the dimension of $S_k(\Gamma_1(N))$ for some sample values of N and k:

N	$S_2(\Gamma_1(N))$	$S_3(\Gamma_1(N))$	$S_4(\Gamma_1(N))$	$S_{24}(\Gamma_1(N))$
1	0	0	0	2
10	0	2	5	65
11	1	5	10	110
100	231	530	830	6830
389	6112	12416	18721	144821
1000	28921	58920	88920	688920
2007	147409	296592	445776	3429456
100000	299792001	599792000	899792000	6899792000

Example 6.8. We compute dimensions of spaces of modular forms for $\Gamma_1(N)$:

```
sage: dimension_cusp_forms(Gamma1(2007),2)
147409
sage: dimension_eis(Gamma1(2007),2)
3551
sage: dimension_modular_forms(Gamma1(2007),2)
150960
```

6.3. Modular Forms with Character

Fix a Dirichlet character ε of modulus N, and let c be the conductor of ε (we do *not* assume that ε is primitive). Assume that $\varepsilon \neq 1$, since otherwise $M_k(N,\varepsilon) = M_k(\Gamma_0(N))$ and the formulas of Section 6.1 apply. Also, assume that $\varepsilon(-1) = (-1)^k$, since otherwise $\dim M_k(\Gamma_0(N)) = 0$. In this section we discuss formulas for computing each of $M_k(N,\varepsilon)$, $S_k(N,\varepsilon)$ and $E_k(N,\varepsilon)$.

In [**CO77**], Cohen and Oesterlé assert (without *published* proof; see Remark 6.11 below) that for any $k \in \mathbb{Z}$ and N, ε as above,

$$\begin{aligned}\dim S_k(N,\varepsilon) &- \dim M_{2-k}(N,\varepsilon)\\ &= \frac{k-1}{12}\cdot\mu_0(N) \;-\; \frac{1}{2}\cdot\prod_{p\mid N}\lambda(p,N,v_p(c))\\ &\quad+\gamma_4(k)\cdot\sum_{x\in A_4(N)}\varepsilon(x) \;+\; \gamma_3(k)\cdot\sum_{x\in A_3(N)}\varepsilon(x)\end{aligned}$$

where $\mu_0(N)$ is as in Section 6.1, $A_4(N) = \{x \in \mathbb{Z}/N\mathbb{Z} : x^2+1 = 0\}$ and $A_3(N) = \{x \in \mathbb{Z}/N\mathbb{Z} : x^2+x+1 = 0\}$, and γ_3, γ_4 are

$$\gamma_4(k) = \begin{cases} -1/4 & \text{if } k \equiv 2 \pmod 4,\\ 1/4 & \text{if } k \equiv 0 \pmod 4,\\ 0 & \text{if } k \text{ is odd.}\end{cases}$$

$$\gamma_3(k) = \begin{cases} -1/3 & \text{if } k \equiv 2 \pmod 3,\\ 1/3 & \text{if } k \equiv 0 \pmod 3,\\ 0 & \text{if } k \equiv 1 \pmod 3.\end{cases}$$

It remains to define λ. Fix a prime divisor $p \mid N$ and let $r = v_p(N)$. Then

$$\lambda(p,N,v_p(c)) = \begin{cases} p^{\frac{r}{2}} + p^{\frac{r}{2}-1} & \text{if } 2\cdot v_p(c) \le r \text{ and } 2 \mid r,\\ 2\cdot p^{\frac{r-1}{2}} & \text{if } 2\cdot v_p(c) \le r \text{ and } 2 \nmid r,\\ 2\cdot p^{r-v_p(c)} & \text{if } 2\cdot v_p(c) > r.\end{cases}$$

This flexible formula can be used to compute the dimension of $M_k(N,\varepsilon)$, $S_k(N,\varepsilon)$, and $E_k(N,\varepsilon)$ for any N, ε, $k \neq 1$, by using that

$$\begin{aligned} \dim S_k(N,\varepsilon) &= 0 \qquad \text{if } k \leq 0,\\ \dim M_k(N,\varepsilon) &= 0 \qquad \text{if } k < 0,\\ \dim M_0(N,\varepsilon) &= 1 \qquad \text{if } k = 0. \end{aligned}$$

One thing that is not straightforward when implementing an algorithm to compute the above dimension formulas is how to efficiently compute the sets $A_4(N)$ and $A_6(N)$. Kevin Buzzard suggested the following two algorithms. Note that if k is odd, then $\gamma_4(k) = 0$, so the sum over $A_4(N)$ is only needed when k is even.

Algorithm 6.9 (Sum over $A_4(N)$). *Given a positive integer N and an even Dirichlet character ε of modulus N, this algorithm computes $\sum_{x \in A_4(N)} \varepsilon(x)$.*

(1) [Factor N] Compute the prime factorization $p_1^{e_1} \cdots p_n^{e_n}$ of N.
(2) [Initialize] Set $t = 1$ and $i = 0$.
(3) [Loop Over Prime Divisors] Set $i = i + 1$. If $i > n$, return t. Otherwise set $p = p_i$ and $e = e_i$.
   (a) If $p \equiv 3 \pmod 4$, return 0.
   (b) If $p = 2$ and $e > 1$, return 0.
   (c) If $p = 2$ and $e = 1$, go to step (3).
   (d) Compute a generator $a \in (\mathbb{Z}/p\mathbb{Z})^*$ using Algorithm 4.4.
   (e) Compute $\omega = a^{(p-1)/4}$.
   (f) Use the Chinese Remainder Theorem to find $x \in \mathbb{Z}/N\mathbb{Z}$ such that $x \equiv a \pmod p$ and $x \equiv 1 \pmod{N/p^e}$.
   (g) Set $x = x^{p^{r-1}}$.
   (h) Set $s = \varepsilon(x)$.
   (i) If $s = 1$, set $t = 2t$ and go to step (3).
   (j) If $s = -1$, set $t = -2t$ and go to step (3).

Proof. Note that $\varepsilon(-x) = \varepsilon(x)$, since ε is even. By the Chinese Remainder Theorem, the set $A_4(N)$ is empty if and only if there is no square root of -1 modulo some prime power divisor of p. If $A_4(N)$ is empty, the algorithm correctly detects this fact in steps (3a)–(3b). Thus assume $A_4(N)$ is nonempty. For each prime power $p_i^{e_i}$ that exactly divides N, let $x_i \in Z/N\mathbb{Z}$ be such that $x_i^2 = -1$ and $x_i \equiv 1 \pmod{p_j^{e_j}}$ for $i \neq j$. This is the value of x computed in steps (3d)–(3g) (as one sees using elementary number theory).

The next key observation is that

$$\prod_i (\varepsilon(x_i) + \varepsilon(-x_i)) = \sum_{x \in A_4(N)} \varepsilon(x), \tag{6.3.1}$$

since by the Chinese Remainder Theorem the elements of $A_4(N)$ are in bijection with the choices for a square root of -1 modulo each prime power divisors of N. The observation (6.3.1) is a huge gain from an efficiency point of view—if N had r prime factors, then $A_4(N)$ would have size 2^r, which could be prohibitive, where the product involves only r factors. To finish the proof, just note that steps (3h)–(3j) compute the local factors $\varepsilon(x_i) + \varepsilon(-x_i) = 2\varepsilon(x_i)$, where again we use that ε is even. Note that a solution of $x^2+1 \equiv 0 \pmod{p}$ lifts uniquely to a solution mod p^n for any n, because the kernel of the natural homomorphism $(\mathbb{Z}/p^n\mathbb{Z})^* \to (\mathbb{Z}/p\mathbb{Z})^*$ is a group of p-power order. □

The algorithm for computing the sum over $A_3(N)$ is similar.

For $k \geq 2$, to compute $\dim S_k(N, \varepsilon)$, use the formula directly and the fact that $\dim M_{2-k}(N, \varepsilon) = 0$, unless $\varepsilon = 1$ and $k = 2$. To compute $\dim M_k(N, \varepsilon)$ for $k \geq 2$, use the fact that the big formula at the beginning of this section is valid for any integer k to replace k by $2 - k$ and that $\dim S_k(N, \varepsilon) = 0$ for $k \leq 0$ to rewrite the formula as

$$\begin{aligned}\dim M_k(N, \varepsilon) &= -(\dim S_{2-k}(N, \varepsilon) - \dim M_k(N, \varepsilon))\\ &= -\Big(\frac{1-k}{12} \cdot \mu_0(N) \;-\; \frac{1}{2} \cdot \prod_{p \mid N} \lambda(p, N, v_p(c))\\ &\qquad + \gamma_4(2-k) \cdot \sum_{x \in A_4(N)} \varepsilon(x) \;+\; \gamma_3(2-k) \cdot \sum_{x \in A_3(N)} \varepsilon(x)\Big).\end{aligned}$$

Note also that for $k = 0$, $\dim E_k(N, \varepsilon) = 1$ if and only if ε is trivial and it equals 0 otherwise. We then also obtain

$$\dim E_k(N, \varepsilon) = \dim M_k(N, \varepsilon) - \dim S_k(N, \varepsilon).$$

We can also compute $\dim E_k(N, \varepsilon)$ when $k = 1$ directly, since

$$\dim S_{2-1}(N, \varepsilon) = \dim S_1(N, \varepsilon).$$

The following table contains the dimension of $S_k(N, \varepsilon)$ for some sample values of N and k. In each case, ε is the product of characters ε_p of maximal order corresponding to the prime power factors of N (i.e., the product of the generators of the group $D(N, \mathbb{C}^*)$ of Dirichlet characters of modulus N).

N	$\dim S_2(N,\varepsilon)$	$\dim S_3(N,\varepsilon)$	$\dim S_4(N,\varepsilon)$	$\dim S_{24}(N,\varepsilon)$
1	0	0	0	2
10	0	1	0	0
11	0	1	0	0
100	13	0	43	343
389	0	64	0	0
1000	148	0	448	3448
2007	222	0	670	5150

Example 6.10. We compute the last line of the above table. First we create the character ε.

```
sage: G = DirichletGroup(2007)
sage: e = prod(G.gens(), G(1))
```

Next we compute the dimension of the four spaces.

```
sage: dimension_cusp_forms(e,2)
222
sage: dimension_cusp_forms(e,3)
0
sage: dimension_cusp_forms(e,4)
670
sage: dimension_cusp_forms(e,24)
5150
```

We can also compute dimensions of the corresponding spaces of Eisenstein series.

```
sage: dimension_eis(e,2)
4
sage: dimension_eis(e,3)
0
sage: dimension_eis(e,4)
4
sage: dimension_eis(e,24)
4
```

Remark 6.11. Cohen and Oesterlé also give dimension formulas for spaces of half-integral weight modular forms, which we do not give in this chapter. Note that [**CO77**] does not contain any *proofs* that their claimed formulas are correct, but instead they say only that “Les formules qui les donnent sont connues de beaucoup de gens et il existe plusieurs méthodes permettant de les obtenir (théorème de Riemann-Roch, application des formules de trace données par Shimura)."[2] Fortunately, in [**Que06**], Jordi Quer derives the (integral weight) formulas of [**CO77**] along with formulas for dimensions of spaces $S_k(G)$ and $M_k(G)$ for more general congruence subgroups.

Let f be the conductor of a Dirichlet character ε of modulus N. Then the dimension of the new subspace of $M_k(N,\varepsilon)$ is

$$\dim S_k(N,\varepsilon)_{\text{new}} = \sum_{M \text{ such that } f|M|N} \overline{\mu}(N/M)\cdot \dim S_k(M,\varepsilon'),$$

where $\overline{\mu}$ is as in the statement of Proposition 6.4, and ε' is the restriction of ε mod M.

Example 6.12. We compute the dimension of $S_2(2007,\varepsilon)_{\text{new}}$ for ε a quadratic character of modulus 2007.

```
sage: G = DirichletGroup(2007, QQ)
sage: e = prod(G.gens(), G(1))
sage: dimension_new_cusp_forms(e,2)
76
```

6.4. Exercises

6.1 Let μ_0 and μ_1 be as in this chapter.
   (a) Prove that $\mu_0(N) = [\mathrm{SL}_2(\mathbb{Z}) : \Gamma_0(N)]$.
   (b) Prove that for $N \geq 3$, $\mu_1(N) = [\mathrm{SL}_2(\mathbb{Z}) : \Gamma_1(N)]/2$, so $\mu_1(N)$ is the index of $\Gamma_1(N)\cdot\{\pm 1\}$ in $\mathrm{PSL}_2(\mathbb{Z}) = \mathrm{SL}_2(\mathbb{Z})/\{\pm 1\}$.

6.2 Use Proposition 6.4 to find a formula for $\dim S_k(\mathrm{SL}_2(\mathbb{Z}))$. Verify that this formula is the same as the one in Corollary 2.16.

6.3 Suppose either that $N = 1$ or that N is prime and $k = 2$. Prove that $M_k(\Gamma_0(N))_{\text{new}} = M_k(\Gamma_0(N))$.

6.4 Fill in the details of the proof of Algorithm 6.9.

6.5 Implement a computer program to compute $\dim S_k(\Gamma_0(N))$ as a function of k and N.

[2]The formulas that we give here are well known and there exist many methods to prove them, e.g., the Riemann-Roch theorem and applications of the trace formula of Shimura.

Chapter 7

Linear Algebra

This chapter is about several algorithms for matrix algebra over the rational numbers and cyclotomic fields. Algorithms for linear algebra over exact fields are necessary in order to implement the modular symbols algorithms that we will describe in Chapter 7. This chapter partly overlaps with [**Coh93**, Sections 2.1–2.4].

Note: We view all matrices as defining linear transformation by acting on row vectors from the right.

7.1. Echelon Forms of Matrices

Definition 7.1 (Reduced Row Echelon Form). A matrix is in (reduced row) *echelon form* if each row in the matrix has more zeros at the beginning than the row above it, the first nonzero entry of every row is 1, and the first nonzero entry of any row is the only nonzero entry in its column.

Given a matrix A, there is another matrix B such that B is obtained from A by left multiplication by an invertible matrix and B is in reduced row echelon form. This matrix B is called the echelon form of A. It is unique.

A *pivot column* of A is a column of A such that the reduced row echelon form of A contains a leading 1.

Example 7.2. The following matrix is not in reduced row echelon form:

$$\begin{pmatrix} 14 & 2 & 7 & 228 & -224 \\ 0 & 0 & 3 & 78 & -70 \\ 0 & 0 & 0 & -405 & 381 \end{pmatrix}.$$

The reduced row echelon form of the above matrix is

$$\begin{pmatrix} 1 & \frac{1}{7} & 0 & 0 & -\frac{1174}{945} \\ 0 & 0 & 1 & 0 & \frac{152}{135} \\ 0 & 0 & 0 & 1 & -\frac{127}{135} \end{pmatrix}.$$

Notice that the entries of the reduced row echelon form can be rationals with large denominators even though the entries of the original matrix A are integers. Another example is the simple looking matrix

$$\begin{pmatrix} -9 & 6 & 7 & 3 & 1 & 0 & 0 & 0 \\ -10 & 3 & 8 & 2 & 0 & 1 & 0 & 0 \\ 3 & -6 & 2 & 8 & 0 & 0 & 1 & 0 \\ -8 & -6 & -8 & 6 & 0 & 0 & 0 & 1 \end{pmatrix}$$

whose echelon form is

$$\begin{pmatrix} 1 & 0 & 0 & 0 & \frac{42}{1025} & -\frac{92}{1025} & \frac{1}{25} & -\frac{9}{205} \\ 0 & 1 & 0 & 0 & \frac{716}{3075} & -\frac{641}{3075} & -\frac{2}{75} & -\frac{7}{615} \\ 0 & 0 & 1 & 0 & -\frac{83}{1025} & \frac{133}{1025} & \frac{1}{25} & -\frac{23}{410} \\ 0 & 0 & 0 & 1 & \frac{184}{1025} & -\frac{159}{1025} & \frac{2}{25} & \frac{9}{410} \end{pmatrix}.$$

A basic fact is that two matrices A and B have the same reduced row echelon form if and only if there is an invertible matrix E such that $EA = B$. Also, many standard operations in linear algebra, e.g., computation of the kernel of a linear map, intersection of subspaces, membership checking, etc., can be encoded as a question about computing the echelon form of a matrix.

The following standard algorithm computes the echelon form of a matrix.

Algorithm 7.3 (Gauss Elimination). *Given an $m \times n$ matrix A over a field, the algorithm outputs the reduced row echelon form of A. Write $a_{i,j}$ for the i, j entry of A, where $0 \leq i \leq m-1$ and $0 \leq j \leq n-1$.*

(1) [Initialize] Set $k = 0$.
(2) [Clear Each Column] For each column $c = 0, 1, \ldots, n-1$, clear the cth column as follows:
   (a) [First Nonzero] Find the smallest r such that $a_{r,c} \neq 0$, or if there is no such r, go to the next column.
   (b) [Rescale] Replace row r of A by $\frac{1}{a_{r,c}}$ times row r.
   (c) [Swap] Swap row r with row k.
   (d) [Clear] For each $i = 0, \ldots, m-1$ with $i \neq k$, if $a_{i,c} \neq 0$, add $-a_{i,c}$ times row k of A to row i to clear the leading entry of the ith row.
   (e) [Increment] Set $k = k + 1$.

This algorithm takes $O(mn^2)$ arithmetic operations in the base field, where A is an $m \times n$ matrix. If the base field is $\mathbb{Q}$, the entries can become huge and arithmetic operations are then very expensive. See Section 7.3 for ways to mitigate this problem.

To conclude this section, we mention how to convert a few standard problems into questions about reduced row echelon forms of matrices. Note that one can also phrase some of these answers in terms of the echelon form, which might be easier to compute, or an LUP decomposition (lower triangular times upper triangular times permutation matrix), which the numerical analysts use.

(1) **Kernel of A:** We explain how to compute the kernel of A acting on column vectors from the right (first transpose to obtain the kernel of A acting on row vectors). Since passing to the reduced row echelon form of A is the same as multiplying on the left by an invertible matrix, the kernel of the reduce row echelon form E of A is the same as the kernel of A. There is a basis vector of $\ker(E)$ that corresponds to each nonpivot column of E. That vector has a 1 at the nonpivot column, 0's at all other nonpivot columns, and for each pivot column, the negative of the entry of A at the nonpivot column in the row with that pivot element.

(2) **Intersection of Subspaces:** Suppose W_1 and W_2 are subspace of a finite-dimensional vector space V. Let A_1 and A_2 be matrices whose columns form a basis for W_1 and W_2, respectively. Let $A = [A_1|A_2]$ be the augmented matrix formed from A_1 and A_2. Let K be the kernel of the linear transformation defined by A. Then K is isomorphic to the desired intersection. To write down the intersection explicitly, suppose that $\dim(W_1) \leq \dim(W_2)$ and do the following: For each b in a basis for K, write down the linear combination of a basis for W_1 obtained by taking the first $\dim(W_1)$ entries of the vector b. The fact that b is in $\text{Ker}(A)$ implies that the vector we just wrote down is also in W_2. This is because a linear relation

$$\sum a_i w_{1,i} + \sum b_j w_{2,j} = 0,$$

i.e., an element of that kernel, is the same as

$$\sum a_i w_{1,i} = \sum -b_j w_{2,j}.$$

For more details, see [**Coh93**, Alg. 2.3.9].

7.2. Rational Reconstruction

Rational reconstruction is a process that allows one to sometimes lift an integer modulo m uniquely to a bounded rational number.

Algorithm 7.4 (Rational Reconstruction). *Given an integer $a \geq 0$ and an integer $m > 1$, this algorithm computes the numerator n and denominator d of the unique rational number n/d, if it exists, with*

$$|n|, d \leq \sqrt{\frac{m}{2}} \qquad \text{and} \qquad n \equiv ad \pmod{m}, \tag{7.2.1}$$

or it reports that there is no such number.

(1) [Reduce mod m] Replace a with the least integer between 0 and $m-1$ that is congruent to a modulo m.
(2) [Trivial Cases] If $a = 0$ or $a = 1$, return a.
(3) [Initialize] Let $b = \sqrt{m/2}$, $u = m$, $v = a$, and set $U = (1, 0, u)$ and $V = (0, 1, v)$. Use the notation U_i and V_i to refer to the ith entries of U, V, for $i = 0, 1, 2$.
(4) [Iterate] Do the following as long as $|V_2| > b$: Set $q = \lfloor U_2/V_2 \rfloor$, set $T = U - qV$, and set $U = V$ and $V = T$.
(5) [Numerator and Denominator] Set $d = |V_1|$ and $n = V_2$.
(6) [Good?] If $d \leq b$ and $\gcd(n, d) = 1$, return n/d; otherwise report that there is no rational number as in (7.2.1).

Algorithm 7.4 for rational reconstruction is described (with proof) in [**Knu**, pgs. 656–657] as the solution to Exercise 51 on page 379 in that book. See, in particular, the paragraph right in the middle of page 657, which describes the algorithm. Knuth attributes this rational reconstruction algorithm to Wang, Kornerup, and Gregory from around 1983.

We now give an indication of why Algorithm 7.4 computes the rational reconstruction of $a \pmod{m}$, leaving the precise details and uniqueness to [**Knu**, pgs. 656–657]. At each step in Algorithm 7.4, the 3-tuple $V = (v_0, v_1, v_2)$ satisfies

$$m \cdot v_0 + a \cdot v_1 = v_2, \tag{7.2.2}$$

and similarly for U. When computing the usual extended gcd, at the end $v_2 = \gcd(a, m)$ and v_0, v_1 give a representation of the v_2 as a $\mathbb{Z}$-linear combination of m and a. In Algorithm 7.4, we are instead interested in finding a rational number n/d such that $n \equiv a \cdot d \pmod{m}$. If we set $n = v_2$ and $d = v_1$ in (7.2.2) and rearrange, we obtain

$$n = a \cdot d + m \cdot v_0.$$

Thus at *every* step of the algorithm we find a rational number n/d such that $n \equiv ad \pmod{m}$. The problem at intermediate steps is that, e.g., v_0 could be 0, or n or d could be too large.

Example 7.5. We compute an example using SAGE.

```
sage: p = 389
sage: k = GF(p)
sage: a = k(7/13); a
210
sage: a.rational_reconstruction()
7/13
```

7.3. Echelon Forms over $\mathbb{Q}$

A difficulty with computation of the echelon form of a matrix over the rational numbers is that arithmetic with large rational numbers is time-consuming; each addition potentially requires a gcd and numerous additions and multiplications of integers. Moreover, the entries of A during intermediate steps of Algorithm 7.3 can be huge even though the entries of A and the answer are small. For example, suppose A is an invertible square matrix. Then the echelon form of A is the identity matrix, but during intermediate steps the numbers involved could be quite large. One technique for mitigating this is to compute the echelon form using a multimodular method.

If A is a matrix with rational entries, let $H(A)$ be the *height* of A, which is the maximum of the absolute values of the numerators and denominators of all entries of A. If x, y are rational numbers and p is a prime, we write $x \equiv y \pmod{p}$ to mean that the denominators of x and y are not divisible by p but the numerator of the rational number $x - y$ (in reduced form) is divisible by p. For example, if $x = 5/7$ and $y = 2/11$, then $x - y = 41/77$, so $x \equiv y \pmod{41}$.

Algorithm 7.6 (Multimodular Echelon Form). *Given an $m \times n$ matrix A with entries in $\mathbb{Q}$, this algorithm computes the reduced row echelon form of A.*

(1) Rescale the input matrix A to have integer entries. This does not change the echelon form and makes reduction modulo many primes easier. We may thus assume A has integer entries.
(2) Let c be a guess for the height of the echelon form.
(3) List successive primes $p_1, p_2, \ldots$ such that the product of the p_i is greater than $n \cdot c \cdot H(A) + 1$, where n is the number of columns of A.
(4) Compute the echelon forms B_i of the reduction $A \pmod{p_i}$ using, e.g., Algorithm 7.3 or any other echelon algorithm.
(5) Discard any B_i whose pivot column list is not maximal among pivot lists of all B_j found so far. (The pivot list associated to B_i is the ordered list of integers k such that the kth column of B_j is a pivot

column. We mean maximal with respect to the following ordering on integer sequences: shorter integer sequences are smaller, and if two sequences have the same length, then order in reverse lexicographic order. Thus $[1, 2]$ is smaller than $[1, 2, 3]$, and $[1, 2, 7]$ is smaller than $[1, 2, 5]$. Think of maximal as "optimal", i.e., best possible pivot columns.)

(6) Use the Chinese Remainder Theorem to find a matrix B with integer entries such that $B \equiv B_i \pmod{p_i}$ for all p_i.

(7) Use Algorithm 7.4 to try to find a matrix C whose coefficients are rational numbers n/r such that $|n|, r \leq \sqrt{M/2}$, where $M = \prod p_i$, and $C \equiv B_i \pmod{p_i}$ for each prime p. If rational reconstruction fails, compute a few more echelon forms mod the next few primes (using the above steps) and attempt rational reconstruction again. Let E be the matrix over $\mathbb{Q}$ so obtained. (A trick here is to keep track of denominators found so far to avoid doing very many rational reconstructions.)

(8) Compute the denominator d of E, i.e., the smallest positive integer such that dE has integer entries. If

$$H(dE) \cdot H(A) \cdot n < \prod p_i, \tag{7.3.1}$$

then E is the reduced row echelon form of A. If not, repeat the above steps with a few more primes.

Proof. We prove that if (7.3.1) is satisfied, then the matrix E computed by the algorithm really is the reduced row echelon form R of A. First note that E is in reduced row echelon form since the set of pivot columns of all matrices B_i used to construct E are the same, so the pivot columns of E are the same as those of any B_i and all other entries in the B_i pivot columns are 0, so the other entries of E in the pivot columns are also 0.

Recall from the end of Section 7.1 that a matrix whose columns are a basis for the kernel of A can be obtained from the reduced row echelon form R. Let K be the matrix whose columns are the vectors in the kernel algorithm applied to E, so $EK = 0$. Since the reduced row echelon form is obtained by left multiplying by an invertible matrix, for each i, there is an invertible matrix V_i mod p_i such that $A \equiv V_i B_i \pmod{p_i}$ so

$$A \cdot dK \equiv V_i B_i \cdot dK \equiv V_i \cdot dE \cdot K \equiv 0 \pmod{p_i}.$$

Since dK and A are integer matrices, the Chinese remainder theorem implies that

$$A \cdot dK \equiv 0 \left(\mathrm{mod}\ \prod p_i\right).$$

The integer entries a of $A \cdot dK$ all satisfy $|a| \leq H(A) \cdot H(dK) \cdot n$, where n is the number of columns of A. Since $H(K) \leq H(E)$, the bound (7.3.1) implies that $A \cdot dK = 0$. Thus $AK = 0$, so $\mathrm{Ker}(E) \subset \mathrm{Ker}(A)$. On the other hand, the rank of E equals the rank of each B_i (since the pivot columns are the same), so

$$\mathrm{rank}(E) = \mathrm{rank}(B_i) = \mathrm{rank}(A \pmod{p_i}) \leq \mathrm{rank}(A).$$

Thus $\dim(\mathrm{Ker}(A)) \leq \dim(\mathrm{Ker}(E))$, and combining this with the bound obtained above, we see that $\mathrm{Ker}(E) = \mathrm{Ker}(A)$. This implies that E is the reduced row echelon form of A, since two matrices have the same kernel if and only if they have the same reduced row echelon form (the echelon form is an invariant of the row space, and the kernel is the orthogonal complement of the row space).

The reason for step (5) is that the matrices B_i need *not* be the reduction of R modulo p_i, and indeed this reduction might not even be defined, e.g., if p_i divides the denominator of some element of R, then this reduction makes no sense. For example, set $p = p_i$ and suppose $A = \left(\begin{smallmatrix} p & 1 \\ 0 & 0 \end{smallmatrix}\right)$. Then $R = \left(\begin{smallmatrix} 1 & 1/p \\ 0 & 0 \end{smallmatrix}\right)$, which has no reduction modulo p; also, the reduction of A modulo B_i is $B_i = \left(\begin{smallmatrix} 0 & 1 \\ 0 & 0 \end{smallmatrix}\right) \pmod{p}$, which is already in reduced row echelon form. However if we were to combine B_i with the echelon form of A modulo another prime, the result could never be lifted using rational reconstruction. Thus the reason we exclude all B_i with nonmaximal pivot column sequence is so that a rational reconstruction will exist. There are only finitely many primes that divide denominators of entries of R, so eventually all B_i will have maximal pivot column sequences, i.e., they are the reduction of the true reduced row echelon form R, so the algorithm terminates. □

Remark 7.7. Algorithm 7.6, with *sparse* matrices seems to work very well in practice. A simple but helpful modification to Algorithm 7.3 in the sparse case is to clear each column using a row with a minimal number of nonzero entries, so as to reduce the amount of "fill in" (denseness) of the matrix. There are much more sophisticated methods along these lines called "intelligent Gauss elimination". (Cryptographers are interested in linear algebra mod p with huge sparse matrices, since they come up in attacks on the discrete log problem and integer factorization.)

One can adapt Algorithm 7.6 to computation of echelon forms of matrices A over cyclotomic fields $\mathbb{Q}(\zeta_n)$. Assume A has denominator 1. Let p be a prime that splits completely in $\mathbb{Q}(\zeta_n)$. Compute the homomorphisms $f_i : \mathbb{Z}_p[\zeta_n] \to \mathbb{F}_p$ by finding the elements of order n in $\mathbb{F}_p^*$. Then compute the mod p matrix $f_i(A)$ for each i, and find its reduced row echelon form. Taken together, the maps f_i together induce an isomorphism $\Psi : \mathbb{F}_p[X]/\Phi_n(X) \cong \mathbb{F}_p^d$, where $\Phi_n(X)$ is the nth cyclotomic polynomial and d is its degree. It is easy to compute $\Psi(f(x))$ by evaluating $f(x)$ at each element of order n in $\mathbb{F}_p$. To compute Ψ^{-1}, simply use linear algebra over $\mathbb{F}_p$

to invert a matrix that represents Ψ. Use Ψ^{-1} to compute the reduced row echelon form of $A \pmod{p}$, where (p) is the nonprime ideal in $\mathbb{Z}[\zeta_n]$ generated by p. Do this for several primes p, and use rational reconstruction on each coefficient of each power of ζ_n, to recover the echelon form of A.

7.4. Echelon Forms via Matrix Multiplication

In this section we explain how to compute echelon forms using matrix multiplication. This is valuable because there are asymptotically fast, i.e., better than $O(n^3)$ field operations, algorithms for matrix multiplication, and implementations of linear algebra libraries often include highly optimized matrix multiplication algorithms. We only sketch the basic ideas behind these asymptotically fast algorithms (following [**Ste**]), since more detail would take us too far from modular forms.

The naive algorithm for multiplying two $m \times m$ matrices requires $O(m^3)$ arithmetic operations in the base ring. In [**Str69**], Strassen described a clever algorithm that computes the product of two $m \times m$ matrices in $O(m^{\log_2(7)}) = O(m^{2.807\ldots})$ arithmetic operations in the base ring. Because of numerical stability issues, Strassen's algorithm is rarely used in numerical analysis. But for matrix arithmetic over exact base rings (e.g., the rational numbers, finite fields, etc.) it is of extreme importance.

In [**Str69**], Strassen also sketched a new algorithm for computing the inverse of a square matrix using matrix multiplication. Using this algorithm, the number of operations to invert an $m \times m$ matrix is (roughly) the same as the number needed to multiply two $m \times m$ matrices. Suppose the input matrix is $2^n \times 2^n$ and we write it in block form as $\left(\begin{smallmatrix} A & B \\ C & D \end{smallmatrix}\right)$ where A, B, C, D are all $2^{n-1} \times 2^{n-1}$ matrices. Assume that any intermediate matrices below that we invert are invertible. Consider the augmented matrix

$$\left(\begin{array}{cc|cc} A & B & I & 0 \\ C & D & 0 & I \end{array}\right).$$

Multiply the top row by A^{-1} to obtain

$$\left(\begin{array}{cc|cc} I & A^{-1}B & A^{-1} & 0 \\ C & D & 0 & I \end{array}\right),$$

and write $E = A^{-1}B$. Subtract C times the first row from the second row to get

$$\left(\begin{array}{cc|cc} I & E & A^{-1} & 0 \\ 0 & D - CE & -CA^{-1} & I \end{array}\right).$$

Set $F = D - CE$ and multiply the bottom row by F^{-1} on the left to obtain

$$\left(\begin{array}{cc|cc} I & E & A^{-1} & 0 \\ 0 & I & -F^{-1}CA^{-1} & F^{-1} \end{array}\right).$$

Set $G = -F^{-1}CA^{-1}$, and subtract E times the second from the first row to arrive at

$$\left(\begin{array}{cc|cc} I & 0 & A^{-1} - EG & -EF^{-1} \\ 0 & I & G & F^{-1} \end{array}\right).$$

The idea listed above can, with significant work, be extended to a general algorithm (as is done in [**Ste06**]).

Next we very briefly sketch how to compute echelon forms of matrices using matrix multiplication and inversion. Its complexity is comparable to the complexity of matrix multiplication.

As motivation, recall the standard algorithm from undergraduate linear algebra for inverting an invertible square matrix A: form the augmented matrix $[A|I]$, and then compute the echelon form of this matrix, which is $[I|A^{-1}]$. If T is the transformation matrix to echelon form, then $T[A|I] = [I|T]$, so $T = A^{-1}$. In particular, we could find the echelon form of $[A|I]$ by multiplying on the left by A^{-1}. Likewise, for any matrix B with the same number of rows as A, we could find the echelon form of $[A|B]$ by multiplying on the left by A^{-1}. Next we extend this idea to give an algorithm to compute echelon forms using only matrix multiplication (and echelon form modulo one prime).

Algorithm 7.8 (Asymptotically Fast Echelon Form). *Given a matrix A over the rational numbers (or a number field), this algorithm computes the echelon form of A.*

(1) [Find Pivots] Choose a random prime p (coprime to the denominator of any entry of A) and compute the echelon form of $A \pmod{p}$, e.g., using Algorithm 7.3. Let $c_0, \ldots, c_{n-1}$ be the pivot columns of $A \pmod{p}$. When computing the echelon form, save the positions $r_0, \ldots, r_{n-1}$ of the rows used to clear each column.
(2) [Extract Submatrix] Extract the $n \times n$ submatrix B of A whose entries are A_{r_i,c_j} for $0 \leq i, j \leq n-1$.
(3) [Compute Inverse] Compute the inverse B^{-1} of B. Note that B must be invertible since its reduction modulo p is invertible.
(4) [Multiply] Let C be the matrix whose rows are the rows $r_0, \ldots, r_{n-1}$ of A. Compute $E = B^{-1}C$. If E is not in echelon form, go to step (1).
(5) [Done?] Write down a matrix D whose columns are a basis for $\ker(E)$ as explained on page 105. Let F be the matrix whose rows are the rows of A other than rows $r_0, \ldots, r_{n-1}$. Compute the product FD. If $FD = 0$, output E, which is the echelon form of A. If $FD \neq 0$, go to step (1) and run the whole algorithm again.

Proof. We prove both that the algorithm terminates and that when it terminates, the matrix E is the echelon form of A.

First we prove that the algorithm terminates. Let E be the echelon form of A. By Exercise 7.3, for all but finitely many primes p (i.e., any prime where $A \pmod{p}$ has the same rank as A) the echelon form of $A \pmod{p}$ equals $E \pmod{p}$. For any such prime p the pivot columns of $E \pmod{p}$ are the pivot columns of E, so the algorithm will terminate for that choice of p.

We next prove that when the algorithm terminates, E is the echelon form of A. By assumption, E is in echelon form and is obtained by multiplying C on the left by an invertible matrix, so E must be *the* echelon form of C. The rows of C are a subset of those of A, so the rows of E are a subset of the rows of the echelon form of A. Thus $\ker(A) \subset \ker(E)$. To show that E equals the echelon form of A, we just need to verify that $\ker(E) \subset \ker(A)$, i.e., that $AD = 0$, where D is as in step (5). Since E is the echelon form of C, we know that $CD = 0$. By step (5) we also know that $FD = 0$. Thus $AD = 0$, since the rows of A are the union of the rows of F and C. □

Example 7.9. Let A be the 4×8 matrix

$$A = \begin{pmatrix} -9 & 6 & 7 & 3 & 1 & 0 & 0 & 0 \\ -10 & 3 & 8 & 2 & 0 & 1 & 0 & 0 \\ 3 & -6 & 2 & 8 & 0 & 0 & 1 & 0 \\ -8 & -6 & -8 & 6 & 0 & 0 & 0 & 1 \end{pmatrix}$$

from Example 7.2.

```
sage: M = MatrixSpace(QQ,4,8)
sage: A = M([[-9,6,7,3,1,0,0,0],[-10,3,8,2,0,1,0,0],
               [3,-6,2,8,0,0,1,0],[-8,-6,-8,6,0,0,0,1]])
```

First choose the "random" prime $p = 41$, which does not divide any of the entries of A, and compute the echelon form of the reduction of A modulo 41.

```
sage: A41 = MatrixSpace(GF(41),4,8)(A)
sage: E41 = A41.echelon_form()
```

The echelon form of $A \pmod{41}$ is

$$\begin{pmatrix} 1 & 0 & 0 & 2 & 0 & 20 & 33 & 18 \\ 0 & 1 & 0 & 40 & 0 & 30 & 7 & 1 \\ 0 & 0 & 1 & 39 & 0 & 19 & 13 & 17 \\ 0 & 0 & 0 & 0 & 1 & 31 & 0 & 37 \end{pmatrix}.$$

Thus we take $c_0 = 0$, $c_1 = 1$, $c_2 = 2$, and $c_3 = 4$. Also $r_i = i$ for $i = 0, 1, 2, 3$. Next extract the submatrix B.

```
sage: B = A.matrix_from_columns([0,1,2,4])
```

The submatrix B is

$$B = \begin{pmatrix} -9 & 6 & 7 & 1 \\ -10 & 3 & 8 & 0 \\ 3 & -6 & 2 & 0 \\ -8 & -6 & -8 & 0 \end{pmatrix}.$$

The inverse of B is

$$B^{-1} = \begin{pmatrix} 0 & -\frac{5}{92} & \frac{1}{46} & -\frac{9}{184} \\ 0 & -\frac{1}{138} & -\frac{3}{23} & -\frac{11}{276} \\ 0 & \frac{11}{184} & \frac{7}{92} & -\frac{17}{368} \\ 1 & -\frac{159}{184} & \frac{41}{92} & \frac{45}{368} \end{pmatrix}.$$

Multiplying by A yields

$$E = B^{-1}A = \begin{pmatrix} 1 & 0 & 0 & -\frac{21}{92} & 0 & -\frac{5}{92} & \frac{1}{46} & -\frac{9}{184} \\ 0 & 1 & 0 & -\frac{179}{138} & 0 & -\frac{1}{138} & -\frac{3}{23} & -\frac{11}{276} \\ 0 & 0 & 1 & \frac{83}{184} & 0 & \frac{11}{184} & \frac{7}{92} & -\frac{17}{368} \\ 0 & 0 & 0 & \frac{1025}{184} & 1 & -\frac{159}{184} & \frac{41}{92} & \frac{45}{368} \end{pmatrix}.$$

```
sage: E = B^(-1)*A
```

This is *not* the echelon form of A. Indeed, it is not even in echelon form, since the last row is not normalized so the leftmost nonzero entry is 1. We thus choose another random prime, say $p = 43$. The echelon form mod 43 has columns $0, 1, 2, 3$ as pivot columns. We thus extract the matrix

$$B = \begin{pmatrix} -9 & 6 & 7 & 3 \\ -10 & 3 & 8 & 2 \\ 3 & -6 & 2 & 8 \\ -8 & -6 & -8 & 6 \end{pmatrix}.$$

```
sage: B = A.matrix_from_columns([0,1,2,3])
```

This matrix has inverse

$$B^{-1} = \begin{pmatrix} \frac{42}{1025} & -\frac{92}{1025} & \frac{1}{25} & -\frac{9}{205} \\ \frac{716}{3075} & -\frac{641}{3075} & -\frac{2}{75} & -\frac{7}{615} \\ -\frac{83}{1025} & \frac{133}{1025} & \frac{1}{25} & -\frac{23}{410} \\ \frac{184}{1025} & -\frac{159}{1025} & \frac{2}{25} & \frac{9}{410} \end{pmatrix}.$$

Finally, the echelon form of A is $E = B^{-1}A$. No further checking is needed since the product so obtained is in echelon form, and the matrix F of the last step of the algorithm has 0 rows.

Remark 7.10. Above we have given only the barest sketch of asymptotically fast "block" algorithms for linear algebra. For optimized algorithms that work in the general case, please see the source code of [**Ste06**].

7.5. Decomposing Spaces under the Action of Matrix

Efficiently solving the following problem is a crucial step in computing a basis of eigenforms for a space of modular forms (see Sections 3.7 and 9.3.2).

Problem 7.11. Suppose T is an $n \times n$ matrix with entries in a field K (typically a number field or finite field) and that the minimal polynomial of T is square-free and has degree n. View T as acting on $V = K^n$. Find a simple module decomposition $W_0 \oplus \cdots \oplus W_m$ of V as a direct sum of simple $K[T]$-modules. Equivalently, find an invertible matrix A such that $A^{-1}TA$ is a block direct sum of matrices $T_0, \ldots, T_m$ such that the minimal polynomial of each T_i is irreducible.

Remark 7.12. A generalization of Problem 7.11 to arbitrary matrices with entries in $\mathbb{Q}$ is finding the *rational Jordan form* (or rational canonical form, or Frobenius form) of T. This is like the usual Jordan form, but the resulting matrix is over $\mathbb{Q}$ and the summands of the matrix corresponding to eigenvalues are replaced by matrices whose minimal polynomials are the minimal polynomials (over $\mathbb{Q}$) of the eigenvalues. The rational Jordan form was extensively studied by Giesbrecht in his Ph.D. thesis and many successive papers, where he analyzes the complexity of his algorithms and observes that the limiting step is factoring polynomials over K. The reason is that given a polynomial $f \in K[x]$, one can easily write down a matrix T such that one can read off the factorization of f from the rational Jordan form of T (see also [**Ste97**]).

7.5.1. Characteristic Polynomials.

The computation of characteristic polynomials of matrices is crucial to modular forms computations. There are many approaches to this problems: compute $\det(xI - A)$ symbolically (bad), compute the traces of the powers of A (bad), or compute the Hessenberg form modulo many primes and use CRT (bad; see for [**Coh93**, §2.2.4] the definition of Hessenberg form and the algorithm). A more sophisticated method is to compute the rational canonical form of A using Giesbrecht's algorithm[1] (see [**GS02**]), which involves computing Krylov subspaces (see Remark 7.13 below), and building up the whole space on which A acts. This

[1] Allan Steel also invented a similar algorithm.

latter method is a generalization of Wiedemann's algorithm for computing minimal polynomials (see Section 7.5.3), but with more structure to handle the case when the characteristic polynomial is not equal to the minimal polynomial.

7.5.2. Polynomial Factorization. Factorization of polynomials in $\mathbb{Q}[X]$ (or over number fields) is an important step in computing an explicit basis of Hecke eigenforms for spaces of modular forms. The best algorithm is the van Hoeij method [**BHKS06**], which uses the LLL lattice basis reduction algorithm [**LLL82**] in a novel way to solve the optimization problems that come up in trying to lift factorizations mod p to $\mathbb{Z}$. It has been generalized by Belebas, van Hoeij, Klüners, and Steel to number fields.

7.5.3. Wiedemann's Minimal Polynomial Algorithm. In this section we describe an algorithm due to Wiedemann for computing the minimal polynomial of an $n \times n$ matrix A over a field.

Choose a random vector v and compute the iterates

$$v_0 = v, \quad v_1 = A(v), \quad v_2 = A^2(v), \quad \dots, \quad v_{2n-1} = A^{2n-1}(v). \tag{7.5.1}$$

If $f = x^m + c_{m-1}x^{m-1} + \dots + c_1 x + c_0$ is the minimal polynomial of A, then

$$A^m + c_{m-1}A^{m-1} + \dots + c_0 I_n = 0,$$

where I_n is the $n \times n$ identity matrix. For any $k \geq 0$, by multiplying both sides on the right by the vector $A^k v$, we see that

$$A^{m+k}v + c_{m-1}A^{m-1+k}v + \dots + c_0 A^k v = 0;$$

hence

$$v_{m+k} + c_{m-1}v_{m-1+k} + \dots + c_0 v_k = 0, \qquad \text{all } k \geq 0.$$

Wiedemann's idea is to observe that any single component of the vectors $v_0, \dots, v_{2n-1}$ satisfies the linear recurrence with coefficients $1, c_{m-1}, \dots, c_0$. The Berlekamp-Massey algorithm (see Algorithm 7.14 below) was introduced in the 1960s in the context of coding theory to find the minimal polynomial of a linear recurrence sequence $\{a_r\}$. The minimal polynomial of this linear recurrence is by definition the unique monic polynomial g, such that if $\{a_r\}$ satisfies a linear recurrence $a_{j+k} + b_{j-1}a_{j-1+k} + \dots + b_0 a_k = 0$ (for all $k \geq 0$), then g divides the polynomial $x^j + \sum_{i=0}^{j-1} b_i x^i$. If we apply Berlekamp-Massey to the top coordinates of the v_i, we obtain a polynomial g_0, which divides f. We then apply it to the second to the top coordinates and find a polynomial g_1 that divides f, etc. Taking the least common multiple of the first few g_i, we find a divisor of the minimal polynomial of f. One can show that with "high probability" one quickly finds f, instead of just a proper divisor of f.

Remark 7.13. In the literature, techniques that involve iterating a vector as in (7.5.1) are often called *Krylov methods*. The subspace generated by the iterates of a vector under a matrix is called a *Krylov subspace*.

Algorithm 7.14 (Berlekamp-Massey). *Suppose $a_0, \dots, a_{2n-1}$ are the first $2n$ terms of a sequence that satisfies a linear recurrence of degree at most n. This algorithm computes the minimal polynomial f of the sequence.*

(1) Let $R_0 = x^{2n}$, $R_1 = \sum_{i=0}^{2n-1} a_i x^i$, $V_0 = 0$, $V_1 = 1$.
(2) While $\deg(R_1) \geq n$, do the following:
    (a) Compute Q and R such that $R_0 = QR_1 + R$.
    (b) Let $(V_0, V_1, R_0, R_1) = (V_1, V_0 - QV_1, R_1, R)$.
(3) Let $d = \max(\deg(V_1), 1 + \deg(R_1))$ and set $P = x^d V_1(1/x)$.
(4) Let c be the leading coefficient of P and output $f = P/c$.

The above description of Berlekamp-Massey is taken from [**ADT04**], which contains some additional ideas for improvements.

Now suppose T is an $n \times n$ matrix as in Problem 7.11. We find the minimal polynomial of T by computing the minimal polynomial of $T \pmod{p}$ using Wiedemann's algorithm, for many primes p, and using the Chinese Remainder Theorem. (One has to bound the number of primes that must be considered; see, e.g., [**Coh93**].)

One can also compute the characteristic polynomial of T directly from the Hessenberg form of T, which can be computed in $O(n^4)$ field operations, as described in [**Coh93**]. This is simple but slow. Also, the T we consider will often be sparse, and Wiedemann is particularly good when T is sparse.

Example 7.15. We compute the minimal polynomial of

$$A = \begin{pmatrix} 3 & 0 & 0 \\ 0 & 0 & 2 \\ -1 & 1/2 & -1 \end{pmatrix}$$

using Wiedemann's algorithm. Let $v = (1,0,0)^t$. Then

$$v = (1,0,0)^t, \quad Av = (3,0,-1)^t, \quad A^2v = (9,-2,-2)^t,$$
$$A^3v = (27,-4,-8)^t, \quad A^4v = (81,-16,-21)^t, \quad A^5v = (243,-42,-68)^t.$$

The linear recurrence sequence coming from the first entries is

$$1, 3, 9, 27, 81, 243.$$

This sequence satisfies the linear recurrence

$$a_{k+1} - 3a_k = 0, \qquad \text{all } k > 0,$$

so its minimal polynomial is $x - 3$. This implies that $x - 3$ divides the minimal polynomial of the matrix A. Next we use the sequence of second coordinates of the iterates of v, which is

$$0, 0, -2, -4, -16, -42.$$

The recurrence that this sequence satisfies is slightly less obvious, so we apply the Berlekamp-Massey algorithm to find it, with $n = 3$.

(1) We have $R_0 = x^6$, $R_1 = -42x^5 - 16x^4 - 4x^3 - 2x^2$, $V_0 = 0, V_1 = 1$.

(2) (a) Dividing R_0 by R_1, we find

$$R_0 = R_1\left(-\frac{1}{42}x + \frac{4}{441}\right) + \left(\frac{22}{441}x^4 - \frac{5}{441}x^3 + \frac{8}{441}x^2\right).$$

(b) The new V_0, V_1, R_0, R_1 are

$$\begin{aligned} V_0 &= 1, \\ V_1 &= \frac{1}{42}x - \frac{4}{441}, \\ R_0 &= -42x^5 - 16x^4 - 4x^3 - 2x^2, \\ R_1 &= \frac{22}{441}x^4 - \frac{5}{441}x^3 + \frac{8}{441}x^2. \end{aligned}$$

Since $\deg(R_1) \geq n = 3$, we do the above three steps again.

(3) We repeat the above three steps.

(a) Dividing R_0 by R_1, we find

$$R_0 = R_1\left(-\frac{9261}{11}x - \frac{123921}{242}\right) + \left(\frac{1323}{242}x^3 + \frac{882}{121}x^2\right).$$

(b) The new V_0, V_1, R_0, R_1 are:

$$\begin{aligned} V_0 &= \frac{1}{42}x - \frac{4}{441}, \\ V_1 &= \frac{441}{22}x^2 + \frac{2205}{484}x + \frac{441}{121}, \\ R_0 &= \frac{22}{441}x^4 - \frac{5}{441}x^3 + \frac{8}{441}x^2, \\ R_1 &= \frac{1323}{242}x^3 + \frac{882}{121}x^2. \end{aligned}$$

(4) We have to repeat the steps yet again:

$$\begin{aligned}
V_0 &= \frac{441}{22}x^2 + \frac{2205}{484}x + \frac{441}{121},\\
V_1 &= -\frac{242}{1323}x^3 + \frac{968}{3969}x^2 + \frac{484}{3969}x - \frac{242}{3969},\\
R_0 &= \frac{1323}{242}x^3 + \frac{882}{121}x^2,\\
R_1 &= \frac{484}{3969}x^2.
\end{aligned}$$

(5) We have $d = 3$, so $P = -\frac{242}{3969}x^3 + \frac{484}{3969}x^2 + \frac{968}{3969}x - \frac{242}{1323}$.

(6) Multiply through by $-3969/242$ and output

$$x^3 - 2x^2 - 4x + 3 = (x-3)(x^2+x-1).$$

The minimal polynomial of T_2 is $(x-3)(x^2+x-1)$, since the minimal polynomial has degree at most 3 and is divisible by $(x-3)(x^2+x-1)$.

7.5.4. p-adic Nullspace. We will use the following algorithm of Dixon [**Dix82**] to compute p-adic approximations to solutions of linear equations over $\mathbb{Q}$. Rational reconstruction modulo p^n then allows us to recover the corresponding solutions over $\mathbb{Q}$.

Algorithm 7.16 (p-adic Nullspace). *Given a matrix A with integer entries and nonzero kernel, this algorithm computes a nonzero element of* $\ker(A)$ *using successive p-adic approximations.*

(1) [Prime] Choose a random prime p.
(2) [Echelon] Compute the echelon form of A modulo p.
(3) [Done?] If A has full rank modulo p, it has full rank, so we terminate the algorithm.
(4) [Setup] Let $b_0 = 0$.
(5) [Iterate] For each $m = 0, 1, 2, \ldots, k$, use the echelon form of A modulo p to find a vector y_m with integer entries such that $Ay_m \equiv b_m \pmod{p}$, and then set

$$b_{m+1} = \frac{b_m - Ay_m}{p}.$$

(If $m = 0$, choose $y_m \neq 0$.)

(6) [p-adic Solution] Let $x = y_0 + y_1p + y_2p^2 + y_3p^3 + \cdots + y_kp^k$.
(7) [Lift] Use rational reconstruction (Algorithm 7.4) to find a vector z with rational entries such that $z \equiv x \pmod{p^{k+1}}$, if such a vector exists. If the vector does not exist, increase k or use a different p. Otherwise, output z.

Proof. We verify the case $k = 2$ only. We have $Ay_0 = 0 \pmod{p}$ and $Ay_1 = -\frac{Ay_0}{p} \pmod{p}$. Thus

$$Ay_0 + pAy_1 \equiv Ay_0 + (-Ay_0) \pmod{p^2}.$$

□

7.5.5. Decomposition Using Kernels. We now know enough to give an algorithm to solve Problem 7.11.

Algorithm 7.17 (Decomposition Using Kernels). *Given an $n \times n$ matrix T over a field K as in Problem 7.11, this algorithm computes the decomposition of V as a direct sum of simple $K[T]$ modules.*

(1) [Minimal Polynomial] Compute the minimal polynomial f of T, e.g., using the multimodular Wiedemann algorithm.
(2) [Factorization] Factor f using the algorithm in Section 7.5.2.
(3) [Compute Kernels] For each irreducible factor g_i of f, compute the following.
   (a) Compute the matrix $A_i = g_i(T)$.
   (b) Compute $W_i = \ker(A_i)$, e.g., using Algorithm 7.16.
(4) [Output Answer] Then $V = \bigoplus W_i$.

Remark 7.18. As mentioned in Remark 7.12, if one can compute such decompositions $V = \bigoplus W_i$, then one can easily factor polynomials f; hence the difficulty of polynomial factorization is a lower bound on the complexity of writing V as a direct sum of simples.

7.6. Exercises

7.1 Given a subspace W of k^n, where k is a field and $n \geq 0$ is an integer, give an algorithm to find a matrix A such that $W = \text{Ker}(A)$.

7.2 If $\text{rref}(A)$ denotes the row reduced echelon form of A and p is a prime not dividing any denominator of any entry of A or of $\text{rref}(A)$, is $\text{rref}(A \pmod{p}) = \text{rref}(A) \pmod{p}$?

7.3 Let A be a matrix with entries in $\mathbb{Q}$. Prove that for all but finitely many primes p we have $\text{rref}(A \pmod{p}) = \text{rref}(A) \pmod{p}$.

7.4 Let

$$A = \begin{pmatrix} 1 & 2 & 3 \\ 4 & 5 & 6 \\ 7 & 8 & 9 \end{pmatrix}.$$

   (a) Compute the echelon form of A using each of Algorithm 7.3, Algorithm 7.6, and Algorithm 7.8.
   (b) Compute the kernel of A.

(c) Find the characteristic polynomial of A using the algorithm of Section 7.5.3.

7.5 The notion of echelon form extends to matrices whose entries come from certain rings other than fields, e.g., Euclidean domains. In the case of matrices over $\mathbb{Z}$ we define a matrix to be in echelon form (or *Hermite normal form*) if it satisfies

- $a_{ij} = 0$, for $i > j$,
- $a_{ii} \geq 0$,
- $a_{ij} < a_{ii}$ for all $j < i$ (unless $a_{ii} = 0$, in which case all $a_{ij} = 0$).

There are algorithms for computing with finitely generated modules over $\mathbb{Z}$ that are analogous to the ones in this chapter for vector spaces, which depend on computation of Hermite forms.

(a) Show that the Hermite form of $\begin{pmatrix} 1 & 2 & 3 \\ 4 & 5 & 6 \\ 7 & 8 & 9 \end{pmatrix}$ is $\begin{pmatrix} 1 & 2 & 3 \\ 0 & 3 & 6 \\ 0 & 0 & 0 \end{pmatrix}$.

(b) Describe an algorithm for transforming an $n \times n$ matrix A with integer entries into Hermite form using row operations and the Euclidean algorithm.

Chapter 8

General Modular Symbols

In this chapter we explain how to generalize the notion of modular symbols given in Chapter 3 to higher weight and more general level. We define Hecke operators on them and their relation to modular forms via the integration pairing. We omit many difficult proofs that modular symbols have certain properties and instead focus on how to compute with modular symbols. For more details see the references given in this section (especially [**Mer94**]) and [**Wie05**].

Modular symbols are a formalism that make it elementary to compute with homology or cohomology related to certain Kuga-Sato varieties (these are $\mathcal{E} \times_X \cdots \times_X \mathcal{E}$, where X is a modular curve and $\mathcal{E}$ is the universal elliptic curve over it). It is not necessary to know anything about these Kuga-Sato varieties in order to compute with modular symbols.

This chapter is about spaces of modular symbols and how to compute with them. It is by far the most important chapter in this book. The algorithms that build on the theory in this chapter are central to all the computations we will do later in the book.

This chapter closely follows Loïc Merel's paper [**Mer94**]. First we define modular symbols of weight $k \geq 2$. Then we define the corresponding Manin symbols and state a theorem of Merel-Shokurov, which gives all relations between Manin symbols. (The proof of the Merel-Shokurov theorem is beyond the scope of this book but is presented nicely in [**Wie05**].) Next we describe how the Hecke operators act on both modular and Manin symbols and how to compute trace and inclusion maps between spaces of modular symbols of different levels.

Not only are modular symbols useful for computation, but they have been used to prove theoretical results about modular forms. For example, certain technical calculations with modular symbols are used in Loïc Merel's proof of the uniform boundedness conjecture for torsion points on elliptic curves over number fields (modular symbols are used to understand linear independence of Hecke operators). Another example is [**Gri05**], which distills hypotheses about Kato's Euler system in K_2 of modular curves to a simple formula involving modular symbols (when the hypotheses are satisfied, one obtains a lower bound on the Shafarevich-Tate group of an elliptic curve).

8.1. Modular Symbols

We recall from Chapter 3 the free abelian group $\mathbb{M}_2$ of modular symbols. We view these as elements of the relative homology of the extended upper half plane $\mathfrak{h}^* = \mathfrak{h} \cup \mathbb{P}^1(\mathbb{Q})$ relative to the cusps. The group $\mathbb{M}_2$ is the free abelian group on symbols $\{\alpha, \beta\}$ with

$$\alpha, \beta \in \mathbb{P}^1(\mathbb{Q}) = \mathbb{Q} \cup \{\infty\}$$

modulo the relations

$$\{\alpha, \beta\} + \{\beta, \gamma\} + \{\gamma, \alpha\} = 0,$$

for all $\alpha, \beta, \gamma \in \mathbb{P}^1(\mathbb{Q})$, and all torsion. More precisely,

$$\mathbb{M}_2 = (F/R)/(F/R)_{\mathrm{tor}},$$

where F is the free abelian group on all pairs (α, β) and R is the subgroup generated by all elements of the form $(\alpha, \beta) + (\beta, \gamma) + (\gamma, \alpha)$. Note that $\mathbb{M}_2$ is a huge free abelian group of countable rank.

For any integer $n \geq 0$, let $\mathbb{Z}[X, Y]_n$ be the abelian group of homogeneous polynomials of degree n in two variables X, Y.

Remark 8.1. Note that $\mathbb{Z}[X, Y]_n$ is isomorphic to $\mathrm{Sym}^n(\mathbb{Z} \times \mathbb{Z})$ as a group, but certain natural actions are different. In [**Mer94**], Merel uses the notation $\mathbb{Z}_n[X, Y]$ for what we denote by $\mathbb{Z}[X, Y]_n$.

Now fix an integer $k \geq 2$. Set

$$\mathbb{M}_k = \mathbb{Z}[X, Y]_{k-2} \otimes_{\mathbb{Z}} \mathbb{M}_2,$$

which is a torsion-free abelian group whose elements are sums of expressions of the form $X^i Y^{k-2-i} \otimes \{\alpha, \beta\}$. For example,

$$X^3 \otimes \{0, 1/2\} - 17XY^2 \otimes \{\infty, 1/7\} \in \mathbb{M}_5.$$

Fix a finite index subgroup G of $\mathrm{SL}_2(\mathbb{Z})$. Define a *left action of G* on $\mathbb{Z}[X,Y]_{k-2}$ as follows. If $g = \left(\begin{smallmatrix} a & b \\ c & d \end{smallmatrix}\right) \in G$ and $P(X,Y) \in \mathbb{Z}[X,Y]_{k-2}$, let

$$(gP)(X,Y) = P(dX - bY, -cX + aY).$$

Note that if we think of $z = (X,Y)$ as a column vector, then

$$(gP)(z) = P(g^{-1}z),$$

since $g^{-1} = \left(\begin{smallmatrix} d & -b \\ -c & a \end{smallmatrix}\right)$. The reason for the inverse is so that this is a left action instead of a right action, e.g., if $g, h \in G$, then

$$((gh)P)(z) = P((gh)^{-1}z) = P(h^{-1}g^{-1}z) = (hP)(g^{-1}z) = (g(hP))(z).$$

Recall that we let G act on the left on $\mathbb{M}_2$ by

$$g\{\alpha, \beta\} = \{g(\alpha), g(\beta)\},$$

where G acts via linear fractional transformations, so if $g = \left(\begin{smallmatrix} a & b \\ c & d \end{smallmatrix}\right)$, then

$$g(\alpha) = \frac{a\alpha + b}{c\alpha + d}.$$

For example, useful special cases to remember are that if $g = \left(\begin{smallmatrix} a & b \\ c & d \end{smallmatrix}\right)$, then

$$g(0) = \frac{b}{d} \qquad \text{and} \qquad g(\infty) = \frac{a}{c}.$$

(Here we view ∞ as $1/0$ in order to describe the action.)

We now combine these two actions to obtain a left action of G on $\mathbb{M}_k$, which is given by

$$g(P \otimes \{\alpha, \beta\}) = (gP) \otimes \{g(\alpha), g(\beta)\}.$$

For example,

$$\begin{aligned}\begin{pmatrix} 1 & 2 \\ -2 & -3 \end{pmatrix}(X^3 \otimes \{0, 1/2\}) &= (-3X - 2Y)^3 \otimes \left\{-\frac{2}{3}, -\frac{5}{8}\right\} \\ &= (-27X^3 - 54X^2Y - 36XY^2 - 8Y^3) \otimes \left\{-\frac{2}{3}, -\frac{5}{8}\right\}.\end{aligned}$$

We will often write $P(X,Y)\{\alpha, \beta\}$ for $P(X,Y) \otimes \{\alpha, \beta\}$.

Definition 8.2 (Modular Symbols)**.** Let $k \geq 2$ be an integer and let G be a finite index subgroup of $\mathrm{SL}_2(\mathbb{Z})$. The space $\mathbb{M}_k(G)$ of *weight k modular symbols for G* is the quotient of $\mathbb{M}_k$ by all relations $gx - x$ for $x \in \mathbb{M}_k$, $g \in G$, and by any torsion.

Note that $\mathbb{M}_k$ is a torsion-free abelian group, and it is a nontrivial fact that $\mathbb{M}_k$ has finite rank. We denote modular symbols for G in exactly the same way we denote elements of $\mathbb{M}_k$; the group G will be clear from context.

The space of *modular symbols over a ring R* is

$$\mathbb{M}_k(G;R) = \mathbb{M}_k(G) \otimes_{\mathbb{Z}} R.$$

8.2. Manin Symbols

Let G be a finite index subgroup of $\mathrm{SL}_2(\mathbb{Z})$ and $k \geq 2$ an integer. Just as in Chapter 3 it is possible to compute $\mathbb{M}_k(G)$ using a computer, despite that, as defined above, $\mathbb{M}_k(G)$ is the quotient of one infinitely generated abelian group by another one. This section is about Manin symbols, which are a distinguished subset of $\mathbb{M}_k(G)$ that lead to a finite presentation for $\mathbb{M}_k(G)$. Formulas written in terms of Manin symbols are frequently much easier to compute using a computer than formulas in terms of modular symbols.

Suppose $P \in \mathbb{Z}[X,Y]_{k-2}$ and $g \in \mathrm{SL}_2(\mathbb{Z})$. Then the *Manin symbol* associated to this pair of elements is

$$[P, g] = g(P\{0, \infty\}) \in \mathbb{M}_k(G).$$

Notice that if $Gg = Gh$, then $[P,g] = [P,h]$, since the symbol $g(P\{0,\infty\})$ is invariant by the action of G on the left (by definition, since it is a modular symbol for G). Thus for a right coset Gg it makes sense to write $[P, Gg]$ for the symbol $[P,h]$ for any $h \in Gg$. Since G has finite index in $\mathrm{SL}_2(\mathbb{Z})$, the abelian group generated by Manin symbols is of finite rank, generated by

$$\left\{[X^{k-2-i}Y^i, Gg_j] \,:\, i = 0, \ldots, k-2 \quad \text{and} \quad j = 0, \ldots, r\right\},$$

where $g_0, \ldots, g_r$ run through representatives for the right cosets $G\backslash \mathrm{SL}_2(\mathbb{Z})$.

We next show that every modular symbol can be written as a $\mathbb{Z}$-linear combination of Manin symbols, so they generate $\mathbb{M}_k(G)$.

Proposition 8.3. *The Manin symbols generate* $\mathbb{M}_k(G)$.

Proof. The proof if very similar to that of Proposition 3.11 except we introduce an extra twist to deal with the polynomial part. Suppose that we are given a modular symbol $P\{\alpha, \beta\}$ and wish to represent it as a sum of Manin symbols. Because

$$P\{a/b, c/d\} = P\{a/b, 0\} + P\{0, c/d\},$$

it suffices to write $P\{0, a/b\}$ in terms of Manin symbols. Let

$$0 = \frac{p_{-2}}{q_{-2}} = \frac{0}{1}, \ \frac{p_{-1}}{q_{-1}} = \frac{1}{0}, \ \frac{p_0}{1} = \frac{p_0}{q_0}, \ \frac{p_1}{q_1}, \ \frac{p_2}{q_2}, \ldots, \frac{p_r}{q_r} = \frac{a}{b}$$

denote the continued fraction convergents of the rational number a/b. Then

$$p_j q_{j-1} - p_{j-1} q_j = (-1)^{j-1} \qquad \text{for } -1 \leq j \leq r.$$

If we let $g_j = \begin{pmatrix} (-1)^{j-1}p_j & p_{j-1} \\ (-1)^{j-1}q_j & q_{j-1} \end{pmatrix}$, then $g_j \in \mathrm{SL}_2(\mathbb{Z})$ and

$$\begin{aligned} P\{0, a/b\} &= P \sum_{j=-1}^{r} \left\{ \frac{p_{j-1}}{q_{j-1}}, \frac{p_j}{q_j} \right\} \\ &= \sum_{j=-1}^{r} g_j((g_j^{-1}P)\{0,\infty\}) \\ &= \sum_{j=-1}^{r} [g_j^{-1}P,\ g_j]. \end{aligned}$$

Since $g_j \in \mathrm{SL}_2(\mathbb{Z})$ and P has integer coefficients, the polynomial $g_j^{-1}P$ also has integer coefficients, so we introduce no denominators. □

Now that we know the Manin symbols generate $\mathbb{M}_k(G)$, we next consider the relations between Manin symbols. Fortunately, the answer is fairly simple (though the proof is not). Let

$$\sigma = \begin{pmatrix} 0 & -1 \\ 1 & 0 \end{pmatrix}, \qquad \tau = \begin{pmatrix} 0 & -1 \\ 1 & -1 \end{pmatrix}, \qquad J = \begin{pmatrix} -1 & 0 \\ 0 & -1 \end{pmatrix}.$$

Define a *right action of* $\mathrm{SL}_2(\mathbb{Z})$ on Manin symbols as follows. If $h \in \mathrm{SL}_2(\mathbb{Z})$, let

$$[P, g]h = [h^{-1}P, gh].$$

This is a right action because both $P \mapsto h^{-1}P$ and $g \mapsto gh$ are right actions.

Theorem 8.4. *If x is a Manin symbol, then*

$$x + x\sigma = 0, \tag{8.2.1}$$

$$x + x\tau + x\tau^2 = 0, \tag{8.2.2}$$

$$x - xJ = 0. \tag{8.2.3}$$

Moreover, these are all the relations between Manin symbols, in the sense that the space $\mathbb{M}_k(G)$ of modular symbols is isomorphic to the quotient of the free abelian group on the finitely many symbols $[X^iY^{k-2-i}, Gg]$ (for $i = 0, \dots, k-2$ and $Gg \in G\backslash \mathrm{SL}_2(\mathbb{Z})$) by the above relations and any torsion.

Proof. First we prove that the Manin symbols satisfy the above relations. We follow Merel's proof (see [**Mer94**, §1.2]). Note that

$$\sigma(0) = \sigma^2(\infty) = \infty \qquad \text{and} \qquad \tau(1) = \tau^2(0) = \infty.$$

Writing $x = [P, g]$, we have

$$\begin{aligned}[P,g] + [P,g]\sigma &= [P,g] + [\sigma^{-1}P, g\sigma]\\ &= g(P\{0,\infty\}) + g\sigma(\sigma^{-1}P\{0,\infty\})\\ &= (gP)\{g(0), g(\infty)\} + (g\sigma)(\sigma^{-1}P)\{g\sigma(0), g\sigma(\infty)\}\\ &= (gP)\{g(0), g(\infty)\} + (gP)\{g(\infty), g(0)\}\\ &= (gP)\{g(0), g(\infty)\} + \{g(\infty), g(0)\})\\ &= 0.\end{aligned}$$

Also,

$$\begin{aligned}&[P,g] + [P,g]\tau + [P,g]\tau^2 = [P,g] + [\tau^{-1}P, g\tau] + [\tau^{-2}P, g\tau^2]\\ &= g(P\{0,\infty\}) + g\tau(\tau^{-1}P\{0,\infty\}) + g\tau^2(\tau^{-2}P\{0,\infty\})\\ &= (gP)\{g(0), g(\infty)\} + (gP)\{g\tau(0), g\tau(\infty)\} + (gP)\{g\tau^2(0), \tau^2(\infty)\}\\ &= (gP)\{g(0), g(\infty)\} + (gP)\{g(1), g(0)\}) + (gP)\{g(\infty), g(1)\}\\ &= (gP)\big(\{g(0), g(\infty)\} + \{g(\infty), g(1)\} + \{g(1), g(0)\}\big)\\ &= 0.\end{aligned}$$

Finally,

$$\begin{aligned}[P,g] + [P,g]J &= g(P\{0,\infty\}) - gJ(J^{-1}P\{gJ(0), gJ(\infty)\}\\ &= (gP)\{g(0), g(\infty)\} - (gP)\{g(0), g(\infty)\}\\ &= 0,\end{aligned}$$

where we use that J acts trivially via linear fractional transformations. This proves that the listed relations are all satisfied.

That the listed relations are all relations is more difficult to prove. One approach is to show (as in [**Mer94**, §1.3]) that the quotient of Manin symbols by the above relations and torsion is isomorphic to a space of Shokurov symbols, which is in turn isomorphic to $\mathbb{M}_k(G)$. A much better approach is to apply some results from group cohomology, as in [**Wie05**]. □

If G is a finite index subgroup and we have an algorithm to enumerate the right cosets $G\backslash \mathrm{SL}_2(\mathbb{Z})$ and to decide which coset an arbitrary element of $\mathrm{SL}_2(\mathbb{Z})$ belongs to, then Theorem 8.4 and the algorithms of Chapter 7 yield an algorithm to compute $\mathbb{M}_k(G;\mathbb{Q})$. Note that if $J \in G$, then the relation $x - xJ = 0$ is automatic.

Remark 8.5. The matrices σ and τ *do not commute*, so in computing $\mathbb{M}_k(G;\mathbb{Q})$, one cannot first quotient out by the two-term σ relations, then quotient out only the remaining free generators by the τ relations, and get the right answer in general.

8.2.1. Coset Representatives and Manin Symbols. The following is analogous to Proposition 3.10:

Proposition 8.6. *The right cosets* $\Gamma_1(N)\backslash \mathrm{SL}_2(\mathbf{Z})$ *are in bijection with pairs* (c, d) *where* $c, d \in \mathbf{Z}/N\mathbf{Z}$ *and* $\gcd(c, d, N) = 1$. *The coset containing a matrix* $\left(\begin{smallmatrix} a & b \\ c & d \end{smallmatrix}\right)$ *corresponds to* (c, d).

Proof. This proof is copied from [**Cre92**, pg. 203], except in that paper Cremona works with the analogue of $\Gamma_1(N)$ in $\mathrm{PSL}_2(\mathbf{Z})$, so his result is slightly different. Suppose $\gamma_i = \left(\begin{smallmatrix} a_i & b_i \\ c_i & d_i \end{smallmatrix}\right) \in \mathrm{SL}_2(\mathbf{Z})$, for $i = 1, 2$. We have

$$\gamma_1\gamma_2^{-1} = \begin{pmatrix} a_1 & b_1 \\ c_1 & d_1 \end{pmatrix}\begin{pmatrix} d_2 & -b_2 \\ -c_2 & a_2 \end{pmatrix} = \begin{pmatrix} a_1d_2 - b_1c_2 & * \\ c_1d_2 - d_1c_2 & a_2d_1 - b_2c_1 \end{pmatrix},$$

which is in $\Gamma_1(N)$ if and only if

$$c_1d_2 - d_1c_2 \equiv 0 \pmod{N} \tag{8.2.4}$$

and

$$a_2d_1 - b_2c_1 \equiv a_1d_2 - b_1c_2 \equiv 1 \pmod{N}. \tag{8.2.5}$$

Since the γ_i have determinant 1, if $(c_1, d_1) = (c_2, d_2) \pmod{N}$, then the congruences (8.2.4)–(8.2.5) hold. Conversely, if (8.2.4)–(8.2.5) hold, then

$$\begin{aligned} c_2 &\equiv a_2d_1c_2 - b_2c_1c_2 \\ &\equiv a_2d_2c_1 - b_2c_2c_1 && \text{since } d_1c_2 \equiv d_2c_1 \pmod{N} \\ &\equiv c_1 && \text{since } a_2d_2 - b_2c_2 = 1, \end{aligned}$$

and likewise

$$d_2 \equiv a_2d_1d_2 - b_2c_1d_2 \equiv a_2d_1d_2 - b_2d_1c_2 \equiv d_1 \pmod{N}.$$

□

Thus we may view weight k Manin symbols for $\Gamma_1(N)$ as triples of integers (i, c, d), where $0 \le i \le k-2$ and $c, d \in \mathbf{Z}/N\mathbf{Z}$ with $\gcd(c, d, N) = 1$. Here (i, c, d) corresponds to the Manin symbol $[X^iY^{k-2-i}, \left(\begin{smallmatrix} a & b \\ c' & d' \end{smallmatrix}\right)]$, where c' and d' are congruent to $c, d \pmod{N}$. The relations of Theorem 8.4 become

$$(i,\, c, d) + (-1)^i(k-2-i,\, d, -c) = 0,$$

$$\begin{aligned} (i,\, c, d) \quad &+ \quad (-1)^{k-2}\sum_{j=0}^{k-2-i}(-1)^j\binom{k-2-i}{j}(j,\, d, -c-d) \\ &+ \quad (-1)^{k-2-i}\sum_{j=0}^{i}(-1)^j\binom{i}{j}(k-2-i+j,\, -c-d, c) = 0, \end{aligned}$$

$$(i,\, c, d) - (-1)^{k-2}(i,\, -c, -d) = 0.$$

Recall that Proposition 3.10 gives a similar description for $\Gamma_0(N)$.

8.2.2. Modular Symbols with Character. Suppose $G = \Gamma_1(N)$. Define an action of *diamond-bracket operators* $\langle d \rangle$o, with $\gcd(d, N) = 1$ on Manin symbols as follows:

$$\langle d \rangle([P, (u, v)]) = [P, (du, dv)].$$

Let

$$\varepsilon : (\mathbb{Z}/N\mathbb{Z})^* \to \mathbb{Q}(\zeta)^*$$

be a Dirichlet character, where ζ is an nth root of unity and n is the order of ε. Let $\mathbb{M}_k(N, \varepsilon)$ be the quotient of $\mathbb{M}_k(\Gamma_1(N); \mathbb{Z}[\zeta])$ by the relations (given in terms of Manin symbols)

$$\langle d \rangle x - \varepsilon(d)x = 0,$$

for all $x \in \mathbb{M}_k(\Gamma_1(N); \mathbb{Z}[\zeta])$, $d \in (\mathbb{Z}/N\mathbb{Z})^*$, and by any $\mathbb{Z}$-torsion. Thus $\mathbb{M}_k(N, \varepsilon)$ is a $\mathbb{Z}[\varepsilon]$-module that has no torsion when viewed as a $\mathbb{Z}$-module.

8.3. Hecke Operators

Suppose Γ is a subgroup of $\mathrm{SL}_2(\mathbb{Z})$ of level N that contains $\Gamma_1(N)$. Just as for modular forms, there is a commutative *Hecke algebra* $\mathbb{T} = \mathbb{Z}[T_1, T_2, \ldots]$, which is the subring of $\mathrm{End}(\mathbb{M}_k(\Gamma))$ generated by all Hecke operators. Let

$$R_p = \left\{ \begin{pmatrix} 1 & r \\ 0 & p \end{pmatrix} : r = 0, 1, \ldots, p-1 \right\} \cup \left\{ \begin{pmatrix} p & 0 \\ 0 & 1 \end{pmatrix} \right\},$$

where we omit $\left(\begin{smallmatrix} p & 0 \\ 0 & 1 \end{smallmatrix}\right)$ if $p \mid N$. Then the *Hecke operator* T_p on $\mathbb{M}_k(\Gamma)$ is given by

$$T_p(x) = \sum_{g \in R_p} gx.$$

Notice when $p \nmid N$ that T_p is defined by summing over $p+1$ matrices that correspond to the $p+1$ subgroups of $\mathbb{Z} \times \mathbb{Z}$ of index p. This is exactly how we defined T_p on modular forms in Definition 2.26.

8.3.1. General Definition of Hecke Operators. Let Γ be a finite index subgroup of $\mathrm{SL}_2(\mathbb{Z})$ and suppose

$$\Delta \subset \mathrm{GL}_2(\mathbb{Q})$$

is a set such that $\Gamma\Delta = \Delta\Gamma = \Delta$ and $\Gamma \backslash \Delta$ is finite. For example, $\Delta = \Gamma$ satisfies this condition. Also, if $\Gamma = \Gamma_1(N)$, then for any positive integer n, the set

$$\Delta_n = \left\{ \begin{pmatrix} a & b \\ c & d \end{pmatrix} \in \mathrm{Mat}_2(\mathbb{Z}) \, : \, ad - bc = n, \quad \begin{pmatrix} a & b \\ c & d \end{pmatrix} \equiv \begin{pmatrix} 1 & * \\ 0 & n \end{pmatrix} \pmod{N} \right\}$$

also satisfies this condition, as we will now prove.

Lemma 8.7. *We have*

$$\Gamma_1(N)\cdot\Delta_n = \Delta_n\cdot\Gamma_1(N) = \Delta_n$$

and

$$\Delta_n = \bigcup_{a,b} \Gamma_1(N)\cdot\sigma_a \begin{pmatrix} a & b \\ 0 & n/a \end{pmatrix},$$

where $\sigma_a \equiv \left(\begin{smallmatrix} 1/a & 0 \\ 0 & a \end{smallmatrix}\right) \pmod{N}$, *the union is disjoint and* $1 \leq a \leq n$ *with* $a \mid n$, $\gcd(a, N) = 1$, *and* $0 \leq b < n/a$. *In particular, the set of cosets* $\Gamma_1(N)\backslash\Delta_n$ *is finite.*

Proof. (This is Lemma 1 of [**Mer94**, §2.3].) If $\gamma \in \Gamma_1(N)$ and $\delta \in \Delta_n$, then

$$\begin{pmatrix} 1 & * \\ 0 & 1 \end{pmatrix}\cdot\begin{pmatrix} 1 & * \\ 0 & n \end{pmatrix} \equiv \begin{pmatrix} 1 & * \\ 0 & n \end{pmatrix}\cdot\begin{pmatrix} 1 & * \\ 0 & 1 \end{pmatrix} \equiv \begin{pmatrix} 1 & * \\ 0 & n \end{pmatrix} \pmod{N}.$$

Thus $\Gamma_1(N)\Delta_n \subset \Delta_n$, and since $\Gamma_1(N)$ is a group, $\Gamma_1(N)\Delta_n = \Delta_n$; likewise $\Delta_n\Gamma_1(N) = \Delta_n$.

For the coset decomposition, we first prove the statement for $N = 1$, i.e., for $\Gamma_1(N) = \mathrm{SL}_2(\mathbb{Z})$. If A is an arbitrary element of $\mathrm{Mat}_2(\mathbb{Z})$ with determinant n, then using row operators on the left with determinant 1, i.e., left multiplication by elements of $\mathrm{SL}_2(\mathbb{Z})$, we can transform A into the form $\left(\begin{smallmatrix} a & b \\ 0 & n/a \end{smallmatrix}\right)$, with $1 \leq a \leq n$ and $0 \leq b < n$. (Just imagine applying the Euclidean algorithm to the two entries in the first column of A. Then a is the gcd of the two entries in the first column, and the lower left entry is 0. Next subtract n/a from b until $0 \leq b < n/a$.)

Next suppose N is arbitrary. Let $g_1, \ldots, g_r$ be such that

$$g_1\Gamma_1(N) \cup \cdots \cup g_r\Gamma_1(N) = \mathrm{SL}_2(\mathbb{Z})$$

is a disjoint union. If $A \in \Delta_n$ is arbitrary, then as we showed above, there is some $\gamma \in \mathrm{SL}_2(\mathbb{Z})$, so that $\gamma\cdot A = \left(\begin{smallmatrix} a & b \\ 0 & n/a \end{smallmatrix}\right)$, with $1 \leq a \leq n$ and $0 \leq b < n/a$, and $a \mid n$. Write $\gamma = g_i \cdot \alpha$, with $\alpha \in \Gamma_1(N)$. Then

$$\alpha\cdot A = g_i^{-1}\cdot\begin{pmatrix} a & b \\ 0 & n/a \end{pmatrix} \equiv \begin{pmatrix} 1 & * \\ 0 & n \end{pmatrix} \pmod{N}.$$

It follows that

$$g_i^{-1} \equiv \begin{pmatrix} 1 & * \\ 0 & n \end{pmatrix}\cdot\begin{pmatrix} a & b \\ 0 & n/a \end{pmatrix}^{-1} \equiv \begin{pmatrix} 1/a & * \\ 0 & a \end{pmatrix} \pmod{N}.$$

Since $\left(\begin{smallmatrix} 1 & 1 \\ 0 & 1 \end{smallmatrix}\right) \in \Gamma_1(N)$ and $\gcd(a, N) = 1$, there is $\gamma' \in \Gamma_1(N)$ such that

$$\gamma' g_i^{-1} \equiv \begin{pmatrix} 1/a & 0 \\ 0 & a \end{pmatrix} \pmod{N}.$$

We may then choose $\sigma_a = \gamma' g_i^{-1}$. Thus every $A \in \Delta_n$ is of the form $\gamma\sigma_a \left(\begin{smallmatrix} a & b \\ 0 & n/a \end{smallmatrix}\right)$, with $\gamma \in \Gamma_1(N)$ and a, b suitably bounded. This proves the second claim. □

Let any element $\delta = \left(\begin{smallmatrix} a & b \\ c & d \end{smallmatrix}\right) \in \mathrm{GL}_2(\mathbb{Q})$ act on the left on modular symbols $\mathbb{M}_k \otimes \mathbb{Q}$ by

$$\delta(P\{\alpha, \beta\}) = P(dX - bY, -cX + aY)\{\delta(\alpha), \delta(\beta)\}.$$

(Until now we had only defined an action of $\mathrm{SL}_2(\mathbb{Z})$ on modular symbols.) For $g = \left(\begin{smallmatrix} a & b \\ c & d \end{smallmatrix}\right) \in \mathrm{GL}_2(\mathbb{Q})$, let

$$\tilde{g} = \begin{pmatrix} d & -b \\ -c & a \end{pmatrix} = \det(g) \cdot g^{-1}. \tag{8.3.1}$$

Note that $\tilde{\tilde{g}} = g$. Also, $\delta P(X, Y) = (P \circ \tilde{g})(X, Y)$, where we set

$$\tilde{g}(X, Y) = (dX - bY, -cX + aY).$$

Suppose Γ and Δ are as above. Fix a finite set R of representatives for $\Gamma \backslash \Delta$. Let

$$T_\Delta : \mathbb{M}_k(\Gamma) \to \mathbb{M}_k(\Gamma)$$

be the linear map

$$T_\Delta(x) = \sum_{\delta \in R} \delta x.$$

This map is well defined because if $\gamma \in \Gamma$ and $x \in \mathbb{M}_k(\Gamma)$, then

$$\sum_{\delta \in R} \delta\gamma x = \sum_{\text{certain } \delta'} \gamma\delta' x = \sum_{\text{certain } \delta'} \delta' x = \sum_{\delta \in R} \delta x,$$

where we have used that $\Delta\Gamma = \Gamma\Delta$, and Γ acts trivially on $\mathbb{M}_k(\Gamma)$.

Let $\Gamma = \Gamma_1(N)$ and $\Delta = \Delta_n$. Then the nth Hecke operator T_n is T_{Δ_n}, and by Lemma 8.7,

$$T_n(x) = \sum_{a,b} \sigma_a \begin{pmatrix} a & b \\ 0 & n/a \end{pmatrix} \cdot x,$$

where a, b are as in Lemma 8.7.

Given this definition, we can compute the Hecke operators on $M_k(\Gamma_1(N))$ as follows. Write x as a modular symbol $P\{\alpha, \beta\}$, compute $T_n(x)$ as a modular symbol, and then convert to Manin symbols using continued fractions expansions. This is extremely inefficient; fortunately Loïc Merel (following [**Maz73**]) found a much better way, which we now describe (see [**Mer94**]).

8.3.2. Hecke Operators on Manin Symbols. If S is a subset of $\mathrm{GL}_2(\mathbb{Q})$, let

$$\tilde{S} = \{\tilde{g} : g \in S\},$$

where $\tilde{g}$ is as in (8.3.1). Also, for any ring R and any subset $S \subset \mathrm{Mat}_2(\mathbb{Z})$, let $R[S]$ denote the free R-module with basis the elements of S, so the elements of $R[S]$ are the finite R-linear combinations of the elements of S.

One of the main theorems of [**Mer94**] is that for any Γ, Δ satisfying the condition at the beginning of Section 8.3.1, if we can find $\sum u_M M \in \mathbb{C}[\mathrm{Mat}_2(\mathbb{Z})]$ and a map

$$\phi : \tilde{\Delta}\,\mathrm{SL}_2(\mathbb{Z}) \to \mathrm{SL}_2(\mathbb{Z})$$

that satisfies certain conditions, then for any Manin symbol $[P, g] \in \mathbb{M}_k(\Gamma)$, we have

$$T_\Delta([P,g]) = \sum_{gM \in \tilde{\Delta}\,\mathrm{SL}_2(\mathbb{Z}) \text{ with } M \in \mathrm{SL}_2(\mathbb{Z})} u_M[\tilde{M} \cdot P,\ \phi(gM)].$$

The paper [**Mer94**] contains many examples of ϕ and $\sum u_M M \in \mathbb{C}[\mathrm{Mat}_2(\mathbb{Z})]$ that satisfy all of the conditions.

When $\Gamma = \Gamma_1(N)$, the complicated list of conditions becomes simpler. Let $\mathrm{Mat}_2(\mathbb{Z})_n$ be the set of 2×2 matrices with determinant n. An element

$$h = \sum u_M[M] \in \mathbb{C}[\mathrm{Mat}_2(\mathbb{Z})_n]$$

satisfies condition C_n if for every $K \in \mathrm{Mat}_2(\mathbb{Z})_n / \mathrm{SL}_2(\mathbb{Z})$, we have that

$$\sum_{M \in K} u_M([M\infty] - [M0]) = [\infty] - [0] \in \mathbb{C}[P^1(\mathbb{Q})]. \tag{8.3.2}$$

If h satisfies condition C_n, then for any Manin symbol $[P, g] \in M_k(\Gamma_1(N))$, Merel proves that

$$T_n([P,(u,v)]) = \sum_M u_M[P(aX+bY,\ cX+dY),(u,v)M]. \tag{8.3.3}$$

Here $(u,v) \in (\mathbb{Z}/N\mathbb{Z})^2$ corresponds via Proposition 8.6 to a coset of $\Gamma_1(N)$ in $\mathrm{SL}_2(\mathbb{Z})$, and if $(u',v') = (u,v)M \in (\mathbb{Z}/N\mathbb{Z})^2$ and $\gcd(u',v',N) \neq 1$, then we omit the corresponding summand.

For example, we will now check directly that the element

$$h_2 = \left[\begin{pmatrix} 2 & 0 \\ 0 & 1 \end{pmatrix}\right] + \left[\begin{pmatrix} 1 & 0 \\ 0 & 2 \end{pmatrix}\right] + \left[\begin{pmatrix} 2 & 1 \\ 0 & 1 \end{pmatrix}\right] + \left[\begin{pmatrix} 1 & 0 \\ 1 & 2 \end{pmatrix}\right]$$

satisfies condition C_2. We have, as in the proof of Lemma 8.7 (but using elementary column operations), that

$$\begin{aligned}\mathrm{Mat}_2(\mathbb{Z})_2/\operatorname{SL}_2(\mathbb{Z}) &= \left\{\begin{pmatrix} a & 0 \\ b & 2/a\end{pmatrix}\operatorname{SL}_2(\mathbb{Z}) : a = 1,2 \text{ and } 0 \leq b < 2/a\right\}\\ &= \left\{\begin{pmatrix} 1 & 0 \\ 0 & 2\end{pmatrix}\operatorname{SL}_2(\mathbb{Z}),\ \begin{pmatrix} 1 & 0 \\ 1 & 2\end{pmatrix}\operatorname{SL}_2(\mathbb{Z}),\ \begin{pmatrix} 2 & 0 \\ 0 & 1\end{pmatrix}\operatorname{SL}_2(\mathbb{Z})\right\}.\end{aligned}$$

To verify condition C_2, we consider in turn each of the three elements of $\mathrm{Mat}_2(\mathbb{Z})_2/\operatorname{SL}_2(\mathbb{Z})$ and check that (8.3.2) holds. We have that

$$\begin{pmatrix} 1 & 0 \\ 0 & 2\end{pmatrix} \in \begin{pmatrix} 1 & 0 \\ 0 & 2\end{pmatrix}\operatorname{SL}_2(\mathbb{Z}),$$

$$\begin{pmatrix} 2 & 1 \\ 0 & 1\end{pmatrix}, \begin{pmatrix} 1 & 0 \\ 1 & 2\end{pmatrix} \in \begin{pmatrix} 1 & 0 \\ 1 & 2\end{pmatrix}\operatorname{SL}_2(\mathbb{Z}),$$

and

$$\begin{pmatrix} 2 & 0 \\ 0 & 1\end{pmatrix} \in \begin{pmatrix} 2 & 0 \\ 0 & 1\end{pmatrix}\operatorname{SL}_2(\mathbb{Z}).$$

Thus if $K = \left(\begin{smallmatrix} 1 & 0 \\ 0 & 2\end{smallmatrix}\right)\operatorname{SL}_2(\mathbb{Z})$, the left sum of (8.3.2) is $[\left(\begin{smallmatrix} 1 & 0 \\ 0 & 2\end{smallmatrix}\right)(\infty)] - [\left(\begin{smallmatrix} 1 & 0 \\ 0 & 2\end{smallmatrix}\right)(0)] = [\infty] - [0]$, as required. If $K = \left(\begin{smallmatrix} 1 & 0 \\ 1 & 2\end{smallmatrix}\right)\operatorname{SL}_2(\mathbb{Z})$, then the left side of (8.3.2) is

$$\begin{aligned}&[\left(\begin{smallmatrix} 2 & 1 \\ 0 & 1\end{smallmatrix}\right)(\infty)] - [\left(\begin{smallmatrix} 2 & 1 \\ 0 & 1\end{smallmatrix}\right)(0)] + [\left(\begin{smallmatrix} 1 & 0 \\ 1 & 2\end{smallmatrix}\right)(\infty)] - [\left(\begin{smallmatrix} 1 & 0 \\ 1 & 2\end{smallmatrix}\right)(0)]\\ &\qquad = [\infty] - [1] + [1] - [0] = [\infty] - [0].\end{aligned}$$

Finally, for $K = \left(\begin{smallmatrix} 2 & 0 \\ 0 & 1\end{smallmatrix}\right)\operatorname{SL}_2(\mathbb{Z})$ we also have $[\left(\begin{smallmatrix} 2 & 0 \\ 0 & 1\end{smallmatrix}\right)(\infty)] - [\left(\begin{smallmatrix} 2 & 0 \\ 0 & 1\end{smallmatrix}\right)(0)] = [\infty] - [0]$, as required. Thus by (8.3.3) we can compute T_2 on *any* Manin symbol, by summing over the action of the four matrices $\left(\begin{smallmatrix} 2 & 0 \\ 0 & 1\end{smallmatrix}\right), \left(\begin{smallmatrix} 1 & 0 \\ 0 & 2\end{smallmatrix}\right), \left(\begin{smallmatrix} 2 & 1 \\ 0 & 1\end{smallmatrix}\right), \left(\begin{smallmatrix} 1 & 0 \\ 1 & 2\end{smallmatrix}\right)$.

Proposition 8.8 (Merel). *The element*

$$\sum_{\substack{a>b\geq 0\\ d>c\geq 0\\ ad-bc=n}} \left[\begin{pmatrix} a & b \\ c & d\end{pmatrix}\right] \in \mathbb{Z}[\mathrm{Mat}_2(\mathbb{Z})_n]$$

satisfies condition C_n.

Merel's two-page proof of Proposition 8.8 is fairly elementary.

Remark 8.9. In [**Cre97a**, §2.4], Cremona discusses the work of Merel and Mazur on Heilbronn matrices in the special cases $\Gamma = \Gamma_0(N)$ and weight 2. He gives a simple proof that the action of T_p on Manin symbols can be computed by summing the action of some set R_p of matrices of determinant p. He then describes the set R_p and gives an efficient continued fractions algorithm for computing it (but he does not prove that his R_p satisfy Merel's hypothesis).

8.3.3. Remarks on Complexity. Merel gives another family $\mathcal{S}_n$ of matrices that satisfy condition C_n, and he proves that as $n \to \infty$,

$$\#\mathcal{S}_n \sim \frac{12\log(2)}{\pi^2} \cdot \sigma_1(n)\log(n),$$

where $\sigma_1(n)$ is the sum of the divisors of n. Thus for a fixed space $\mathbb{M}_k(\Gamma)$ of modular symbols, one can compute T_n using $O(\sigma_1(n)\log(n))$ arithmetic operations. Note that we have fixed $\mathbb{M}_k(\Gamma)$, so we ignore the linear algebra involved in computation of a presentation; also, adding elements takes a bounded number of field operations when the space is fixed. Thus, using Manin symbols the complexity of computing T_p, for p prime, is $O((p+1)\log(p))$ field operations, which is *exponential* in the number of digits of p.

8.3.4. Basmaji's Trick. There is a trick of Basmaji (see [**Bas96**]) for computing a matrix of T_n on $\mathbb{M}_k(\Gamma)$, when n is very large, and it is more efficient than one might naively expect. Basmaji's trick does not improve the big-oh complexity for a fixed space, but it does improve the complexity by a constant factor of the dimension of $\mathbb{M}_k(\Gamma;\mathbb{Q})$. Suppose we are interested in computing the matrix for T_n for some massive integer n and that $\mathbb{M}_k(\Gamma;\mathbb{Q})$ has large dimension. The trick is as follows. Choose a list

$$x_1 = [P_1, g_1], \ldots, x_r = [P_r, g_r] \in V = \mathbb{M}_k(\Gamma;\mathbb{Q})$$

of Manin symbols such that the map $\Psi : \mathbb{T} \to V^r$ given by

$$t \mapsto (tx_1, \ldots, tx_r)$$

is injective. In practice, it is often possible to do this with r "very small". Also, we emphasize that V^r is a $\mathbb{Q}$-vector space of dimension $r \cdot \dim(V)$.

Next find Hecke operators T_i, with i small, whose images form a basis for the image of Ψ. Now with the above data precomputed, which only required working with Hecke operators T_i for small i, we are ready to compute T_n with n huge. Compute $y_i = T_n(x_i)$, for each $i = 1, \ldots, r$, which we can compute using Heilbronn matrices since each $x_i = [P_i, g_i]$ is a Manin symbol. We thus obtain $\Psi(T_n) \in V^r$. Since we have precomputed Hecke operators T_j such that $\Psi(T_j)$ generate V^r, we can find a_j such that $\sum a_j\Psi(T_j) = \Psi(T_n)$. Then since Ψ is injective, we have $T_n = \sum a_j T_j$, which gives the full matrix of T_n on $M_k(\Gamma;\mathbb{Q})$.

8.4. Cuspidal Modular Symbols

Let $\mathbb{B}$ be the free abelian group on symbols $\{\alpha\}$, for $\alpha \in \mathbb{P}^1(\mathbb{Q})$, and set

$$\mathbb{B}_k = \mathbb{Z}[X,Y]_{k-2} \otimes \mathbb{B}.$$

Define a *left action of* $\mathrm{SL}_2(\mathbb{Z})$ on $\mathbb{B}_k$ by

$$g(P\{\alpha\}) = (gP)\{g(\alpha)\},$$

for $g \in \mathrm{SL}_2(\mathbb{Z})$. For any finite index subgroup $\Gamma \subset \mathrm{SL}_2(\mathbb{Z})$, let $\mathbb{B}_k(\Gamma)$ be the quotient of $\mathbb{B}_k$ by the relations $x - gx$ for all $g \in \Gamma$ and by any torsion. Thus $\mathbb{B}_k(\Gamma)$ is a torsion-free abelian group.

The *boundary map* is the map

$$b : \mathbb{M}_k(\Gamma) \to \mathbb{B}_k(\Gamma)$$

given by extending the map

$$b(P\{\alpha, \beta\}) = P\{\beta\} - P\{\alpha\}$$

linearly. The space $\mathbb{S}_k(\Gamma)$ of *cuspidal modular symbols* is the kernel

$$\mathbb{S}_k(\Gamma) = \ker(\mathbb{M}_k(\Gamma) \to \mathbb{B}_k(\Gamma)),$$

so we have an exact sequence

$$0 \to \mathbb{S}_k(\Gamma) \to \mathbb{M}_k(\Gamma) \to \mathbb{B}_k(\Gamma).$$

One can prove that when $k > 2$, this sequence is exact on the right.

Next we give a presentation of $\mathbb{B}_k(\Gamma)$ in terms of "boundary Manin symbols".

8.4.1. Boundary Manin Symbols. We give an explicit description of the boundary map in terms of Manin symbols for $\Gamma = \Gamma_1(N)$, then describe an efficient way to compute the boundary map.

Let $\mathcal{R}$ be the equivalence relation on $\Gamma\backslash\mathbb{Q}^2$ given by

$$[\Gamma \left(\begin{smallmatrix} \lambda u \\ \lambda v \end{smallmatrix}\right)] \sim \mathrm{sign}(\lambda)^k [\Gamma \left(\begin{smallmatrix} u \\ v \end{smallmatrix}\right)],$$

for any $\lambda \in \mathbb{Q}^*$. Denote by $B_k(\Gamma)$ the finite-dimensional $\mathbb{Q}$-vector space with basis the equivalence classes $(\Gamma\backslash\mathbb{Q}^2)/\mathcal{R}$. The following two propositions are proved in [**Mer94**].

Proposition 8.10. *The map*

$$\mu : \mathbb{B}_k(\Gamma) \to B_k(\Gamma), \qquad P\left\{\frac{u}{v}\right\} \mapsto P(u, v)\left[\Gamma \begin{pmatrix} u \\ v \end{pmatrix}\right]$$

is well defined and injective. Here u and v are assumed coprime.

Thus the kernel of $\delta : \mathbb{S}_k(\Gamma) \to \mathbb{B}_k(\Gamma)$ is the same as the kernel of $\mu \circ \delta$.

Proposition 8.11. *Let $P \in V_{k-2}$ and $g = \left(\begin{smallmatrix} a & b \\ c & d \end{smallmatrix}\right) \in \mathrm{SL}_2(\mathbb{Z})$. We have*

$$\mu \circ \delta([P, (c, d)]) = P(1, 0)[\Gamma \left(\begin{smallmatrix} a \\ c \end{smallmatrix}\right)] - P(0, 1)[\Gamma \left(\begin{smallmatrix} b \\ d \end{smallmatrix}\right)].$$

We next describe how to explicitly compute $\mu \circ \delta : \mathbb{M}_k(N, \varepsilon) \to B_k(N, \varepsilon)$ by generalizing the algorithm of [**Cre97a**, §2.2]. To compute the image of $[P, (c, d)]$, with $g = \left(\begin{smallmatrix} a & b \\ c & d \end{smallmatrix}\right) \in \mathrm{SL}_2(\mathbb{Z})$, we must compute the class of $[\left(\begin{smallmatrix} a \\ c \end{smallmatrix}\right)]$ and of $[\left(\begin{smallmatrix} b \\ d \end{smallmatrix}\right)]$. Instead of finding a canonical form for cusps, we use a quick test for equivalence modulo scalars. In the following algorithm, by the ith symbol we mean the ith basis vector for a basis of $B_k(N,\varepsilon)$. This basis is constructed as the algorithm is called successively. We first give the algorithm, and then prove the facts used by the algorithm in testing equivalence.

Algorithm 8.12 (Cusp Representation). *Given a boundary Manin symbol* $[(\begin{smallmatrix} u \\ v \end{smallmatrix})]$, *this algorithm outputs an integer* i *and a scalar* α *such that* $[(\begin{smallmatrix} u \\ v \end{smallmatrix})]$ *is equivalent to* α *times the* i*th symbol found so far. (We call this algorithm repeatedly and maintain a running list of cusps seen so far.)*

(1) Use Proposition 3.21 to check whether or not $[(\begin{smallmatrix} u \\ v \end{smallmatrix})]$ is equivalent, modulo scalars, to any cusp found. If so, return the representative, its index, and the scalar. If not, record $(\begin{smallmatrix} u \\ v \end{smallmatrix})$ in the representative list.
(2) Using Proposition 8.16, check whether or not $[(\begin{smallmatrix} u \\ v \end{smallmatrix})]$ is forced to equal 0 by the relations. If it does not equal 0, return its position in the list and the scalar 1. If it equals 0, return the scalar 0 and the position 1; keep $(\begin{smallmatrix} u \\ v \end{smallmatrix})$ in the list, and record that it is equivalent to 0.

The case considered in Cremona's book [**Cre97a**] only involve the trivial character, so no cusp classes are forced to vanish. Cremona gives the following two criteria for equivalence.

Proposition 8.13 (Cremona). *Consider* $(\begin{smallmatrix} u_i \\ v_i \end{smallmatrix})$, $i=1,2$, *with* u_i, v_i *integers such that* $\gcd(u_i,v_i)=1$ *for each* i.

(1) *There exists* $g\in\Gamma_0(N)$ *such that* $g(\begin{smallmatrix} u_1 \\ v_1 \end{smallmatrix}) = (\begin{smallmatrix} u_2 \\ v_2 \end{smallmatrix})$ *if and only if*

$$s_1v_2 \equiv s_2v_1 \pmod{\gcd(v_1v_2,N)}, \text{ where } s_j \text{ satisfies } u_js_j\equiv 1 \pmod{v_j}.$$

(2) *There exists* $g\in\Gamma_1(N)$ *such that* $g(\begin{smallmatrix} u_1 \\ v_1 \end{smallmatrix}) = (\begin{smallmatrix} u_2 \\ v_2 \end{smallmatrix})$ *if and only if*

$$v_2\equiv v_1 \pmod{N} \text{ and } u_2\equiv u_1 \pmod{\gcd(v_1,N)}.$$

Proof. The first statement is [**Cre97a**, Prop. 2.2.3], and the second is [**Cre92**, Lem. 3.2]. □

Algorithm 8.14 (Explicit Cusp Equivalence). *Suppose* $(\begin{smallmatrix} u_1 \\ v_1 \end{smallmatrix})$ *and* $(\begin{smallmatrix} u_2 \\ v_2 \end{smallmatrix})$ *are equivalent modulo* $\Gamma_0(N)$. *This algorithm computes a matrix* $g\in\Gamma_0(N)$ *such that* $g(\begin{smallmatrix} u_1 \\ v_1 \end{smallmatrix}) = (\begin{smallmatrix} u_2 \\ v_2 \end{smallmatrix})$.

(1) Let s_1,s_2,r_1,r_2 be solutions to $s_1u_1-r_1v_1=1$ and $s_2u_2-r_2v_2=1$.
(2) Find integers x_0 and y_0 such that $x_0v_1v_2+y_0N=1$.
(3) Let $x=-x_0(s_1v_2-s_2v_1)/(v_1v_2,N)$ and $s_1'=s_1+xv_1$.
(4) Output $g=(\begin{smallmatrix} u_2 & r_2 \\ v_2 & s_2 \end{smallmatrix})\cdot\begin{pmatrix} u_1 & r_1 \\ v_1 & s_1' \end{pmatrix}^{-1}$, which sends $(\begin{smallmatrix} u_1 \\ v_1 \end{smallmatrix})$ to $(\begin{smallmatrix} u_2 \\ v_2 \end{smallmatrix})$.

Proof. See the proof of [**Cre97a**, Prop. 8.13]. □

The ε relations can make the situation more complicated, since it is possible that $\varepsilon(\alpha) \neq \varepsilon(\beta)$ but

$$\varepsilon(\alpha)\left[\begin{pmatrix}u\\v\end{pmatrix}\right] = \left[\gamma_\alpha\begin{pmatrix}u\\v\end{pmatrix}\right] = \left[\gamma_\beta\begin{pmatrix}u\\v\end{pmatrix}\right] = \varepsilon(\beta)\left[\begin{pmatrix}u\\v\end{pmatrix}\right].$$

One way out of this difficulty is to construct the cusp classes for $\Gamma_1(N)$, and then quotient out by the additional ε relations using Gaussian elimination. This is far too inefficient to be useful in practice because the number of $\Gamma_1(N)$ cusp classes can be unreasonably large. Instead, we give a quick test to determine whether or not a cusp vanishes modulo the ε-relations.

Lemma 8.15. *Suppose α and α' are integers such that $\gcd(\alpha, \alpha', N) = 1$. Then there exist integers β and β', congruent to α and α' modulo N, respectively, such that $\gcd(\beta, \beta') = 1$.*

Proof. By Exercise 8.2 the map $\mathrm{SL}_2(\mathbb{Z}) \to \mathrm{SL}_2(\mathbb{Z}/N\mathbb{Z})$ is surjective. By the Euclidean algorithm, there exist integers x, y and z such that $x\alpha + y\alpha' + zN = 1$. Consider the matrix $\left(\begin{smallmatrix}y & -x\\ \alpha & \alpha'\end{smallmatrix}\right) \in \mathrm{SL}_2(\mathbb{Z}/N\mathbb{Z})$ and take β, β' to be the bottom row of a lift of this matrix to $\mathrm{SL}_2(\mathbb{Z})$. □

Proposition 8.16. *Let N be a positive integer and ε a Dirichlet character of modulus N. Suppose $\left(\begin{smallmatrix}u\\v\end{smallmatrix}\right)$ is a cusp with u and v coprime. Then $\left(\begin{smallmatrix}u\\v\end{smallmatrix}\right)$ vanishes modulo the relations*

$$\left[\gamma\left(\begin{smallmatrix}u\\v\end{smallmatrix}\right)\right] = \varepsilon(\gamma)\left[\left(\begin{smallmatrix}u\\v\end{smallmatrix}\right)\right], \qquad \textit{all } \gamma \in \Gamma_0(N),$$

if and only if there exists $\alpha \in (\mathbb{Z}/N\mathbb{Z})^*$, with $\varepsilon(\alpha) \neq 1$, such that

$$\begin{aligned} v &\equiv \alpha v \pmod{N},\\ u &\equiv \alpha u \pmod{\gcd(v, N)}. \end{aligned}$$

Proof. First suppose such an α exists. By Lemma 8.15 there exists $\beta, \beta' \in \mathbb{Z}$ lifting α, α^{-1} such that $\gcd(\beta, \beta') = 1$. The cusp $\left(\begin{smallmatrix}\beta u\\ \beta' v\end{smallmatrix}\right)$ has coprime coordinates so, by Proposition 8.13 and our congruence conditions on α, the cusps $\left(\begin{smallmatrix}\beta u\\ \beta' v\end{smallmatrix}\right)$ and $\left(\begin{smallmatrix}u\\v\end{smallmatrix}\right)$ are equivalent by an element of $\Gamma_1(N)$. This implies that $\left[\left(\begin{smallmatrix}\beta u\\ \beta' v\end{smallmatrix}\right)\right] = \left[\left(\begin{smallmatrix}u\\v\end{smallmatrix}\right)\right]$. Since $\left[\left(\begin{smallmatrix}\beta u\\ \beta' v\end{smallmatrix}\right)\right] = \varepsilon(\alpha)\left[\left(\begin{smallmatrix}u\\v\end{smallmatrix}\right)\right]$ and $\varepsilon(\alpha) \neq 1$, we have $\left[\left(\begin{smallmatrix}u\\v\end{smallmatrix}\right)\right] = 0$.

Conversely, suppose $\left[\left(\begin{smallmatrix}u\\v\end{smallmatrix}\right)\right] = 0$. Because all relations are two-term relations and the $\Gamma_1(N)$-relations identify $\Gamma_1(N)$-orbits, there must exists α and β with

$$\left[\gamma_\alpha\begin{pmatrix}u\\v\end{pmatrix}\right] = \left[\gamma_\beta\begin{pmatrix}u\\v\end{pmatrix}\right] \qquad \text{and } \varepsilon(\alpha) \neq \varepsilon(\beta).$$

Indeed, if this did not occur, then we could mod out by the ε relations by writing each $[\gamma_\alpha \left(\begin{smallmatrix} u \\ v \end{smallmatrix}\right)]$ in terms of $[\left(\begin{smallmatrix} u \\ v \end{smallmatrix}\right)]$, and there would be no further relations left to kill $[\left(\begin{smallmatrix} u \\ v \end{smallmatrix}\right)]$. Next observe that

$$\begin{aligned}\left[\gamma_{\beta^{-1}\alpha}\begin{pmatrix} u \\ v \end{pmatrix}\right] &= \left[\gamma_{\beta^{-1}}\gamma_\alpha\begin{pmatrix} u \\ v \end{pmatrix}\right] \\ &= \varepsilon(\beta^{-1})\left[\gamma_\alpha\begin{pmatrix} u \\ v \end{pmatrix}\right] = \varepsilon(\beta^{-1})\left[\gamma_\beta\begin{pmatrix} u \\ v \end{pmatrix}\right] = \left[\begin{pmatrix} u \\ v \end{pmatrix}\right].\end{aligned}$$

Applying Proposition 8.13 and noting that $\varepsilon(\beta^{-1}\alpha) \neq 1$ shows that $\beta^{-1}\alpha$ satisfies the properties of the "α" in the statement of the proposition. □

We enumerate the possible α appearing in Proposition 8.16 as follows. Let $g = (v, N)$ and list the $\alpha = v \cdot \frac{N}{g} \cdot a + 1$, for $a = 0, \ldots, g-1$, such that $\varepsilon(\alpha) \neq 0$.

8.5. Pairing Modular Symbols and Modular Forms

In this section we define a pairing between modular symbols and modular forms that the Hecke operators respect. We also define an involution on modular symbols and study its relationship with the pairing. This pairing is crucial in much that follows, because it gives rise to period maps from modular symbols to certain complex vector spaces.

Fix an integer weight $k \geq 2$ and a finite index subgroup Γ of $\mathrm{SL}_2(\mathbb{Z})$. Let $M_k(\Gamma)$ denote the space of holomorphic modular forms of weight k for Γ, and let $S_k(\Gamma)$ denote its cuspidal subspace. Following [**Mer94**, §1.5], let

$$\overline{S}_k(\Gamma) = \{\overline{f} : f \in S_k(\Gamma)\} \tag{8.5.1}$$

denote the space of *antiholomorphic* cusp forms. Here $\overline{f}$ is the function on $\mathfrak{h}^*$ given by $\overline{f}(z) = \overline{f(z)}$.

Define a pairing

$$(S_k(\Gamma) \oplus \overline{S}_k(\Gamma)) \times \mathbb{M}_k(\Gamma) \to \mathbb{C} \tag{8.5.2}$$

by letting

$$\langle (f_1, f_2), P\{\alpha, \beta\}\rangle = \int_\alpha^\beta f_1(z)P(z,1)dz + \int_\alpha^\beta f_2(z)P(\overline{z},1)d\overline{z}$$

and extending linearly. Here the integral is a complex path integral along a simple path from α to β in $\mathfrak{h}$ (so, e.g., write $z(t) = x(t) + iy(t)$, where $(x(t), y(t))$ traces out the path and consider two real integrals).

Proposition 8.17. *The integration pairing is well defined, i.e., if we replace $P\{\alpha,\beta\}$ by an equivalent modular symbol (equivalent modulo the left action of Γ), then the integral is the same.*

Proof. This follows from the change of variables formulas for integration and the fact that $f_1 \in S_k(\Gamma)$ and $f_2 \in \overline{S}_k(\Gamma)$. For example, if $k = 2$, $g \in \Gamma$ and $f \in S_k(\Gamma)$, then

$$\begin{aligned}\langle f, g\{\alpha,\beta\}\rangle &= \langle f, \{g(\alpha), g(\beta)\}\rangle \\ &= \int_{g(\alpha)}^{g(\beta)} f(z)dz \\ &= \int_{\alpha}^{\beta} f(g(z))dg(z) \\ &= \int_{\alpha}^{\beta} f(z)dz = \langle f, \{\alpha,\beta\}\rangle,\end{aligned}$$

where $f(g(z))dg(z) = f(z)dz$ because f is a weight 2 modular form. For the case of arbitrary weight $k > 2$, see Exercise 8.4. □

The integration pairing is very relevant to the study of special values of L-functions. The L-function of a cusp form $f = \sum a_n q^n \in S_k(\Gamma_1(N))$ is

$$L(f,s) = (2\pi)^s \Gamma(s)^{-1} \int_0^\infty f(it)t^s \frac{dt}{t} \tag{8.5.3}$$

$$= \sum_{n=1}^{\infty} \frac{a_n}{n^s} \qquad \text{for } \mathrm{Re}(s) \gg 0. \tag{8.5.4}$$

The equality of the integral and the Dirichlet series follows by switching the order of summation and integration, which is justified using standard estimates on $|a_n|$ (see, e.g., [**Kna92**, Section VIII.5]).

For each integer j with $1 \leq j \leq k-1$, we have, setting $s = j$ and making the change of variables $t \mapsto -it$ in (8.5.3), that

$$L(f,j) = \frac{(-2\pi i)^j}{(j-1)!} \cdot \left\langle f,\ X^{j-1}Y^{k-2-(j-1)}\{0,\infty\}\right\rangle. \tag{8.5.5}$$

The integers j as above are called *critical integers*. When f is an eigenform, they have deep conjectural significance (see [**BK90, Sch90**]). One can approximate $L(f,j)$ to any desired precision by computing the above pairing explicitly using the method described in Chapter 10. Alternatively, [**Dok04**] contains methods for computing special values of very general L-functions, which can be used for approximating $L(f,s)$ for arbitrary s, in addition to just the critical integers $1,\ldots,k-1$.

Theorem 8.18 (Shokoruv). *The pairing*

$$\langle \cdot, \cdot \rangle : (S_k(\Gamma) \oplus \overline{S}_k(\Gamma)) \times \mathbb{S}_k(\Gamma, \mathbb{C}) \to \mathbb{C}$$

is a nondegenerate pairing of complex vector spaces.

Proof. This is [**Sho80b**, Thm. 0.2] and [**Mer94**, §1.5]. □

Corollary 8.19. *We have*

$$\dim_{\mathbb{C}} \mathbb{S}_k(\Gamma, \mathbb{C}) = 2 \dim_{\mathbb{C}} S_2(\Gamma).$$

The pairing is also compatible with Hecke operators. Before proving this, we define an *action of Hecke operators* on $M_k(\Gamma_1(N))$ and on $\overline{S}_k(\Gamma_1(N))$. The definition is similar to the one we gave in Section 2.4 for modular forms of level 1. For a positive integer n, let R_n be a set of coset representatives for $\Gamma_1(N)\backslash\Delta_n$ from Lemma 8.7. For any $\gamma = \left(\begin{smallmatrix} a & b \\ c & d \end{smallmatrix}\right) \in \mathrm{GL}_2(\mathbb{Q})$ and $f \in M_k(\Gamma_1(N))$ set

$$f^{[\gamma]_k} = \det(\gamma)^{k-1}(cz+d)^{-k} f(\gamma(z)).$$

Also, for $f \in \overline{S}_k(\Gamma_1(N))$, set

$$f^{[\gamma]'_k} = \det(\gamma)^{k-1}(c\overline{z}+d)^{-k} f(\gamma(z)).$$

Then for $f \in M_k(\Gamma_1(N))$,

$$T_n(f) = \sum_{\gamma \in R_n} f^{[\gamma]_k}$$

and for $f \in \overline{S}_k(\Gamma_1(N))$,

$$T_n(f) = \sum_{\gamma \in R_n} f^{[\gamma]'_k}.$$

This agrees with the definition from Section 2.4 when $N = 1$.

Remark 8.20. If Γ is an arbitrary finite index subgroup of $\mathrm{SL}_2(\mathbb{Z})$, then we can define operators T_Δ on $M_k(\Gamma)$ for any Δ with $\Delta\Gamma = \Gamma\Delta = \Delta$ and $\Gamma\backslash\Delta$ finite. For concreteness we do not do the general case here or in the theorem below, but the proof is exactly the same (see [**Mer94**, §1.5]).

Finally we prove the promised Hecke compatibility of the pairing. This proof should convince you that the definition of modular symbols is sensible, in that they are natural objects to integrate against modular forms.

Theorem 8.21. *If*

$$f = (f_1, f_2) \in S_k(\Gamma_1(N)) \oplus \overline{S}_k(\Gamma_1(N))$$

and $x \in \mathbb{M}_k(\Gamma_1(N))$, *then for any* n,

$$\langle T_n(f), x\rangle = \langle f, T_n(x)\rangle.$$

Proof. We follow [**Mer94**, §2.1] (but with more details) and will only prove the theorem when $f = f_1 \in S_k(\Gamma_1(N))$, the proof in the general case being the same.

Let $\alpha, \beta \in \mathbb{P}^1(\mathbb{Q})$, $P \in \mathbb{Z}[X, Y]_{k-2}$, and for $g = \left(\begin{smallmatrix} a & b \\ c & d \end{smallmatrix}\right) \in \mathrm{GL}_2(\mathbb{Q})$, set

$$j(g, z) = cz + d.$$

Let n be any positive integer, and let R_n be a set of coset representatives for $\Gamma_1(N)\backslash\Delta_n$ from Lemma 8.7.

We have

$$\begin{aligned}\langle T_n(f), P\{\alpha,\beta\}\rangle &= \int_\alpha^\beta T_n(f)P(z,1)dz\\ &= \sum_{\delta\in R}\int_\alpha^\beta \det(\delta)^{k-1}f(\delta(z))j(\delta,z)^{-k}P(z,1)dz.\end{aligned}$$

Now for each summand corresponding to the $\delta\in R$, make the change of variables $u=\delta z$. Thus we make $\#R$ change of variables. Also, we will use the notation

$$\tilde{g} = \begin{pmatrix} d & -b\\ -c & a\end{pmatrix} = \det(g)\cdot g^{-1}$$

for $g\in \mathrm{GL}_2(\mathbb{Q})$. We have

$$\langle T_n(f), P\{\alpha,\beta\}\rangle =$$
$$\sum_{\delta\in R}\int_{\delta(\alpha)}^{\delta(\beta)} \det(\delta)^{k-1}f(u)j(\delta,\delta^{-1}(u))^{-k}P(\delta^{-1}(u),1)d(\delta^{-1}(u)).$$

We have $\delta^{-1}(u)=\tilde{\delta}(u)$, since a linear fractional transformation is unchanged by a nonzero rescaling of a matrix that induces it. Thus by the quotient rule, using that $\tilde{\delta}$ has determinant $\det(\delta)$, we see that

$$d(\delta^{-1}(u)) = d(\tilde{\delta}(u)) = \frac{\det(\delta)du}{j(\tilde{\delta},u)^2}.$$

We next show that

$$j(\delta,\delta^{-1}(u))^{-k}P(\delta^{-1}(u),1) = j(\tilde{\delta},u)^k\det(\delta)^{-k}P(\tilde{\delta}(u),1). \tag{8.5.6}$$

From the definitions, and again using that $\delta^{-1}(u)=\tilde{\delta}(u)$, we see that

$$j(\delta,\delta^{-1}(u)) = \frac{\det(\delta)}{j(\tilde{\delta},u)},$$

which proves that (8.5.6) holds. Thus

$$\langle T_n(f), P\{\alpha,\beta\}\rangle =$$
$$\sum_{\delta\in R}\int_{\delta(\alpha)}^{\delta(\beta)} \det(\delta)^{k-1}f(u)j(\tilde{\delta},u)^k\det(\delta)^{-k}P(\tilde{\delta}(u),1)\frac{\det(\delta)du}{j(\tilde{\delta},u)^2}.$$

Next we use that

$$(\delta P)(u,1) = j(\tilde{\delta},u)^{k-2}P(\tilde{\delta}(u),1).$$

To see this, note that $P(X,Y) = P(X/Y,1)\cdot Y^{k-2}$. Using this we see that

$$\begin{aligned}(\delta P)(X,Y) &= (P\circ\tilde{\delta})(X,Y)\\ &= P\left(\tilde{\delta}\left(\frac{X}{Y}\right),1\right)\left(-c\cdot\frac{X}{Y}+a\right)^{k-2}\cdot Y^{k-2}.\end{aligned}$$

Now substituting $(u,1)$ for $(X,1)$, we see that

$$(\delta P)(u,1) = P(\tilde{\delta}(u),1)\cdot(-cu+a)^{k-2},$$

as required. Thus finally

$$\begin{aligned}\langle T_n(f), P\{\alpha,\beta\}\rangle &= \sum_{\delta\in R}\int_{\delta(\alpha)}^{\delta(\beta)} f(u)j(\tilde{\delta},u)^{k-2}P(\tilde{\delta}(u),1)du\\ &= \sum_{\delta\in R}\int_{\delta(\alpha)}^{\delta(\beta)} f(u)\cdot((\delta P)(u,1))du\\ &= \langle f,\, T_n(P\{\alpha,\beta\})\rangle.\end{aligned}$$

□

Suppose that Γ is a finite index subgroup of $\mathrm{SL}_2(\mathbb{Z})$ such that if $\eta = \left(\begin{smallmatrix}-1&0\\0&1\end{smallmatrix}\right)$, then

$$\eta\Gamma\eta = \Gamma.$$

For example, $\Gamma = \Gamma_1(N)$ satisfies this condition. There is an involution ι^* on $\mathbb{M}_k(\Gamma)$ given by

$$\iota^*(P(X,Y)\{\alpha,\beta\}) = -P(X,-Y)\{-\alpha,-\beta\}, \tag{8.5.7}$$

which we call the *star involution*. On Manin symbols, ι^* is

$$\iota^*[P,(u,v)] = -[P(-X,Y),(-u,v)]. \tag{8.5.8}$$

Let $\mathbb{S}_k(\Gamma)^+$ be the $+1$ eigenspace for ι^* on $\mathbb{S}_k(\Gamma)$, and let $\mathbb{S}_k(\Gamma)^-$ be the -1 eigenspace. There is also a map ι on modular forms, which is adjoint to ι^*.

Remark 8.22. Notice the minus sign in front of $-P(X,-Y)\{-\alpha,-\beta\}$ in (8.5.7). This sign is missing in [**Cre97a**], which is a potential source of confusion (this is not a mistake, but a different choice of convention).

We now state the final result about the pairing, which explains how modular symbols and modular forms are related.

Theorem 8.23. *The integration pairing $\langle\cdot\,,\,\cdot\rangle$ induces nondegenerate Hecke compatible bilinear pairings*

$$\mathbb{S}_k(\Gamma)^+\times S_k(\Gamma)\to\mathbb{C}\qquad\textit{and}\qquad \mathbb{S}_k(\Gamma)^-\times \overline{S}_k(\Gamma)\to\mathbb{C}.$$

Remark 8.24. We make some remarks about computing the boundary map of Section 8.4.1 when working in the ±1 quotient. Let s be a sign, either $+1$ or -1. To compute $\mathbb{S}_k(N,\varepsilon)_s$, it is necessary to replace $B_k(N,\varepsilon)$ by its quotient modulo the additional relations $[(\begin{smallmatrix}-u\\v\end{smallmatrix})] = s\,[(\begin{smallmatrix}u\\v\end{smallmatrix})]$ for all cusps $(\begin{smallmatrix}u\\v\end{smallmatrix})$. Algorithm 8.12 can be modified to deal with this situation as follows. Given a cusp $x = (\begin{smallmatrix}u\\v\end{smallmatrix})$, proceed as in Algorithm 8.12 and check if either $(\begin{smallmatrix}u\\v\end{smallmatrix})$ or $(\begin{smallmatrix}-u\\v\end{smallmatrix})$ is equivalent (modulo scalars) to any cusp seen so far. If not, use the following trick to determine whether the ε and s-relations kill the class of $(\begin{smallmatrix}u\\v\end{smallmatrix})$: use the unmodified Algorithm 8.12 to compute the scalars α_1, α_2 and indices i_1, i_2 associated to $(\begin{smallmatrix}u\\v\end{smallmatrix})$ and $(\begin{smallmatrix}-u\\v\end{smallmatrix})$, respectively. The s-relation kills the class of $(\begin{smallmatrix}u\\v\end{smallmatrix})$ if and only if $i_1 = i_2$ but $\alpha_1 \neq s\alpha_2$.

8.6. Degeneracy Maps

In this section, we describe natural maps between spaces of modular symbols with character of different levels. We consider spaces with character, since they are so important in applications.

Fix a positive integer N and a Dirichlet character $\varepsilon : (\mathbb{Z}/N\mathbb{Z})^* \to \mathbb{C}^*$. Let M be a positive divisor of N that is divisible by the conductor of ε, in the sense that ε factors through $(\mathbb{Z}/M\mathbb{Z})^*$ via the natural map $(\mathbb{Z}/N\mathbb{Z})^* \to (\mathbb{Z}/M\mathbb{Z})^*$ composed with some uniquely defined character $\varepsilon' : (\mathbb{Z}/M\mathbb{Z})^* \to \mathbb{C}^*$. For any positive divisor t of N/M, let $T = (\begin{smallmatrix}1&0\\0&t\end{smallmatrix})$ and fix a choice $D_t = \{T\gamma_i : i = 1,\ldots,n\}$ of coset representatives for $\Gamma_0(N)\backslash T\Gamma_0(M)$.

Remark 8.25. Note that [**Mer94**, §2.6] contains a typo: The quotient "$\Gamma_1(N)\backslash\Gamma_1(M)T$" should be replaced by "$\Gamma_1(N)\backslash T\Gamma_1(M)$".

Proposition 8.26. *For each divisor t of N/M there are well-defined linear maps*

$$\alpha_t : \mathbb{M}_k(N,\varepsilon) \to \mathbb{M}_k(M,\varepsilon'), \qquad \alpha_t(x) = (tT^{-1})x = \begin{pmatrix} t & 0\\ 0 & 1\end{pmatrix} x,$$

$$\beta_t : \mathbb{M}_k(M,\varepsilon') \to \mathbb{M}_k(N,\varepsilon), \qquad \beta_t(x) = \sum_{T\gamma_i \in D_t} \varepsilon'(\gamma_i)^{-1} T\gamma_i x.$$

Furthermore, $\alpha_t \circ \beta_t$ is multiplication by $t^{k-2}\cdot[\Gamma_0(M) : \Gamma_0(N)]$.

Proof. To show that α_t is well defined, we must show that for each $x \in \mathbb{M}_k(N,\varepsilon)$ and $\gamma = (\begin{smallmatrix}a&b\\c&d\end{smallmatrix}) \in \Gamma_0(N)$, we have

$$\alpha_t(\gamma x - \varepsilon(\gamma)x) = 0 \in \mathbb{M}_k(M,\varepsilon').$$

We have

$$\alpha_t(\gamma x) = \begin{pmatrix} t & 0\\ 0 & 1\end{pmatrix}\gamma x = \begin{pmatrix} a & tb\\ c/t & d\end{pmatrix}\begin{pmatrix} t & 0\\ 0 & 1\end{pmatrix} x = \varepsilon'(a)\begin{pmatrix} t & 0\\ 0 & 1\end{pmatrix} x,$$

so

$$\alpha_t(\gamma x - \varepsilon(\gamma)x) = \varepsilon'(a)\alpha_t(x) - \varepsilon(\gamma)\alpha_t(x) = 0.$$

We next verify that β_t is well defined. Suppose that $x \in \mathbb{M}_k(M, \varepsilon')$ and $\gamma \in \Gamma_0(M)$; then $\varepsilon'(\gamma)^{-1}\gamma x = x$, so

$$\begin{aligned}\beta_t(x) &= \sum_{T\gamma_i \in D_t} \varepsilon'(\gamma_i)^{-1} T\gamma_i \varepsilon'(\gamma)^{-1}\gamma x \\ &= \sum_{T\gamma_i\gamma \in D_t} \varepsilon'(\gamma_i\gamma)^{-1} T\gamma_i\gamma x.\end{aligned}$$

This computation shows that the definition of β_t does not depend on the choice D_t of coset representatives. To finish the proof that β_t is well defined, we must show that, for $\gamma \in \Gamma_0(M)$, we have $\beta_t(\gamma x) = \varepsilon'(\gamma)\beta_t(x)$ so that β_t respects the relations that define $\mathbb{M}_k(M, \varepsilon)$. Using that β_t does not depend on the choice of coset representative, we find that for $\gamma \in \Gamma_0(M)$,

$$\begin{aligned}\beta_t(\gamma x) &= \sum_{T\gamma_i \in D_t} \varepsilon'(\gamma_i)^{-1} T\gamma_i\gamma x \\ &= \sum_{T\gamma_i\gamma^{-1} \in D_t} \varepsilon'(\gamma_i\gamma^{-1})^{-1} T\gamma_i\gamma^{-1}\gamma x \\ &= \varepsilon'(\gamma)\beta_t(x).\end{aligned}$$

To compute $\alpha_t \circ \beta_t$, we use that $\#D_t = [\Gamma_C(N) : \Gamma_0(M)]$:

$$\begin{aligned}\alpha_t(\beta_t(x)) &= \alpha_t\left(\sum_{T\gamma_i} \varepsilon'(\gamma_i)^{-1} T\gamma_i x\right) \\ &= \sum_{T\gamma_i} \varepsilon'(\gamma_i)^{-1}(tT^{-1})T\gamma_i x \\ &= t^{k-2}\sum_{T\gamma_i} \varepsilon'(\gamma_i)^{-1}\gamma_i x \\ &= t^{k-2}\sum_{T\gamma_i} x \\ &= t^{k-2} \cdot [\Gamma_0(N) : \Gamma_0(M)] \cdot x.\end{aligned}$$

The scalar factor of t^{k-2} appears instead of t, because t is acting on x as an element of $\mathrm{GL}_2(\mathbb{Q})$ and *not* as an an element of $\mathbb{Q}$. □

Definition 8.27 (New and Old Modular Symbols). The space $\mathbb{M}_k(N, \varepsilon)_{\mathrm{new}}$ of *new modular symbols* is the intersection of the kernels of the α_t as t runs through all positive divisors of N/M and M runs through positive divisors of M strictly less than N and divisible by the conductor of ε. The subspace

$\mathbb{M}_k(N,\varepsilon)_{\text{old}}$ of *old modular symbols* is the subspace generated by the images of the β_t where t runs through all positive divisors of N/M and M runs through positive divisors of M strictly less than N and divisible by the conductor of ε. The new and old subspaces of cuspidal modular symbols are the intersections of the above spaces with $\mathbb{S}_k(N,\varepsilon)$.

Example 8.28. The new and old subspaces need not be disjoint, as the following example illustrates! (This contradicts [**Mer94**, pg. 80].) Consider, for example, the case $N = 6$, $k = 2$, and trivial character. The spaces $\mathbb{M}_2(\Gamma_0(2))$ and $\mathbb{M}_2(\Gamma_0(3))$ are each of dimension 1, and each is generated by the modular symbol $\{\infty, 0\}$. The space $\mathbb{M}_2(\Gamma_0(6))$ is of dimension 3 and is generated by the three modular symbols $\{\infty, 0\}$, $\{-1/4, 0\}$, and $\{-1/2, -1/3\}$. The space generated by the two images of $\mathbb{M}_2(\Gamma_0(2))$ under the two degeneracy maps has dimension 2, and likewise for $\mathbb{M}_2(\Gamma_0(3))$. Together these images generate $\mathbb{M}_2(\Gamma_0(6))$, so $\mathbb{M}_2(\Gamma_0(6))$ is equal to its old subspace. However, the new subspace is nontrivial because the two degeneracy maps $\mathbb{M}_2(\Gamma_0(6)) \to \mathbb{M}_2(\Gamma_0(2))$ are equal, as are the two degeneracy maps

$$\mathbb{M}_2(\Gamma_0(6)) \to \mathbb{M}_2(\Gamma_0(3)).$$

In particular, the intersection of the kernels of the degeneracy maps has dimension at least 1 (in fact, it equals 1). We verify some of the above claims using SAGE.

```
sage: M = ModularSymbols(Gamma0(6)); M
Modular Symbols space of dimension 3 for Gamma_0(6)
of weight 2 with sign 0 over Rational Field
sage: M.new_subspace()
Modular Symbols subspace of dimension 1 of Modular
Symbols space of dimension 3 for Gamma_0(6) of weight
2 with sign 0 over Rational Field
sage: M.old_subspace()
Modular Symbols subspace of dimension 3 of Modular
Symbols space of dimension 3 for Gamma_0(6) of weight
2 with sign 0 over Rational Field
```

8.7. Explicitly Computing $\mathbb{M}_k(\Gamma_0(N))$

In this section we explicitly compute $\mathbb{M}_k(\Gamma_0(N))$ for various k and N. We represent Manin symbols for $\Gamma_0(N)$ as triples of integers (i, u, v), where $(u, v) \in \mathbb{P}^1(\mathbb{Z}/N\mathbb{Z})$, and (i, u, v) corresponds to $[X^iY^{k-2-i}, (u, v)]$ in the usual notation. Also, recall from Proposition 3.10 that (u, v) corresponds to the right coset of $\Gamma_0(N)$ that contains a matrix $\left(\begin{smallmatrix} a & b \\ c & d \end{smallmatrix}\right)$ with $(u,v) \equiv (c,d)$ as elements of $\mathbb{P}^1(\mathbb{Z}/N\mathbb{Z})$, i.e., up to rescaling by an element of $(\mathbb{Z}/N\mathbb{Z})^*$.

8.7.1. Computing $\mathbb{P}^1(\mathbb{Z}/N\mathbb{Z})$.

In this section we give an algorithm to compute a canonical representative for each element of $\mathbb{P}^1(\mathbb{Z}/N\mathbb{Z})$. This algorithm is extremely important because modular symbols implementations use it a huge number of times. A more naive approach would be to store all pairs $(u,v) \in (\mathbb{Z}/N\mathbb{Z})^2$ and a fixed reduced representative, but this wastes a huge amount of memory. For example, if $N = 10^5$, we would store an array of

$$2 \cdot 10^5 \cdot 10^5 = 20 \text{ billion integers.}$$

Another approach to enumerating $\mathbb{P}^1(\mathbb{Z}/N\mathbb{Z})$ is described at the end of [**Cre97a**, §2.2]. It uses the fact that is easy to test whether two pairs $(u_0,v_0),(u_1,v_1)$ define the same element of $\mathbb{P}^1(\mathbb{Z}/N\mathbb{Z})$; they do if and only if we have equality of cross terms $u_0v_1 = v_0u_1 \pmod N$ (see [**Cre97a**, Prop. 2.2.1]). So we consider the 0-based list of elements

$$(1,0),(1,1),\ldots,(1,N-1),(0,1) \tag{8.7.1}$$

concatenated with the list of nonequivalent elements (d,a) for $d \mid N$ and $a = 1,\ldots,N-1$, checking each time we add a new element to our list (of (d,a)) whether we have already seen it.

Given a random pair (u,v), the problem is then to find the index of the element of our list of the equivalent representative in $\mathbb{P}^1(\mathbb{Z}/N\mathbb{Z})$. We use the following algorithm, which finds a canonical representative for each element of $\mathbb{P}^1(\mathbb{Z}/N\mathbb{Z})$. Given an arbitrary (u,v), we first find the canonical equivalent elements (u',v'). If $u' = 1$, then the index is v'. If $u' \neq 1$, we find the corresponding element in an explicit sorted list, e.g., using binary search.

In the following algorithm, $a \pmod N$ denotes the residue of a modulo N that satisfies $0 \le a < N$. Note that we *never* create and store the list (8.7.1) itself in memory.

Algorithm 8.29 (Reduction in $\mathbb{P}^1(\mathbb{Z}/N\mathbb{Z})$ to Canonical Form). *Given u and v and a positive integer N, this algorithm outputs a pair u_0, v_0 such that $(u,v) \equiv (u_0,v_0)$ as elements of $\mathbb{P}^1(\mathbb{Z}/N\mathbb{Z})$ and $s \in \mathbb{Z}$ such that $(u,v) = (su_0, sv_0) \pmod{\mathbb{Z}/n\mathbb{Z}}$. Moreover, the element (u_0,v_0) does not depend on the class of (u,v), i.e., for any s with $\gcd(N,s) = 1$ the input (su,sv) also outputs (u_0,v_0). If (u,v) is not in $\mathbb{P}^1(\mathbb{Z}/N\mathbb{Z})$, this algorithm outputs $(0,0), 0$.*

(1) [Reduce] Reduce both u and v modulo N.
(2) [Easy $(0,1)$ Case] If $u = 0$, check that $\gcd(v,N) = 1$. If so, return $s = 1$ and $(0,1)$; otherwise return 0.
(3) [GCD] Compute $g = \gcd(u,N)$ and $s,t \in \mathbb{Z}$ such that $g = su + tN$.

(4) [Not in P^1?] We have $\gcd(u, v, N) = \gcd(g, v)$, so if $\gcd(g, v) > 1$, then $(u, v) \notin \mathbb{P}^1(\mathbb{Z}/N\mathbb{Z})$, and we return 0.

(5) [Pseudo-Inverse] Now $g = su+tN$, so we may think of s as "pseudo-inverse" of $u \pmod N$, in the sense that su is as close as possible to being 1 modulo N. Note that since $g \mid u$, changing s modulo N/g does not change $su \pmod N$. We can adjust s modulo N/g so it is coprime to N (by adding multiples of N/g to s). (This is because $1 = su/g + tN/g$, so s is a unit mod N/g, and the map $(\mathbb{Z}/N\mathbb{Z})^* \to (\mathbb{Z}/(N/g)\mathbb{Z})^*$ is surjective, e.g., as we saw in the proof of Algorithm 4.28.)

(6) [Multiply by s] Multiply (u, v) by s, and replace (u, v) by the equivalent element (g, sv) of $\mathbb{P}^1(\mathbb{Z}/N\mathbb{Z})$.

(7) [Normalize] Compute the unique pair (g, v') equivalent to (g, v) that minimizes v, as follows:
   (a) [Easy Case] If $g = 1$, this pair is $(1, v)$.
   (b) [Enumerate and Find Best] Otherwise, note that if $1 \neq t \in (\mathbb{Z}/N\mathbb{Z})^*$ and $tg \equiv g \pmod N$, then $(t-1)g \equiv 0 \pmod N$, so $t - 1 = kN/g$ for some k with $1 \leq k \leq g - 1$. Then for $t = 1 + kN/g$ coprime to N, we have $(gt, vt) = (g, v + kvN/g)$. So we compute all pairs $(g, v + kvN/g)$ and pick out the one that minimizes the least nonnegative residue of vt modulo N.
   (c) [Invert s and Output] The s that we computed in the above steps multiplies the input (u, v) to give the output (u_0, v_0). Thus we invert it, since the scalar we output multiplies (u_0, v_0) to give (u, v).

Remark 8.30. In the above algorithm, there are many gcd's with N so one should create a table of the gcd's of $0, 1, 2, \ldots, N - 1$ with N.

Remark 8.31. Another approach is to instead use that

$$\mathbb{P}^1(\mathbb{Z}/N\mathbb{Z}) \cong \prod_{p \mid N} \mathbb{P}^1(\mathbb{Z}/p^{\nu_p}\mathbb{Z}),$$

where $\nu_p = \mathrm{ord}_p(N)$, and that it is relatively easy to enumerate the elements of $\mathbb{P}^1(\mathbb{Z}/p^n\mathbb{Z})$ for a prime power p^n.

Algorithm 8.32 (List $\mathbb{P}^1(\mathbb{Z}/N\mathbb{Z})$). *Given an integer $N > 1$, this algorithm makes a sorted list of the distinct representatives (c, d) of $\mathbb{P}^1(\mathbb{Z}/N\mathbb{Z})$ with $c \neq 0, 1$, as output by Algorithm 8.29.*

(1) For each $c = 1, \ldots, N - 1$ with $g = \gcd(c, N) > 1$ do the following:
   (a) Use Algorithm 8.29 to compute the canonical representative (u', v') equivalent to $(c, 1)$, and include it in the list.

(b) If $g = c$, for each $d = 2, \dots, N-1$ with $\gcd(d, N) > 1$ and $\gcd(c, d) = 1$, append the normalized representative of (c, d) to the list.

(2) Sort the list.

(3) Pass through the sorted list and delete any duplicates.

8.8. Explicit Examples

Explicit detailed examples are crucial when implementing modular symbols algorithms from scratch. This section contains a number of such examples.

8.8.1. Examples of Computation of $\mathbb{M}_k(\Gamma_0(N))$. In this section, we compute $\mathbb{M}_k(\Gamma_0(N))$ explicitly in a few cases.

Example 8.33. We compute $V = \mathbb{M}_4(\Gamma_0(1))$. Because $S_k(\Gamma_0(1)) = 0$ and $M_k(\Gamma_0(1)) = \mathbb{C}E_4$, we expect V to have dimension 1 and for each integer n the Hecke operator T_n to have eigenvalue the sum $\sigma_3(n)$ of the cubes of positive divisors of n.

The Manin symbols are

$$x_0 = (0,0,0), \quad x_1 = (1,0,0), \quad x_2 = (2,0,0).$$

The relation matrix is

$$\left(\begin{array}{ccc} 1 & 0 & 1 \\ 0 & 0 & 0 \\ \hline 2 & -2 & 2 \\ 1 & -1 & 1 \\ 2 & -2 & 2 \end{array}\right),$$

where the first two rows correspond to S-relations and the second three to T-relations. Note that we do not include all S-relations, since it is obvious that some are redundant, e.g., $x + xS = 0$ and $(xS) + (xS)S = xS + x = 0$ are the same since S has order 2.

The echelon form of the relation matrix is

$$\begin{pmatrix} 1 & 0 & 1 \\ 0 & 1 & 0 \end{pmatrix},$$

where we have deleted the zero rows from the bottom. Thus we may replace the above complicated list of relations with the following simpler list of relations:

$$\begin{aligned} x_0 + x_2 &= 0, \\ x_1 &= 0 \end{aligned}$$

from which we immediately read off that the second generator x_1 is 0 and $x_0 = -x_2$. Thus $\mathbb{M}_4(\Gamma_0(1))$ has dimension 1, with basis the equivalence class of x_2 (or of x_0).

Next we compute the Hecke operator T_2 on $\mathbb{M}_4(\Gamma_0(1))$. The Heilbronn matrices of determinant 2 from Proposition 8.8 are

$$\begin{aligned} h_0 &= \begin{pmatrix} 1 & 0 \\ 0 & 2 \end{pmatrix}, \\ h_1 &= \begin{pmatrix} 1 & 0 \\ 1 & 2 \end{pmatrix}, \\ h_2 &= \begin{pmatrix} 2 & 0 \\ 0 & 1 \end{pmatrix}, \\ h_3 &= \begin{pmatrix} 2 & 1 \\ 0 & 1 \end{pmatrix}. \end{aligned}$$

To compute T_2, we apply each of these matrices to x_0, then reduce modulo the relations. We have

$$\begin{aligned} x_2 \begin{pmatrix} 1 & 0 \\ 0 & 2 \end{pmatrix} &= [X^2, (0,0)] \begin{pmatrix} 1 & 0 \\ 0 & 2 \end{pmatrix} x_2, \\ x_2 \begin{pmatrix} 1 & 0 \\ 1 & 2 \end{pmatrix} &= [X^2, (0,0)] = x_2, \\ x_2 \begin{pmatrix} 2 & 0 \\ 0 & 1 \end{pmatrix} &= [(2X)^2, (0,0)] = 4x_2, \\ x_2 \begin{pmatrix} 2 & 1 \\ 0 & 1 \end{pmatrix} &= [(2X+1)^2, (0,0)] = x_0 + 4x_1 + 4x_2 \sim 3x_2. \end{aligned}$$

Summing we see that $T_2(x_2) \sim 9x_2$ in $\mathbb{M}_4(\Gamma_0(1))$. Notice that

$$9 = 1^3 + 2^3 = \sigma_3(2).$$

The Heilbronn matrices of determinant 3 from Proposition 8.8 are

$$\begin{aligned} h_0 &= \begin{pmatrix} 1 & 0 \\ 0 & 3 \end{pmatrix}, & h_1 &= \begin{pmatrix} 1 & 0 \\ 1 & 3 \end{pmatrix}, \\ h_2 &= \begin{pmatrix} 1 & 0 \\ 2 & 3 \end{pmatrix}, & h_3 &= \begin{pmatrix} 2 & 1 \\ 1 & 2 \end{pmatrix}, \\ h_4 &= \begin{pmatrix} 3 & 0 \\ 0 & 1 \end{pmatrix}, & h_5 &= \begin{pmatrix} 3 & 1 \\ 0 & 1 \end{pmatrix}, \\ h_6 &= \begin{pmatrix} 3 & 2 \\ 0 & 1 \end{pmatrix}. \end{aligned}$$

We have

$$\begin{aligned}
x_2\begin{pmatrix}1&0\\0&3\end{pmatrix} &= [X^2,(0,0)]\begin{pmatrix}1&0\\0&3\end{pmatrix} = x_2,\\
x_2\begin{pmatrix}1&0\\1&3\end{pmatrix} &= [X^2,(0,0)] = x_2,\\
x_2\begin{pmatrix}1&0\\2&3\end{pmatrix} &= [X^2,(0,0)] = x_2,\\
x_2\begin{pmatrix}2&1\\2&2\end{pmatrix} &= [(2X+1)^2,(0,0)] = x_0+4x_1+4x_2 \sim 3x_2,\\
x_2\begin{pmatrix}3&0\\0&1\end{pmatrix} &= [(3X)^2,(0,0)] = 9x_2,\\
x_2\begin{pmatrix}3&1\\0&1\end{pmatrix} &= [(3X+1)^2,(0,0)] = x_0+6x_1+9x_2 \sim 8x_2,\\
x_2\begin{pmatrix}3&2\\0&1\end{pmatrix} &= [(3X+2)^2,(0,0)] = 4x_0+12x_1+9x_2 \sim 5x_2.
\end{aligned}$$

Summing we see that

$$T_3(x_2) \sim x_2+x_2+x_2+3x_2+9x_2+8x_2+5x_2 = 28x_2.$$

Notice that

$$28 = 1^3+3^3 = \sigma_3(3).$$

Example 8.34. Next we compute $\mathbb{M}_2(\Gamma_0(11))$ explicitly. The Manin symbol generators are

$$x_0=(0,1),\ x_1=(1,0),\ x_2=(1,1),\ x_3=(1,2),\ x_4=(1,3),\ x_5=(1,4),$$

$$x_6=(1,5),\ x_7=(1,6),\ x_8=(1,7),\ x_9=(1,8),\ x_{10}=(1,9),\ x_{11}=(1,10).$$

The relation matrix is as follows, where the S-relations are above the line and the T-relations are below it:

$$\left(\begin{array}{cccccccccccc}
1&1&0&0&0&0&0&0&0&0&0&0\\
0&0&1&0&0&0&0&0&0&0&0&1\\
0&0&0&1&0&0&1&0&0&0&0&0\\
0&0&0&0&1&0&0&0&1&0&0&0\\
0&0&0&0&0&1&0&0&0&1&0&0\\
0&0&0&0&0&0&0&1&0&0&1&0\\
\hline
1&1&0&0&0&0&0&0&0&0&0&1\\
0&0&1&0&0&0&1&0&0&0&1&0\\
0&0&0&1&0&1&0&0&1&0&0&0\\
0&0&0&0&1&0&0&1&0&1&0&0
\end{array}\right).$$

In weight 2, two out of three T-relations are redundant, so we do not include them. The reduced row echelon form of the relation matrix is

$$\begin{pmatrix} 1 & 1 & 0 & 0 & 0 & 0 & 0 & 0 & 0 & 0 & 0 & 0 \\ 0 & 0 & 1 & 0 & 0 & 0 & 0 & 0 & 0 & 0 & 0 & 0 \\ 0 & 0 & 0 & 1 & 0 & 0 & 0 & 0 & 0 & 0 & -1 & 0 \\ 0 & 0 & 0 & 0 & 1 & 0 & 0 & 0 & 0 & 1 & -1 & 0 \\ 0 & 0 & 0 & 0 & 0 & 1 & 0 & 0 & 0 & 1 & 0 & 0 \\ 0 & 0 & 0 & 0 & 0 & 0 & 1 & 0 & 0 & 0 & 1 & 0 \\ 0 & 0 & 0 & 0 & 0 & 0 & 0 & 1 & 0 & 0 & 1 & 0 \\ 0 & 0 & 0 & 0 & 0 & 0 & 0 & 0 & 1 & -1 & 1 & 0 \\ 0 & 0 & 0 & 0 & 0 & 0 & 0 & 0 & 0 & 0 & 0 & 1 \\ 0 & 0 & 0 & 0 & 0 & 0 & 0 & 0 & 0 & 0 & 0 & 0 \end{pmatrix}.$$

From the echelon form we see that every symbol is equivalent to a combination of $x_1 = (1,0)$, $x_9 = (1,8)$, and $x_{10} = (1,9)$. (Notice that columns 1, 9, and 10 are the pivot columns, where we index columns starting at 0.)

To compute T_2, we apply each of the Heilbronn matrices of determinant 2 from Proposition 8.8 to x_1, then to x_9, and finally to x_{10}. The matrices are as in Example 8.33 above. We have

$$T_2(x_1) = 3(1,0) + (1,6) \sim 3x_1 - x_{10}.$$

Applying T_2 to $x_9 = (1,8)$, we get

$$T_2(x_9) = (1,3) + (1,4) + (1,5) + (1,10) \sim -2x_9.$$

Applying T_2 to $x_{10} = (1,9)$, we get

$$T_2(x_{10}) = (1,4) + (1,5) + (1,7) + (1,10) \sim -x_1 - 2x_{10}.$$

Thus the matrix of T_2 with respect to this basis is

$$T_2 = \begin{pmatrix} 3 & 0 & 0 \\ 0 & -2 & 0 \\ -1 & 0 & -2 \end{pmatrix},$$

where we write the matrix as an operator on the left on vectors written in terms of x_1, x_9, and x_{10}. The matrix T_2 has characteristic polynomial

$$(x-3)(x+2)^2.$$

The $(x-3)$ factor corresponds to the weight 2 Eisenstein series, and the $x+2$ factor corresponds to the elliptic curve $E = X_0(11)$, which has

$$a_2 = -2 = 2 + 1 - \#E(\mathbb{F}_2).$$

Example 8.35. In this example, we compute $\mathbb{M}_6(\Gamma_0(3))$, which illustrates both weight greater than 2 and level greater than 1. We have the following generating Manin symbols:

$$
\begin{aligned}
x_0 &= [XY^4, (0,1)], & x_1 &= [XY^4, (1,0)], \\
x_2 &= [XY^4, (1,1)], & x_3 &= [XY^4, (1,2)], \\
x_4 &= [XY^3, (0,1)], & x_5 &= [XY^3, (1,0)], \\
x_6 &= [XY^3, (1,1)], & x_7 &= [XY^3, (1,2)], \\
x_8 &= [X^2Y^2, (0,1)], & x_9 &= [X^2Y^2, (1,0)], \\
x_{10} &= [X^2Y^2, (1,1)], & x_{11} &= [X^2Y^2, (1,2)], \\
x_{12} &= [X^3Y, (0,1)], & x_{13} &= [X^3Y, (1,0)], \\
x_{14} &= [X^3Y, (1,1)], & x_{15} &= [X^3Y, (1,2)], \\
x_{16} &= [X^4Y, (0,1)], & x_{17} &= [X^4Y, (1,0)], \\
x_{18} &= [X^4Y, (1,1)], & x_{19} &= [X^4Y, (1,2)].
\end{aligned}
$$

The relation matrix is already very large for $\mathbb{M}_6(\Gamma_0(3))$. It is as follows, where the S-relations are before the line and the T-relations after it:

$$
\left(\begin{array}{cccccccccccccccccccc}
1 & 0 & 0 & 0 & 0 & 0 & 0 & 0 & 0 & 0 & 0 & 0 & 0 & 0 & 0 & 0 & 0 & 1 & 0 & 0 \\
0 & 1 & 0 & 0 & 0 & 0 & 0 & 0 & 0 & 0 & 0 & 0 & 0 & 0 & 0 & 0 & 1 & 0 & 0 & 0 \\
0 & 0 & 1 & 0 & 0 & 0 & 0 & 0 & 0 & 0 & 0 & 0 & 0 & 0 & 0 & 0 & 0 & 0 & 0 & 1 \\
0 & 0 & 0 & 1 & 0 & 0 & 0 & 0 & 0 & 0 & 0 & 0 & 0 & 0 & 0 & 0 & 0 & 0 & 1 & 0 \\
0 & 0 & 0 & 0 & 1 & 0 & 0 & 0 & 0 & 0 & 0 & 0 & 0 & -1 & 0 & 0 & 0 & 0 & 0 & 0 \\
0 & 0 & 0 & 0 & 0 & 1 & 0 & 0 & 0 & 0 & 0 & 0 & -1 & 0 & 0 & 0 & 0 & 0 & 0 & 0 \\
0 & 0 & 0 & 0 & 0 & 0 & 1 & 0 & 0 & 0 & 0 & 0 & 0 & 0 & 0 & -1 & 0 & 0 & 0 & 0 \\
0 & 0 & 0 & 0 & 0 & 0 & 0 & 1 & 0 & 0 & 0 & 0 & 0 & 0 & -1 & 0 & 0 & 0 & 0 & 0 \\
0 & 0 & 0 & 0 & 0 & 0 & 0 & 0 & 1 & 1 & 0 & 0 & 0 & 0 & 0 & 0 & 0 & 0 & 0 & 0 \\
0 & 0 & 0 & 0 & 0 & 0 & 0 & 0 & 0 & 0 & 1 & 1 & 0 & 0 & 0 & 0 & 0 & 0 & 0 & 0 \\
\hline
1 & 0 & 0 & 1 & 0 & 0 & 0 & -4 & 0 & 0 & 0 & 6 & 0 & 0 & 0 & -4 & 0 & 1 & 0 & 1 \\
1 & 1 & 0 & 0 & -4 & 0 & 0 & 0 & 6 & 0 & 0 & 0 & -4 & 0 & 0 & 0 & 1 & 0 & 0 & 1 \\
0 & 0 & 2 & 0 & 0 & 0 & -4 & 0 & 0 & 0 & 6 & 0 & 0 & 0 & -4 & 0 & 0 & 0 & 2 & 0 \\
0 & 1 & 0 & 1 & 0 & -4 & 0 & 0 & 0 & 6 & 0 & 0 & 0 & -4 & 0 & 0 & 1 & 1 & 0 & 0 \\
0 & 0 & 0 & 1 & 1 & 0 & 0 & -3 & 0 & 0 & 0 & 3 & 0 & -1 & 0 & -1 & 0 & 1 & 0 & 0 \\
1 & 0 & 0 & 0 & -3 & 1 & 0 & 0 & 3 & 0 & 0 & 0 & -1 & 0 & 0 & -1 & 0 & 0 & 0 & 1 \\
0 & 0 & 1 & 0 & 0 & 0 & -2 & 0 & 0 & 0 & 3 & 0 & 0 & 0 & -2 & 0 & 0 & 0 & 1 & 0 \\
0 & 1 & 0 & 0 & 0 & -3 & 0 & 1 & 0 & 3 & 0 & 0 & -1 & -1 & 0 & 0 & 1 & 0 & 0 & 0 \\
0 & 0 & 0 & 1 & 0 & 0 & 0 & -2 & 1 & 1 & 0 & 1 & 0 & -2 & 0 & 0 & 0 & 1 & 0 & 0 \\
1 & 0 & 0 & 0 & -2 & 0 & 0 & 0 & 1 & 1 & 0 & 1 & 0 & 0 & 0 & -2 & 0 & 0 & 0 & 1 \\
0 & 0 & 1 & 0 & 0 & 0 & -2 & 0 & 0 & 0 & 3 & 0 & 0 & 0 & -2 & 0 & 0 & 0 & 1 & 0 \\
0 & 1 & 0 & 0 & 0 & -2 & 0 & 0 & 1 & 1 & 0 & 1 & -2 & 0 & 0 & 0 & 1 & 0 & 0 & 0 \\
0 & 0 & 0 & 1 & 0 & -1 & 0 & -1 & 0 & 3 & 0 & 0 & 1 & -3 & 0 & 0 & 0 & 1 & 0 & 0 \\
1 & 0 & 0 & 0 & -1 & 0 & 0 & -1 & 0 & 0 & 0 & 3 & 0 & 1 & 0 & -3 & 0 & 0 & 0 & 1 \\
0 & 0 & 1 & 0 & 0 & 0 & -2 & 0 & 0 & 0 & 3 & 0 & 0 & 0 & -2 & 0 & 0 & 0 & 1 & 0 \\
0 & 1 & 0 & 0 & -1 & -1 & 0 & 0 & 3 & 0 & 0 & 0 & -3 & 0 & 0 & 1 & 1 & 0 & 0 & 0 \\
0 & 1 & 0 & 1 & 0 & -4 & 0 & 0 & 0 & 6 & 0 & 0 & 0 & -4 & 0 & 0 & 1 & 1 & 0 & 0 \\
1 & 0 & 0 & 1 & 0 & 0 & 0 & -4 & 0 & 0 & 0 & 6 & 0 & 0 & 0 & -4 & 0 & 1 & 0 & 1 \\
0 & 0 & 2 & 0 & 0 & 0 & -4 & 0 & 0 & 0 & 6 & 0 & 0 & 0 & -4 & 0 & 0 & 0 & 2 & 0 \\
1 & 1 & 0 & 0 & -4 & 0 & 0 & 0 & 6 & 0 & 0 & 0 & -4 & 0 & 0 & 0 & 1 & 0 & 0 & 1
\end{array}\right).
$$

The reduced row echelon form of the relations matrix, with zero rows removed is

$$\left(\begin{array}{cccccccccccccccccccc}
1&0&0&0&0&0&0&0&0&0&0&0&0&0&0&0&0&1&0&0\\
0&1&0&0&0&0&0&0&0&0&0&0&0&0&0&0&1&0&0&0\\
0&0&1&0&0&0&0&0&0&0&0&0&0&0&0&0&0&0&0&1\\
0&0&0&1&0&0&0&0&0&0&0&0&0&0&0&0&0&0&1&0\\
0&0&0&0&1&0&0&0&0&0&0&0&0&0&0&0&0&0&3/16&-3/16\\
0&0&0&0&0&1&0&0&0&0&0&0&0&0&0&0&0&0&-1/16&1/16\\
0&0&0&0&0&0&1&0&0&0&0&0&0&0&0&0&0&1/2&-5/16&-3/16\\
0&0&0&0&0&0&0&1&0&0&0&0&0&0&0&0&0&-1/2&3/16&5/16\\
0&0&0&0&0&0&0&0&1&0&0&0&0&0&0&0&0&-1/6&1/12&1/12\\
0&0&0&0&0&0&0&0&0&1&0&0&0&0&0&0&0&1/6&-1/12&-1/12\\
0&0&0&0&0&0&0&0&0&0&1&0&0&0&0&0&0&0&1/4&-1/4\\
0&0&0&0&0&0&0&0&0&0&0&1&0&0&0&0&0&0&-1/4&1/4\\
0&0&0&0&0&0&0&0&0&0&0&0&1&0&0&0&0&0&-1/16&1/16\\
0&0&0&0&0&0&0&0&0&0&0&0&0&1&0&0&0&0&3/16&-3/16\\
0&0&0&0&0&0&0&0&0&0&0&0&0&0&1&0&0&-1/2&3/16&5/16\\
0&0&0&0&0&0&0&0&0&0&0&0&0&0&0&1&0&1/2&-5/16&-3/16
\end{array}\right).$$

Since these relations are equivalent to the original relations, we see how $x_0, \ldots, x_{15}$ can be expressed in terms of x_{16}, x_{17}, x_{18}, and x_{19}. Thus $\mathbb{M}_6(\Gamma_0(3))$ has dimension 4. For example,

$$x_{15} \sim \frac{1}{2}x_{17} - \frac{5}{16}x_{18} - \frac{3}{16}x_{19}.$$

Notice that the number of relations is already quite large. It is perhaps surprising how complicated the presentation is already for $\mathbb{M}_6(\Gamma_0(3))$. Because there are denominators in the relations, the above calculation is only a computation of $\mathbb{M}_6(\Gamma_0(3); \mathbb{Q})$. Computing $\mathbb{M}_6(\Gamma_0(3); \mathbb{Z})$ involves finding a $\mathbb{Z}$-basis for the kernel of the relation matrix (see Exercise 7.5).

As before, we find that with respect to the basis x_{16}, x_{17}, x_{18}, and x_{19}

$$T_2 = \begin{pmatrix} 33 & 0 & 0 & 0 \\ 3 & 6 & 12 & 12 \\ -3/2 & 27/2 & 15/2 & 27/2 \\ -3/2 & 27/2 & 27/2 & 15/2 \end{pmatrix}.$$

Notice that there are denominators in the matrix for T_2 with respect to this basis. It is clear from the definition of T_2 acting on Manin symbols that T_2 preserves the $\mathbb{Z}$-module $\mathbb{M}_6(\Gamma_0(3))$, so there is some basis for $\mathbb{M}_6(\Gamma_0(3))$ such that T_2 is given by an integer matrix. Thus the characteristic polynomial f_2 of T_2 will have integer coefficients; indeed,

$$f_2 = (x - 33)^2 \cdot (x + 6)^2.$$

Note the factor $(x-33)^2$, which comes from the two images of the Eisenstein series E_4 of level 1. The factor $x + 6$ comes from the cusp form

$$g = q - 6q^2 + \cdots \in S_6(\Gamma_0(3)).$$

By computing more Hecke operators T_n, we can find more coefficients of g. For example, the charpoly of T_3 is $(x - 1)(x - 243)(x - 9)^2$, and the matrix of T_5 is

$$T_5 = \begin{pmatrix} 3126 & 0 & 0 & 0 \\ 240 & 966 & 960 & 960 \\ -120 & 1080 & 1086 & 1080 \\ -120 & 1080 & 1080 & 1086 \end{pmatrix},$$

which has characteristic polynomial

$$f_5 = (x - 3126)^2(x - 6)^2.$$

The matrix of T_7 is

$$T_7 = \begin{pmatrix} 16808 & 0 & 0 & 0 \\ 1296 & 5144 & 5184 & 5184 \\ -648 & 5832 & 5792 & 5832 \\ -648 & 5832 & 5832 & 5792 \end{pmatrix},$$

with characteristic polynomial

$$f_7 = (x - 16808)^2(x + 40)^2.$$

One can put this information together to deduce that

$$g = q - 6q^2 + 9q^3 + 4q^4 + 6q^5 - 54q^6 - 40q^7 + \cdots .$$

Example 8.36. Consider $\mathbb{M}_2(\Gamma_0(43))$, which has dimension 7. With respect to the symbols

$$x_1 = (1, 0), \quad x_{32} = (1, 31), \quad x_{33} = (1, 32),$$
$$x_{39} = (1, 38), \quad x_{40} = (1, 39), \quad x_{41} = (1, 40), \quad x_{42} = (1, 41),$$

the matrix of T_2 is

$$T_2 = \begin{pmatrix} 3 & 0 & 0 & 0 & 0 & 0 & 0 \\ 0 & -2 & -1 & -1 & -1 & 0 & 0 \\ 0 & 1 & 1 & 0 & 0 & -2 & -1 \\ 0 & 0 & 1 & -1 & 1 & 0 & 0 \\ 0 & 0 & 0 & 2 & 1 & 2 & 1 \\ 0 & 0 & -1 & -1 & -1 & -2 & 0 \\ -1 & 0 & 0 & 1 & 1 & 1 & -1 \end{pmatrix},$$

which has characteristic polynomial

$$(x - 3)(x + 2)^2(x^2 - 2)^2.$$

There is one Eisenstein series and there are three cusp forms with $a_2 = -2$ and $a_2 = \pm\sqrt{2}$.

Example 8.37. To compute $\mathbb{M}_2(\Gamma_0(2004); \mathbb{Q})$, we first make a list of the

$$4032 = (2^2 + 2) \cdot (3 + 1) \cdot (167 + 1)$$

elements $(a, b) \in \mathbb{P}^1(\mathbb{Z}/2004\mathbb{Z})$ using Algorithm 8.29. The list looks like this:

$$(0, 1), (1, 0), (1, 1), (1, 2), \ldots, (668, 1), (668, 3), (668, 5), (1002, 1).$$

For each of the symbols x_i, we consider the S-relations and T-relations. Ignoring the redundant relations, we find 2016 S-relations and 1344 T-relations. It is simple to quotient out by the S-relations, e.g., by identifying x_i with $-x_iS = -x_j$ for some j (or setting $x_i = 0$ if $x_iS = x_i$). Once we have taken the quotient by the S-relations, we take the *image* of all 1344 of the T-relations modulo the S-relations and quotient out by those relations. Because S and T do not commutate, we cannot only quotient out by T-relations $x_i + x_iT + x_iT^2 = 0$ where the x_i are the basis after quotienting out by the S-relations. The relation matrix has rank 3359, so $\mathbb{M}_2(\Gamma_0(2004); \mathbb{Q})$ has dimension 673.

If we instead compute the quotient $\mathbb{M}_2(\Gamma_0(2004); \mathbb{Q})^+$ of $\mathbb{M}_2(\Gamma_0(2004); \mathbb{Q})$ by the subspace of elements $x - \eta^*(x)$, we include relations $x_i + x_iI = 0$, where $I = \left(\begin{smallmatrix}-1 & 0\\ 0 & 1\end{smallmatrix}\right)$. There are now 2016 S-relations, 2024 I-relations, and 1344 T-relations. Again, it is relatively easy to quotient out by the S-relations by identifying x_i and $-x_iS$. We then take the image of all 2024 I-relations modulo the S-relations, and again directly quotient out by the I-relations by identifying $[x_i]$ with $-[x_iI] = -[x_j]$ for some j, where by $[x_i]$ we mean the class of x_i modulo the S-relations. Finally, we quotient out by the 1344 T-relations, which involves sparse Gauss elimination on a matrix with 1344 rows and at most three nonzero entries per row. The dimension of $\mathbb{M}_2(\Gamma_0(2004); \mathbb{Q})^+$ is 331.

8.9. Refined Algorithm for the Presentation

Algorithm 8.38 (Modular Symbols Presentation). *This is an algorithm to compute $\mathbb{M}_k(\Gamma_0(N); \mathbb{Q})$ or $\mathbb{M}_k(\Gamma_0(N); \mathbb{Q})^{\pm}$, which only requires doing generic sparse linear algebra to deal with the three term T-relations.*

(1) Let $x_0, \ldots, x_n$ by a list of all Manin symbols.
(2) Quotient out the two-term S-relations and if the $\pm$ quotient is desired, by the two-term η-relations. (Note that this is more subtle than just "identifying symbols in pairs", since complicated relations can cause generators to surprisingly equal 0.) Let $[x_i]$ denote the class of x_i after this quotienting process.
(3) Create a sparse matrix A with m columns, whose rows encode the relations
$$[x_i] + [x_iT] + [x_iT^2] = 0.$$
For example, there are about $n/3$ such rows when $k = 2$. The number of nonzero entries per row is at most $3(k-1)$. Note that we must include rows for *all* i, since even if $[x_i] = [x_j]$, it need not be the case that $[x_iT] = [x_jT]$, since the matrices S and T do not commute. However, we have an a priori formula for the dimension of the quotient by all these relations, so we could omit

many relations and just check that there are enough at the end—if there are not, we add in more.

(4) Compute the reduced row echelon form of A using Algorithm 7.6. For $k = 2$, this is the echelon form of a matrix with size about $n/3 \times n/4$.

(5) Use what we have done above to read off a sparse matrix R that expresses each of the n Manin symbols in terms of a basis of Manin symbols, modulo the relations.

8.10. Applications

8.10.1. Later in This Book. We sketch some of the ways in which we will apply the modular symbols algorithms of this chapter later in this book.

Cuspidal modular symbols are in Hecke-equivariant duality with cuspidal modular forms, and as such we can compute modular forms by computing systems of eigenvalues for the Hecke operators acting on modular symbols. By the Atkin-Lehner-Li theory of newforms (see, e.g., Theorem 9.4), we can construct $S_k(N, \varepsilon)$ for any N, any ε, and $k \geq 2$ using this method. See Chapter 1 for more details.

Once we can compute spaces of modular symbols, we move to computing the corresponding modular forms. We define inclusion and trace maps from modular symbols of one level N to modular symbols of level a multiple or divisor of N. Using these, we compute the quotient V of the new subspace of cuspidal modular symbols on which a "star involution" acts as $+1$. The Hecke operators act by diagonalizable commuting matrices on this space, and computing the systems of Hecke eigenvalues is equivalent to computing newforms $\sum a_n q^n$. In this way, we obtain a list of *all* newforms (normalized eigenforms) in $S_k(N, \varepsilon)$ for any N, ε, and $k \geq 2$.

In Chapter 10, we compute with the period mapping from modular symbols to $\mathbb{C}$ attached to a newform $f \in S_k(N, \varepsilon)$. When $k = 2, \varepsilon = 1$ and f has rational Fourier coefficients, this gives a method to compute the period lattice associated to a modular elliptic curve attached to a newform (see Section 10.7). In general, computation of this map is important when finding equations for modular $\mathbb{Q}$-curves, CM curves, and curves with a given modular Jacobian. It is also important for computing special values of the L-function $L(f, s)$ at integer points in the critical strip.

8.10.2. Discussion of the Literature and Research. Modular symbols were introduced by Birch [**Bir71**] for computations in support of the Birch and Swinnerton-Dyer conjecture. Manin [**Man72**] used modular symbols to prove rationality results about special values of L-functions.

Merel's paper [**Mer94**] builds on work of Shokurov (mainly [**Sho80a**]), which develops a higher-weight generalization of Manin's work partly to understand rationality properties of special values of L-functions. Cremona's book [**Cre97a**] discusses how to compute the space of weight 2 modular symbols for $\Gamma_0(N)$, in connection with the problem of enumerating all elliptic curves of given conductor, and his article [**Cre92**] discusses the $\Gamma_1(N)$ case and computation of modular symbols with character.

There have been several Ph.D. theses about modular symbols. Basmaji's thesis [**Bas96**] contains tricks to efficiently compute Hecke operators T_p, with p very large (see Section 8.3.4), and also discusses how to compute spaces of half integral weight modular forms building on what one can get from modular symbols of integral weight. The author's Ph.D. thesis [**Ste00**] discusses higher-weight modular symbols and applies modular symbols to study Shafarevich-Tate groups (see also [**Aga00**]). Martin's thesis [**Mar01**] is about an attempt to study an analogue of analytic modular symbols for weight 1. Gabor Wiese's thesis [**Wie05**] uses modular symbols methods to study weight 1 modular forms modulo p. Lemelin's thesis [**Lem01**] discusses modular symbols for quadratic imaginary fields in the context of p-adic analogues of the Birch and Swinnerton-Dyer conjecture. See also the survey paper [**FM99**], which discusses computation with weight 2 modular symbols in the context of modular abelian varieties.

The appendix of this book is about analogues of modular symbols for groups besides finite index subgroups of $\mathrm{SL}_2(\mathbb{Z})$, e.g., for subgroup of higher rank groups such as $\mathrm{SL}_3(\mathbb{Z})$. There has also been work on computing Hilbert modular forms, e.g., by Lassina Dembelé [**Dem05**] Hilbert modular forms are functions on a product of copies of $\mathfrak{h}$, and $\mathrm{SL}_2(\mathbb{Z})$ is replaced by a group of matrices with entries in a totally real field.

Glenn Stevens, Robert Pollack and Henri Darmon (see [**DP04**]) have worked for many years to develop an analogue of modular symbols in a rigid analytic context, which is helpful for questions about computing with overconvergent p-adic modular forms or proving results about p-adic L-functions.

Finally we mention that Barry Mazur and some other authors use the term "modular symbol" in a different way than we do. They use the term in a way that is dual to our usage; for example, they attach a "modular symbol" to a modular form or elliptic curve. See [**MTT86**] for an extensive discussion of modular symbols from this point of view, where they are used to construct p-adic L-functions.

8.11. Exercises

8.1 Suppose M is an integer multiple of N. Prove that the natural map $(\mathbb{Z}/M\mathbb{Z})^* \to (\mathbb{Z}/N\mathbb{Z})^*$ is surjective.

8.2 Prove that $\mathrm{SL}_2(\mathbf{Z}) \to \mathrm{SL}_2(\mathbf{Z}/N\mathbf{Z})$ is surjective (see Lemma 8.15).

8.3 Compute $\mathbb{M}_3(\Gamma_1(3))$. List each Manin symbol the relations they satisfy, compute the quotient, etc. Find the matrix of T_2. (Check: The dimension of $\mathbb{M}_3(\Gamma_1(3))$ is 2, and the characteristic polynomial of T_2 is $(x-3)(x+3)$.)

8.4 Finish the proof of Proposition 8.17.

8.5 (a) Show that if $\eta = \left(\begin{smallmatrix} -1 & 0 \\ 0 & 1 \end{smallmatrix}\right)$, then $\eta\Gamma\eta = \Gamma$ for $\Gamma = \Gamma_0(N)$ and $\Gamma = \Gamma_1(N)$.

(b) (*) Give an example of a finite index subgroup Γ such that $\eta\Gamma\eta \neq \Gamma$.

Chapter 9

Computing with Newforms

In this chapter we pull together results and algorithms from Chapter 3, 4, 7, and 8 and explain how to use linear algebra techniques to compute cusp forms and eigenforms using modular symbols.

We first discuss in Section 9.1 how to decompose $M_k(\Gamma_1(N))$ as a direct sum of subspaces corresponding to Dirichlet characters. Next in Section 9.2 we state the main theorems of Atkin-Lehner-Li theory, which decomposes $S_k(\Gamma_1(N))$ into subspaces on which the Hecke operators act diagonalizably with "multiplicity one". In Section 9.3 we describe two algorithms for computing modular forms. One algorithm finds a basis of q-expansions, and the other computes eigenvalues of newforms.

9.1. Dirichlet Character Decomposition

The group $(\mathbb{Z}/N\mathbb{Z})^*$ acts on $M_k(\Gamma_1(N))$ through *diamond-bracket operators* $\langle d\rangle$, as follows. For $d \in (\mathbb{Z}/N\mathbb{Z})^*$, define

$$f|\langle d\rangle = f^{\left[\left(\begin{smallmatrix} a & b \\ c & d' \end{smallmatrix}\right)\right]_k},$$

where $\left(\begin{smallmatrix} a & b \\ c & d' \end{smallmatrix}\right) \in \mathrm{SL}_2(\mathbb{Z})$ is congruent to $\left(\begin{smallmatrix} d^{-1} & 0 \\ 0 & d \end{smallmatrix}\right) \pmod N$. Note that the map $\mathrm{SL}_2(\mathbb{Z}) \to \mathrm{SL}_2(\mathbb{Z}/N\mathbb{Z})$ is surjective (see Exercise 8.2), so the matrix $\left(\begin{smallmatrix} a & b \\ c & d' \end{smallmatrix}\right)$ exists. To prove that $\langle d\rangle$ preserves $M_k(\Gamma_1(N))$, we prove the more general fact that $\Gamma_1(N)$ is a normal subgroup of

$$\Gamma_0(N) = \left\{\begin{pmatrix} a & b \\ c & d \end{pmatrix} \in \mathrm{SL}_2(\mathbb{Z}) : \begin{pmatrix} a & b \\ c & d \end{pmatrix} \equiv \begin{pmatrix} * & * \\ 0 & * \end{pmatrix} \pmod N\right\}.$$

This will imply that $\langle d\rangle$ preserves $M_k(\Gamma_1(N))$ since $\left(\begin{smallmatrix} a & b \\ c & d' \end{smallmatrix}\right) \in \Gamma_0(N)$.

Lemma 9.1. *The group $\Gamma_1(N)$ is a normal subgroup of $\Gamma_0(N)$, and the quotient $\Gamma_0(N)/\Gamma_1(N)$ is isomorphic to $(\mathbb{Z}/N\mathbb{Z})^*$.*

Proof. See Exercise 9.1. □

Alternatively, one can prove that $\langle d\rangle$ preserves $M_k(\Gamma_1(N))$ by showing that $\langle d\rangle \in \mathbb{T}$ and noting that $M_k(\Gamma_1(N))$ is preserved by $\mathbb{T}$ (see Remark 9.11).

The *diamond-bracket action* is the action of $\Gamma_0(N)/\Gamma_1(N) \cong (\mathbb{Z}/N\mathbb{Z})^*$ on $M_k(\Gamma_1(N))$. Since $M_k(\Gamma_1(N))$ is a finite-dimensional vector space over $\mathbb{C}$, the $\langle d\rangle$ action breaks $M_k(\Gamma_1(N))$ up as a direct sum of factors corresponding to the Dirichlet characters $D(N,\mathbb{C})$ of modulus N.

Proposition 9.2. *We have*

$$M_k(\Gamma_1(N)) = \bigoplus_{\varepsilon\in D(N,\mathbb{C})} M_k(N,\varepsilon),$$

where

$$M_k(N,\varepsilon) = \left\{f \in \mathbb{M}_k(\Gamma_1(N)) : f|\langle d\rangle = \varepsilon(d)f, \text{ all } d \in (\mathbb{Z}/N\mathbb{Z})^*\right\}.$$

Proof. The linear transformations $\langle d\rangle$, for the $d \in (\mathbb{Z}/N\mathbb{Z})^*$, all commute, since $\langle d\rangle$ acts through the abelian group $\Gamma_0(N)/\Gamma_1(N)$. Also, if e is the exponent of $(\mathbb{Z}/N\mathbb{Z})^*$, then $\langle d\rangle^e = \langle d^e\rangle = \langle 1\rangle = 1$, so the matrix of $\langle d\rangle$ is diagonalizable. It is a standard fact from linear algebra that any commuting family of diagonalizable linear transformations is simultaneously diagonalizable (see Exercise 5.1), so there is a basis $f_1,\ldots,f_n$ for $M_k(\Gamma_1(N))$ such that all $\langle d\rangle$ act by diagonal matrices. The system of eigenvalues of the action of $(\mathbb{Z}/N\mathbb{Z})^*$ on a fixed f_i defines a Dirichlet character, i.e., each f_i has the property that $f_i|\langle d\rangle = \varepsilon_i(d)$, for all $d \in (\mathbb{Z}/N\mathbb{Z})^*$ and some Dirichlet character ε_i. The f_i for a given ε then span $M_k(N,\varepsilon)$, and taken together the $M_k(N,\varepsilon)$ must span $M_k(\Gamma_1(N))$. □

Definition 9.3 (Character of Modular Form). If $f \in M_k(N,\varepsilon)$, we say that ε is the *character of the modular form* f.

The spaces $M_k(N,\varepsilon)$ are a direct sum of subspaces $S_k(N,\varepsilon)$ and $E_k(N,\varepsilon)$, where $S_k(N,\varepsilon)$ is the subspace of cusp forms, i.e., forms that vanish at *all* cusps (elements of $\mathbb{Q}\cup\{\infty\}$), and $E_k(N,\varepsilon)$ is the subspace of Eisenstein series, which is the unique subspace of $M_k(N,\varepsilon)$ that is invariant under all Hecke operators and is such that $M_k(N,\varepsilon) = S_k(N,\varepsilon)\oplus E_k(N,\varepsilon)$. The space $E_k(N,\varepsilon)$ can also be defined as the space spanned by all Eisenstein series of weight k and level N, as defined in Chapter 5. The space $E_k(N,\varepsilon)$ can be defined in a third way using the Petersson inner product (see [**Lan95**, §VII.5]).

The diamond-bracket operators preserve cusp forms, so the isomorphism of Proposition 9.2 restricts to an isomorphism of the corresponding cuspidal subspaces. We illustrate how to use SAGE to make a table of dimension of $M_k(\Gamma_1(N))$ and $M_k(N, \varepsilon)$ for $N = 13$.

```
sage: G = DirichletGroup(13)
sage: G
Group of Dirichlet characters of modulus 13 over
Cyclotomic Field of order 12 and degree 4
sage: dimension_modular_forms(Gamma1(13),2)
13
sage: [dimension_modular_forms(e,2) for e in G]
[1, 0, 3, 0, 2, 0, 2, 0, 2, 0, 3, 0]
```

Next we do the same for $N = 100$.

```
sage: G = DirichletGroup(100)
sage: G
Group of Dirichlet characters of modulus 100 over
Cyclotomic Field of order 20 and degree 8
sage: dimension_modular_forms(Gamma1(100),2)
370
sage: v = [dimension_modular_forms(e,2) for e in G]; v
[24, 0, 0, 17, 18, 0, 0, 17, 18, 0, 0, 21, 18, 0, 0, 17,
  18, 0, 0, 17, 24, 0, 0, 17, 18, 0, 0, 17, 18, 0, 0, 21,
  18, 0, 0, 17, 18, 0, 0, 17]
sage: sum(v)
370
```

9.2. Atkin-Lehner-Li Theory

In Section 8.6 we defined maps between modular symbols spaces of different level. There are similar maps between spaces of cusp forms. Suppose N and M are positive integers with $M \mid N$ and that t is a divisor of N/M. Let

$$\alpha_t : S_k(\Gamma_1(M)) \to S_k(\Gamma_1(N)) \tag{9.2.1}$$

be the *degeneracy map*, which is given by $f(q) \mapsto f(q^t)$. There are also maps β_t in the other direction; see [**Lan95**, Ch. VIII].

The *old subspace* of $S_k(\Gamma_1(N))$, denoted $S_k(\Gamma_1(N))_{\mathrm{old}}$, is the sum of the images of all maps α_t with M a proper divisor of N and t any divisor of N/M (note that α_t depends on t, N, and M, so there is a slight abuse of notation). The *new subspace* of $S_k(\Gamma_1(N))$, which we denote by $S_k(\Gamma_1(N))_{\mathrm{new}}$, is the intersection of the kernel of all maps β_t with M a proper divisor of N. One can use the Petersson inner product to show that

$$S_k(\Gamma_1(N)) = S_k(\Gamma_1(N))_{\mathrm{new}} \oplus S_k(\Gamma_1(N))_{\mathrm{old}}.$$

Moroever, the new and old subspaces are preserved by all Hecke operators.

Let $\mathbb{T} = \mathbb{Z}[T_1, T_2, \ldots]$ be the commutative polynomial ring in infinitely many indeterminates T_n. This ring acts (via T_n acting as the nth Hecke operator) on $S_k(\Gamma_1(N))$ for every integer N. Let $\mathbb{T}^{(N)}$ be the subring of $\mathbb{T}$ generated by the T_n with $\gcd(n, N) = 1$.

Theorem 9.4 (Atkin, Lehner, Li). *We have a decomposition*

$$S_k(\Gamma_1(N)) = \bigoplus_{M \mid N} \bigoplus_{d \mid N/M} \alpha_d(S_k(\Gamma_1(M))_{\mathrm{new}}). \tag{9.2.2}$$

Each space $S_k(\Gamma_1(M))_{\mathrm{new}}$ is a direct sum of distinct (nonisomorphic) simple $\mathbb{T}^{(N)}_{\mathbb{C}}$-modules.

Proof. The complete proof is in [**Li75**]. See also [**DS05**, Ch. 5] for a beautiful modern treatement of this and related results. □

The analogue of Theorem 9.4 with Γ_1 replaced by Γ_0 is also true (this is what was proved in [**AL70**]). The analogue for $S_k(N, \varepsilon)$ is also valid, as long as we omit the spaces $S_k(\Gamma_1(M), \varepsilon)$ for which $\mathrm{cond}(\varepsilon) \nmid M$.

Example 9.5. If N is prime and $k \leq 11$, then $S_k(\Gamma_1(N))_{\mathrm{new}} = S_k(\Gamma_1(N))$, since $S_k(\Gamma_1(1)) = 0$.

One can prove using the Petersson inner product that the operators T_n on $S_k(\Gamma_1(N))$, with $\gcd(n, N) = 1$, are diagonalizable. Another result of Atkin-Lehner-Li theory is that the ring of endomorphisms of $S_k(\Gamma_1(N))_{\mathrm{new}}$ generated by all Hecke operators equals the ring generated by the Hecke operators T_n with $(n, N) = 1$. This statement need not be true if we do not restrict to the new subspace, as the following example shows.

Example 9.6. We have

$$S_2(\Gamma_0(22)) = S_2(\Gamma_0(11)) \oplus \alpha_2(S_2(\Gamma_0(11))),$$

where each of the spaces $S_2(\Gamma_0(11))$ has dimension 1. Thus $S_2(\Gamma_0(22))_{\mathrm{new}} = 0$. The Hecke operator T_2 on $S_2(\Gamma_0(22))$ has characteristic polynomial $x^2 + 2x + 2$, which is irreducible. Since α_2 commutes with all Hecke operators T_n, with $\gcd(n, 2) = 1$, the subring $\mathbb{T}^{(22)}$ of the Hecke algebra generated by operators T_n with n odd is isomorphic to $\mathbb{Z}$ (the 2×2 scalar matrices). Thus on the full space $S_2(\Gamma_0(22))$, we do not have $\mathbb{T}^{(22)} = \mathbb{T}$. However, on the new subspace we do have this equality, since the new subspace has dimension 0.

Example 9.7. The space $S_2(\Gamma_0(45))$ has dimension 3 and basis

$$\begin{aligned} f_0 &= q - q^4 - q^{10} - 2q^{13} - q^{16} + 4q^{19} + \cdots, \\ f_1 &= q^2 - q^5 - 3q^8 + 4q^{11} - 2q^{17} + \cdots, \\ f_2 &= q^3 - q^6 - q^9 - q^{12} + q^{15} + q^{18} + \cdots. \end{aligned}$$

The new subspace $S_2(\Gamma_0(45))_{\text{new}}$ is spanned by the single cusp form

$$q + q^2 - q^4 - q^5 - 3q^8 - q^{10} + 4q^{11} - 2q^{13} + \cdots.$$

We have $S_2(\Gamma_0(45/5)) = 0$ and $S_2(\Gamma_0(15))$ has dimension 1 with basis

$$q - q^2 - q^3 - q^4 + q^5 + q^6 + 3q^8 + q^9 - q^{10} - 4q^{11} + q^{12} - 2q^{13} + \cdots.$$

We use SAGE to verify the above assertions.

```
sage: S = CuspForms(Gamma0(45), 2, prec=14); S
Cuspidal subspace of dimension 3 of Modular Forms space
of dimension 10 for Congruence Subgroup Gamma0(45) of
weight 2 over Rational Field
sage: S.basis()
[
q - q^4 - q^10 - 2*q^13 + O(q^14),
q^2 - q^5 - 3*q^8 + 4*q^11 + O(q^14),
q^3 - q^6 - q^9 - q^12 + O(q^14)
]
sage: S.new_subspace().basis()
(q - q^4 - q^10 - 2*q^13 + O(q^14),)
sage: CuspForms(Gamma0(9),2)
Cuspidal subspace of dimension 0 of Modular Forms space
of dimension 3 for Congruence Subgroup Gamma0(9) of
weight 2 over Rational Field
sage: CuspForms(Gamma0(15),2, prec=10).basis()
[
q - q^2 - q^3 - q^4 + q^5 + q^6 + 3*q^8 + q^9 + O(q^10)
]
```

Example 9.8. This example is similar to Example 9.6, except that there are newforms. We have

$$S_2(\Gamma_0(55)) = S_2(\Gamma_0(11)) \oplus \alpha_5(S_2(\Gamma_0(11))) \oplus S_2(\Gamma_0(55))_{\text{new}},$$

where $S_2(\Gamma_0(11))$ has dimension 1 and $S_2(\Gamma_0(55))_{\mathrm{new}}$ has dimension 3. The Hecke operator T_5 on $S_2(\Gamma_0(55))_{\mathrm{new}}$ acts via the matrix

$$\begin{pmatrix} -2 & 2 & -1 \\ -1 & 1 & -1 \\ 1 & -2 & 0 \end{pmatrix}$$

with respect to some basis. This matrix has eigenvalues 1 and -1. Atkin-Lehner theory asserts that T_5 must be a linear combination of T_n, with $\gcd(n, 55) = 1$. Upon computing the matrix for T_2, we find by simple linear algebra that $T_5 = 2T_2 - T_4$.

Definition 9.9 (Newform). A *newform* is a $\mathbb{T}$-eigenform $f \in S_k(\Gamma_1(N))_{\mathrm{new}}$ that is normalized so that the coefficient of q is 1.

We now motivate this definition by explaining why any $\mathbb{T}$-eigenform can be normalized so that the coefficient of q is 1 and how such an eigenform has the property that its Fourier coefficients are exactly the Hecke eigenvalues.

Proposition 9.10. *If $f = \sum_{n=0}^{\infty} a_n q^n \in M_k(N, \varepsilon)$ is an eigenvector for all Hecke operators T_n normalized so that $a_1 = 1$, then $T_n(f) = a_n f$.*

Proof. If $\varepsilon = 1$, then $f \in M_k(\Gamma_0(N))$ and this is Lemma 3.22. However, we have not yet considered Hecke operators on q-expansions for more general spaces of modular forms.

The Hecke operators T_p, for p prime, act on $S_k(N, \varepsilon)$ by

$$T_p\left(\sum_{n=0}^{\infty} a_n q^n\right) = \sum_{n=0}^{\infty}\left(a_{np}q^n + \varepsilon(p)p^{k-1}a_n q^{np}\right),$$

and there is a similar formula for T_m with m composite. If $f = \sum_{n=0}^{\infty} a_n q^n$ is an eigenform for all T_p, with eigenvalues λ_p, then by the above formula

$$\lambda_p f = \lambda_p a_1 q + \lambda_p a_2 q^2 + \cdots = T_p(f) = a_p q + \text{higher terms}. \tag{9.2.3}$$

Equating coefficients of q, we see that if $a_1 = 0$, then $a_p = 0$ for all p; hence $a_n = 0$ for all n, because of the multiplicativity of Fourier coefficients and the recurrence

$$a_{p^r} = a_{p^{r-1}}a_p - \varepsilon(p)p^{k-1}a_{p^{r-2}}.$$

This would mean that $f = 0$, a contradiction. Thus $a_1 \neq 0$, and it makes sense to normalize f so that $a_1 = 1$. With this normalization, (9.2.3) implies that $\lambda_p = a_p$, as desired. □

Remark 9.11. The Hecke algebra $\mathbb{T}_{\mathbb{Q}}$ on $M_k(\Gamma_1(N))$ contains the operators $\langle d\rangle$, since they satisfy the relation $T_{p^2} = T_p^2 - \langle p\rangle p^{k-1}$. Thus any $\mathbb{T}$-eigenform in $M_k(\Gamma_1(N))$ lies in a subspace $M_k(N, \varepsilon)$ for some Dirichlet character ε. Also, one can even prove that $\langle d\rangle \in \mathbb{Z}[\ldots, T_n, \ldots]$ (see Exercise 9.2).

9.3. Computing Cusp Forms

Let $\mathbb{S}_k(N, \varepsilon; \mathbb{C})$ be the space of cuspidal modular symbols as in Chapter 8. Let ι^* be the map of (8.5.8), and let $\mathbb{S}_k(N, \varepsilon; \mathbb{C})^+$ be the *plus one quotient* of cuspidal modular symbols, i.e., the quotient of $\mathbb{S}_k(N, \varepsilon; \mathbb{C})$ by the image of $\iota^* - 1$. It follows from Theorem 8.23 and compatibility of the degeneracy maps (for modular symbols they are defined in Section 8.6) that the $\mathbb{T}$-modules $S_k(N, \varepsilon)_{\text{new}}$ and $\mathbb{S}_k(N, \varepsilon; \mathbb{C})^+_{\text{new}}$ are dual as $\mathbb{T}$-modules. Thus finding the systems of $\mathbb{T}$-eigenvalues on cusp forms is the same as finding the systems of $\mathbb{T}$-eigenvalues on cuspidal modular symbols.

Our strategy to compute $S_k(N, \varepsilon)$ is to first compute spaces $S_k(N, \varepsilon)_{\text{new}}$ using the Atkin-Lehner-Li decomposition (9.2.2). To compute $S_k(N, \varepsilon)_{\text{new}}$ to a given precision, we compute the systems of eigenvalues of the Hecke operators T_p on the space $V = \mathbb{S}_k(N, \varepsilon; \mathbb{C})^+_{\text{new}}$, which we will define below. Using Proposition 9.10, we then recover a basis of q-expansions for newforms. Note that we only need to compute Hecke eigenvalues T_p, for p prime, not the T_n for n composite, since the a_n can be quickly recovered in terms of the a_p using multiplicativity and the recurrence.

For some problems, e.g., construction of models for modular curves, having a basis of q-expansions is enough. For many other problems, e.g., enumeration of modular abelian varieties, one is really interested in the newforms. We next discuss algorithms aimed at each of these problems.

9.3.1. A Basis of q-Expansions.

The following algorithm generalizes Algorithm 3.26. It computes $S_k(N, \varepsilon)$ without finding any eigenspaces.

Algorithm 9.12 (Merel's Algorithm for Computing a Basis). *Given integers m, N and k and a Dirichlet character ε with modulus N, this algorithm computes a basis of q-expansions for $S_k(N, \varepsilon)$ to precision $O(q^{m+1})$.*

(1) [Compute Modular Symbols] Use Algorithm 8.38 to compute

$$V = \mathbb{S}_k(N, \varepsilon)^- \otimes \mathbb{Q}(\varepsilon),$$

viewed as a $K = \mathbb{Q}(\varepsilon)$ vector space, with an action of the T_n.

(2) [Basis for Linear Dual] Write down a basis for $V^* = \text{Hom}(V, \mathbb{Q}(\varepsilon))$. E.g., if we identify V with K^n viewed as column vectors, then V^* is the space of row vectors of length n, and the pairing is the row $\times$ column product.

(3) [Find Generator] Find $x \in V$ such that $\mathbb{T}x = V$ by choosing random x until we find one that generates. The set of x that fail to generate lie in a union of a finite number of proper subspaces.

(4) [Compute Basis] The set of power series

$$f_i = \sum_{n=1}^{m} \psi_i(T_n(x))q^n + O(q^{m+1})$$

forms a basis for $S_k(N, \varepsilon)$ to precision m.

In practice Algorithm 9.12 seems slower than the eigenspace algorithm that we will describe in the rest of this chapter. The theoretical complexity of Algorithm 9.12 *may* be better, because it is not necessary to factor any polynomials. Polynomial factorization is difficult from the worst-case complexity point of view, though it is usually fast in practice. The eigenvalue algorithm only requires computing a few images $T_p(x)$ for p prime and x a Manin symbol on which T_p can easily be computed. The Merel algorithm involves computing $T_n(x)$ for all n and for a fairly easy x, which is potentially more work.

Remark 9.13. By "easy x", I mean that computing $T_n(x)$ is easier on x than on a completely random element of $\mathbb{S}_k(N, \varepsilon)^+$, e.g., x could be a Manin symbol.

9.3.2. Newforms: Systems of Eigenvalues. In this section we describe an algorithm for computing the system of Hecke eigenvalues associated to a simple subspace of a space of modular symbols. This algorithm is better than doing linear algebra directly over the number field generated by the eigenvalues. It only involves linear algebra over the base field and also yields a compact representation for the answer, which is better than writing the eigenvalues in terms of a power basis for a number field. In order to use this algorithm, it is necessary to decompose the space of cuspidal modular symbols as a direct sum of simples, e.g., using Algorithm 7.17.

Fix N and a Dirichlet character ε of modulus N, and let

$$V = \mathbb{M}_k(N, \varepsilon)^+$$

be the $+1$ quotient of modular symbols (see equation (8.5.8)).

Algorithm 9.14 (System of Eigenvalues). *Given a $\mathbb{T}$-simple subspace $W \subset V$ of modular symbols, this algorithm outputs maps ψ and e, where $\psi : \mathbb{T}_K \to W$ is a K-linear map and $e : W \cong L$ is an isomorphism of W with a number field L, such that $a_n = e(\psi(T_n))$ is the eigenvalue of the nth Hecke operator acting on a fixed $\mathbb{T}$-eigenvector in $W \otimes \overline{\mathbb{Q}}$. (Thus $f = \sum_{n=1}^{\infty} e(\psi(T_n))q^n$ is a newform.)*

(1) [Compute Projection] Let $\varphi : V \to W'$ be any surjective linear map such that $\ker(\varphi)$ equals the kernel of the $\mathbb{T}$-invariant projection onto W. For example, compute φ by finding a simple submodule

of $V^* = \mathrm{Hom}(V, K)$ that is isomorphic to W, e.g., by applying Algorithm 7.17 to V^* with T replaced by the transpose of T.

(2) [Choose v] Choose a nonzero element $v \in V$ such that $\pi(v) \neq 0$ and computation of $T_n(v)$ is "easy", e.g., choose v to be a Manin symbol.

(3) [Map from Hecke Ring] Let ψ be the map $\mathbb{T} \to W'$, given by $\psi(t) = \pi(tv)$. Note that computation of ψ is relatively easy, because v was chosen so that tv is relatively easy to compute. In particular, if $t = T_p$, we do not need to compute the full matrix of T_p on V; instead we just compute $T_p(v)$.

(4) [Find Generator] Find a random $T \in \mathbb{T}$ such that the iterates

$$\psi(T^0), \quad \psi(T), \quad \psi(T^2), \quad \ldots, \quad \psi(T^{d-1})$$

are a basis for W', where W has dimension d.

(5) [Characteristic Polynomial] Compute the characteristic polynomial f of $T|_W$, and let $L = K[x]/(f)$. Because of how we chose T in step (4), the minimal and characteristic polynomials of $T|_W$ are equal, and both are irreducible, so L is an extension of K of degree $d = \dim(W)$.

(6) [Field Structure] In this step we endow W' with a field structure. Let $e : W' \to L$ be the unique K-linear isomorphism such that

$$e(\psi(T^i)) \equiv x^i \pmod{f}$$

for $i = 0, 1, 2, ..., \deg(f) - 1$. The map e is uniquely determined since the $\psi(T^i)$ are a basis for W'. To compute e, we compute the change of basis matrix from the standard basis for W' to the basis $\{\psi(T^i)\}$. This change of basis matrix is the inverse of the matrix whose rows are the $\psi(T^i)$ for $i = 0, ..., \deg(f) - 1$.

(7) [Hecke Eigenvalues] Finally for each integer $n \geq 1$, we have

$$a_n = e(\psi(T_n)) = e(\pi(T_n(v))),$$

where a_n is the eigenvalue of T_n. Output the maps ψ and e and terminate.

One reason we separate ψ and e is that when $\dim(W)$ is large, the values $\psi(T_n)$ take less space to store and are easier to compute, whereas each one of the values $e(\psi(n))$ is huge.[1] The function e typically involves large numbers if $\dim(W)$ is large, since e is obtained from the iterates of a single vector. For many applications, e.g., databases, it is better to store a matrix that defines e and the images under ψ of many T_n.

[1] John Cremona initially suggested to me the idea of separating these two maps.

Example 9.15. The space $S_2(\Gamma_0(23))$ of cusp forms has dimension 2 and is spanned by two $\mathrm{Gal}(\overline{\mathbb{Q}}/\mathbb{Q})$-conjugate newforms, one of which is

$$f = q + aq^2 + (-2a-1)q^3 + (-a-1)q^4 + 2aq^5 + \cdots,$$

where $a = (-1+\sqrt{5})/2$. We will use Algorithm 9.14 to compute a few of these coefficients.

The space $\mathbb{M}_2(\Gamma_0(23))^+$ of modular symbols has dimension 3. It has the following basis of Manin symbols:

$$[(0,0)], \quad [(1,0)], \quad [(0,1)],$$

where we use square brackets to differentiate Manin symbols from vectors. The Hecke operator

$$T_2 = \begin{pmatrix} 3 & 0 & 0 \\ 0 & 0 & 2 \\ -1 & 1/2 & -1 \end{pmatrix}$$

has characteristic polynomial $(x-3)(x^2+x-1)$. The kernel of $T_2 - 3$ corresponds to the span of the Eisenstein series of level 23 and weight 2, and the kernel V of $T_2^2 + T_2 - 1$ corresponds to $S_2(\Gamma_0(23))$. (We could also have computed V as the kernel of the boundary map $\mathbb{M}_2(\Gamma_0(23))^+ \to \mathbb{B}_2(\Gamma_0(23))^+$.) Each of the following steps corresponds to the step of Algorithm 9.14 with the same number.

(1) [Compute Projection] We compute projection onto V (this will suffice to give us a map ϕ as in the algorithm). The matrix whose first two columns are the echelon basis for V and whose last column is the echelon basis for the Eisenstein subspace is

$$\begin{pmatrix} 0 & 0 & 1 \\ 1 & 0 & -2/11 \\ 0 & 1 & -3/11 \end{pmatrix}$$

and

$$B^{-1} = \begin{pmatrix} 2/11 & 1 & 0 \\ 3/11 & 0 & 1 \\ 1 & 0 & 0 \end{pmatrix},$$

so projection onto V is given by the first two rows:

$$\pi = \begin{pmatrix} 2/11 & 1 & 0 \\ 3/11 & 0 & 1 \end{pmatrix}.$$

(2) [Choose v] Let $v = (0,1,0)^t$. Notice that $\pi(v) = (1,0)^t \neq 0$, and $v = [(1,0)]$ is a sum of only one Manin symbol.

(3) [Map from Hecke Ring] This step is purely conceptual, since no actual work needs to be done. We illustrate it by computing $\psi(T_1)$ and $\psi(T_2)$. We have

$$\psi(T_1) = \pi(v) = (1,0)^t$$

and

$$\psi(T_2) = \pi(T_2(v)) = \pi((0,0,1/2)^t) = (0,1/2)^t.$$

(4) [Find Generator] We have

$$\psi(T_2^0) = \psi(T_1) = (1,0)^t,$$

which is clearly independent from $\psi(T_2) = (0,1/2)^t$. Thus we find that the image of the powers of $T = T_2$ generate V.

(5) [Characteristic Polynomial] The matrix of $T_2|_V$ is $\left(\begin{smallmatrix} 0 & 2 \\ 1/2 & -1 \end{smallmatrix}\right)$, which has characteristic polynomial $f = x^2 + x - 1$. Of course, we already knew this because we computed V as the kernel of $T_2^2 + T_2 - 1$.

(6) [Field Structure] We have

$$\psi(T_2^0) = \pi(v) = (1,0)^t \text{ and } \psi(T_2) = (0,1/2).$$

The matrix with rows the $\psi(T_2^i)$ is $\left(\begin{smallmatrix} 1 & 0 \\ 0 & 1/2 \end{smallmatrix}\right)$, which has inverse $e = \left(\begin{smallmatrix} 1 & 0 \\ 0 & 2 \end{smallmatrix}\right)$. The matrix e defines an isomorphism between V and the field

$$L = \mathbf{Q}[x]/(f) = \mathbf{Q}((-1+\sqrt{5})/2).$$

I.e., $e((1,0)) = 1$ and $e((0,1)) = 2x$, where $x = (-1+\sqrt{5})/2$.

(7) [Hecke Eigenvalues] We have $a_n = e(\Psi(T_n))$. For example,

$$\begin{aligned}
a_1 &= e(\Psi(T_1)) = e((1,0)) = 1,\\
a_2 &= e(\Psi(T_2)) = e((0,1/2)) = x,\\
a_3 &= e(\Psi(T_3)) = e(\pi(T_3(v))) = e(\pi((0,-1,-1)^t))\\
&= e((-1,-1)^t) = -1-2x,\\
a_4 &= e(\Psi(T_4)) = e(\pi((0,-1,-1/2)^t)) = e((-1,-1/2)^t) = -1-x,\\
a_5 &= e(\Psi(T_5)) = e(\pi((0,0,1)^t)) = e((0,1)^t) = 2x,\\
a_{23} &= e(\Psi(T_{23})) = e(\pi((0,1,0)^t)) = e((1,0)^t) = 1,\\
a_{97} &= e(\Psi(T_{23})) = e(\pi((0,14,3)^t)) = e((14,3)^t) = 14+6x.
\end{aligned}$$

Example 9.16. It is easier to appreciate Algorithm 9.14 after seeing how big the coefficients of the power series expansion of a newform typically are, when the newform is defined over a large field. For example, there is a newform

$$f = \sum_{n=1}^{\infty} a_n q^n \in S_2(\Gamma_0(389))$$

such that if $\alpha = a_2$, then

$$\begin{aligned}1097385680 \cdot a_3(f) = {} & -20146763x^{19} + 102331615x^{18} + 479539092x^{17} \\ & - 3014444212x^{16} - 3813583550x^{15} + 36114755350x^{14} \\ & + 6349339639x^{13} - 227515736964x^{12} + 71555185319x^{11} \\ & + 816654992625x^{10} - 446376673498x^{9} - 1698789732650x^{8} \\ & + 1063778499268x^{7} + 1996558922610x^{6} - 1167579836501x^{5} \\ & - 1238356001958x^{4} + 523532113822x^{3} + 352838824320x^{2} \\ & - 58584308844x - 25674258672.\end{aligned}$$

In contrast, if we take $v = \{0, \infty\} = (0, 1) \in \mathbb{M}_2(\Gamma_0(389))^+$, then

$$T_3(v) = -4(1,0) + 2(1,291) - 2(1,294) - 2(1,310) + 2(1,313) + 2(1,383).$$

Storing $T_3(v), T_5(v), \ldots$ as vectors is more compact than storing $a_3(f)$, $a_5(f)$, $\ldots$ directly as polynomials in a_2!

9.4. Congruences between Newforms

This section is about congruences between modular forms. Understanding congruences is crucial for studying Serre's conjectures, Galois representations, and explicit construction of Hecke algebras. We assume more background in algebraic number theory here than elsewhere in this book.

9.4.1. Congruences between Modular Forms. Let Γ be an arbitrary congruence subgroup of $\mathrm{SL}_2(\mathbb{Z})$, and suppose $f \in M_k(\Gamma)$ is a modular form of integer weight k for Γ. Since $\left(\begin{smallmatrix} 1 & N \\ 0 & 1 \end{smallmatrix}\right) \in \Gamma$ for some integer N, the form f has a Fourier expansion in nonnegative powers of $q^{1/N}$. For a rational number n, let $a_n(f)$ be the coefficient of q^n in the Fourier expansion of f. Put

$$\mathrm{ord}_q(f) = \min\{n \in \mathbb{Q} : a_n \neq 0\},$$

where by convention we take $\min \emptyset = +\infty$, so $\mathrm{ord}_q(0) = +\infty$.

9.4.1.1. *The j-invariant.* Let

$$j = \frac{1}{q} + 744 + 196884q + \cdots$$

be the j-function, which is a weight 0 modular function that is holomorphic except for a simple pole at ∞ and has integer Fourier coefficients (see, e.g., [**Ser73**, Section VIII.3.3]).

Lemma 9.17. *Suppose g is a weight 0 level 1 modular function that is holomorphic except possibly with a pole of order n at ∞. Then g is a polynomial in j of degree at most n. Moreover, the coefficients of this polynomial lie in the ideal I generated by the coefficients $a_m(g)$ with $m \leq 0$.*

Proof. If $n = 0$, then $g \in M_0(\mathrm{SL}_2(\mathbb{Z})) = \mathbb{C}$, so g is constant with constant term in I, so the statement is true. Next suppose $n > 0$ and the lemma has been proved for all functions with smaller order poles. Let $\alpha = a_n(g)$, and note that

$$\mathrm{ord}_q(g - \alpha j^n) = \mathrm{ord}_q\left(g - \alpha \cdot \left(\frac{1}{q} + 744 + 196884q + \cdots\right)^n\right) > -n.$$

Thus by induction $h = g - \alpha j^n$ is a polynomial in j of degree $< n$ with coefficients in the ideal generated by the coefficients $a_m(g)$ with $m < 0$. It follows that $g = \alpha \cdot j^n - h$ satisfies the conclusion of the lemma. $\square$

9.4.1.2. *Sturm's Theorem.* If $\mathcal{O}$ is the ring of integers of a number field, $\mathfrak{m}$ is a maximal ideal of $\mathcal{O}$, and $f = \sum a_n q^n \in \mathcal{O}[[q^{1/N}]]$ for some integer N, let

$$\mathrm{ord}_{\mathfrak{m}}(f) = \mathrm{ord}_q(f \mod \mathfrak{m}) = \min\{n \in \mathbb{Q} : a_n \notin \mathfrak{m}\}.$$

Note that $\mathrm{ord}_{\mathfrak{m}}(fg) = \mathrm{ord}_{\mathfrak{m}}(f) + \mathrm{ord}_{\mathfrak{m}}(g)$. The following theorem was first proved in [**Stu87**].

Theorem 9.18 (Sturm). *Let $\mathfrak{m}$ be a prime ideal in the ring of integers $\mathcal{O}$ of a number field K, and let Γ be a congruence subgroup of $\mathrm{SL}_2(\mathbb{Z})$ of index m and level N. Suppose $f \in M_k(\Gamma, \mathcal{O})$ is a modular form and*

$$\mathrm{ord}_{\mathfrak{m}}(f) > \frac{km}{12}$$

or $f \in S_k(\Gamma, \mathcal{O})$ is a cusp form and

$$\mathrm{ord}_{\mathfrak{m}}(f) > \frac{km}{12} - \frac{m-1}{N}.$$

Then $f \equiv 0 \pmod{\mathfrak{m}}$.

Proof. Case 1: First we assume $\Gamma = \mathrm{SL}_2(\mathbb{Z})$.
Let

$$\Delta = q + 24q^2 + \cdots \in S_{12}(\mathrm{SL}_2(\mathbb{Z}), \mathbb{Z})$$

be the Δ function. Since $\mathrm{ord}_{\mathfrak{m}}(f) > k/12$, we have $\mathrm{ord}_{\mathfrak{m}}(f^{12}) > k$. We have

$$\mathrm{ord}_q(f^{12} \cdot \Delta^{-k}) = 12 \cdot \mathrm{ord}_q(f) - k \cdot \mathrm{ord}_q(\Delta) \geq -k, \tag{9.4.1}$$

since f is holomorphic at infinity and Δ has a zero of order 1. Also

$$\mathrm{ord}_{\mathfrak{m}}(f^{12} \cdot \Delta^{-k}) = \mathrm{ord}_{\mathfrak{m}}(f^{12}) - k \cdot \mathrm{ord}_{\mathfrak{m}}(\Delta) > k - k = 0. \tag{9.4.2}$$

Combining (9.4.1) and (9.4.2), we see that

$$f^{12} \cdot \Delta^{-k} = \sum_{n \geq -k} b_n q^n,$$

with $b_n \in \mathcal{O}$ and $b_n \in \mathfrak{m}$ if $n \leq 0$.

By Lemma 9.17,

$$f^{12} \cdot \Delta^{-k} \in \mathfrak{m}[j]$$

is a polynomial in j of degree at most k with coefficients in $\mathfrak{m}$. Thus

$$f^{12} \in \mathfrak{m}[j] \cdot \Delta^k,$$

so since the coefficients of Δ are integers, every coefficient of f^{12} is in $\mathfrak{m}$. Thus $\mathrm{ord}_{\mathfrak{m}}(f^{12}) = +\infty$, hence $\mathrm{ord}_{\mathfrak{m}}(f) = +\infty$, so $f = 0$, as claimed.

Case 2: Γ Arbitrary

Let N be such that $\Gamma(N) \subset \Gamma$, so also $f \in M_k(\Gamma(N))$. If $g \in M_k(\Gamma(N))$ is arbitrary, then because $\Gamma(N)$ is a normal subgroup of $\mathrm{SL}_2(\mathbb{Z})$, we have that for any $\gamma \in \Gamma(N)$ and $\delta \in \mathrm{SL}_2(\mathbb{Z})$,

$$(g^{[\delta]_k})^{[\gamma]_k} = g^{[\delta\gamma]_k} = g^{[\gamma'\delta]_k} = (g^{[\gamma']_k})^{[\delta]_k} = g^{[\delta]_k},$$

where $\gamma' \in \mathrm{SL}_2(\mathbb{Z})$. Thus for any $\delta \in \mathrm{SL}_2(\mathbb{Z})$, we have that $g^{[\delta]_k} \in M_k(\Gamma(N))$, so $\mathrm{SL}_2(\mathbb{Z})$ acts on $M_k(\Gamma(N))$.

It is a standard (but nontrivial) fact about modular forms, which comes from the geometry of the modular curve $X(N)$ over $\mathbb{Q}(\zeta_N)$ and $\mathbb{Z}[\zeta_N]$, that $M_k(\Gamma(N))$ has a basis with Fourier expansions in $\mathbb{Z}[\zeta_N][[q^{1/N}]]$ and that the action of $\mathrm{SL}_2(\mathbb{Z})$ on $M_k(\Gamma(N))$ preserves

$$M_k(\Gamma(N), \mathbb{Q}(\zeta_N)) = M_k(\Gamma(N)) \cap (\mathbb{Q}(\zeta_N)[[q^{1/N}]])$$

and the cuspidal subspace $S_k(\Gamma(N), \mathbb{Q}(\zeta_N))$. In particular, for any $\gamma \in \mathrm{SL}_2(\mathbb{Z})$,

$$f^{[\gamma]_k} \in M_k(\Gamma(N), K(\zeta_N))$$

Moreover, the denominators of $f^{[\gamma]_k}$ are bounded, since f is an $\mathcal{O}[\zeta_N]$-linear combination of a basis for $M_k(\Gamma(N), \mathbb{Z}[\zeta_N])$, and the denominators of $f^{[\gamma]_k}$ divide the product of the denominators of the images of each of these basis vectors under $[\gamma]_k$.

Let $L = K(\zeta_N)$. Let $\mathfrak{M}$ be a prime of $\mathcal{O}_L$ that divides $\mathfrak{m}\mathcal{O}_L$. We will now show that for each $\gamma \in \mathrm{SL}_2(\mathbb{Z})$, the Chinese Remainder Theorem implies that there is an element $A_\gamma \in L^*$ such that

$$(9.4.3) \qquad A_\gamma \cdot f^{[\gamma]_k} \in M_k(\Gamma(N), \mathcal{O}_L) \qquad \text{and} \qquad \mathrm{ord}_{\mathfrak{M}}(A_\gamma \cdot f^{[\gamma]_k}) < \infty.$$

First find $A \in L^*$ such that $A \cdot f^{[\gamma]_k}$ has coefficients in $\mathcal{O}_L$. Choose $\alpha \in \mathfrak{M}$ with $\alpha \notin \mathfrak{M}^2$, and find a negative power α^t such that $\alpha^t \cdot A \cdot f^{[\gamma]_k}$ has $\mathfrak{M}$-integral coefficients and finite valuation. This is possible because we assumed that f is nonzero. Use the Chinese Remainder Theorem to find $\beta \in \mathcal{O}_L$ such that $\beta \equiv 1 \pmod{\mathfrak{M}}$ and $\beta \equiv 0 \pmod{\wp}$ fcr each prime $\wp \neq \mathfrak{M}$ that divides (α). Then for some s we have

$$\beta^s \cdot \alpha^t \cdot A \cdot f^{[\gamma]_k} = A_\gamma \cdot f^{[\gamma]_k} \in M_k(\Gamma(N), \mathcal{O}_L)$$

and $\mathrm{ord}_{\mathfrak{M}}(A_\gamma \cdot f^{[\gamma]_k}) < \infty$.

Write

$$\mathrm{SL}_2(\mathbf{Z}) = \bigcup_{i=1}^{m} \Gamma \gamma_i$$

with $\gamma_1 = \left(\begin{smallmatrix} 1 & 0 \\ 0 & 1 \end{smallmatrix}\right)$, and let

$$F = f \cdot \prod_{i=2}^{m} A_{\gamma_i} \cdot f^{[\gamma_i]_k}.$$

Then $F \in M_{km}(\mathrm{SL}_2(\mathbf{Z}))$ and since $\mathfrak{M} \cap \mathcal{O}_K = \mathfrak{m}$, we have $\mathrm{ord}_{\mathfrak{M}}(f) = \mathrm{ord}_{\mathfrak{m}}(f)$, so

$$\mathrm{ord}_{\mathfrak{M}}(F) \geq \mathrm{ord}_{\mathfrak{M}}(f) = \mathrm{ord}_{\mathfrak{m}}(f) > \frac{km}{12}.$$

Thus we can apply Case 1 to conclude that

$$\mathrm{ord}_{\mathfrak{M}}(F) = +\infty.$$

Thus

$$\infty = \mathrm{ord}_{\mathfrak{M}}(F) = \mathrm{ord}_{\mathfrak{m}}(f) + \sum_{i=2}^{m} \mathrm{ord}_{\mathfrak{M}}(A_{\gamma_i} f^{[\gamma]_k}), \tag{9.4.4}$$

so $\mathrm{ord}_{\mathfrak{m}}(f) = +\infty$, because of (9.4.3).

We next obtain a better bound when f is a cusp form. Since $[\gamma]_k$ preserves cusp forms, $\mathrm{ord}_{\mathfrak{M}}(A_{\gamma_i} f^{[\gamma]_k}) \geq \frac{1}{N}$ for each i. Thus

$$\mathrm{ord}_{\mathfrak{M}}(F) \geq \mathrm{ord}_{\mathfrak{M}}(f) + \frac{m-1}{N} = \mathrm{ord}_{\mathfrak{m}}(f) + \frac{m-1}{N} > \frac{km}{12},$$

since now we are merely assuming that

$$\mathrm{ord}_{\mathfrak{m}}(f) > \frac{km}{12} - \frac{m-1}{N}.$$

Thus we again apply Case 1 to conclude that $\mathrm{ord}_{\mathfrak{M}}(F) = +\infty$, and using (9.4.4), conclude that $\mathrm{ord}_{\mathfrak{m}}(f) = +\infty$. □

Corollary 9.19. *Let $\mathfrak{m}$ be a prime ideal in the ring of integers $\mathcal{O}$ of a number field. Suppose $f, g \in M_k(\Gamma, \mathcal{O})$ are modular forms and*

$$a_n(f) \equiv a_n(g) \pmod{\mathfrak{m}}$$

for all

$$n \leq \begin{cases} \dfrac{km}{12} - \dfrac{m-1}{N} & \text{if } f - g \in S_k(\Gamma, \mathcal{O}), \\ \dfrac{km}{12} & \text{otherwise}, \end{cases}$$

where $m = [\mathrm{SL}_2(\mathbb{Z}) : \Gamma]$. *Then* $f \equiv g \pmod{\mathfrak{m}}$.

Buzzard proved the following corollary, which is extremely useful in practical computations. It asserts that the Sturm bound for modular forms with character is the same as the Sturm bound for $\Gamma_0(N)$.

Corollary 9.20 (Buzzard). *Let* $\mathfrak{m}$ *be a prime ideal in the ring of integers* $\mathcal{O}$ *of a number field. Suppose* $f, g \in M_k(N, \varepsilon, \mathcal{O})$ *are modular forms with Dirichlet character* $\varepsilon : (\mathbb{Z}/N\mathbb{Z})^* \to \mathbb{C}^*$ *and assume that*

$$a_n(f) \equiv a_n(g) \pmod{\mathfrak{m}} \qquad \text{for all} \qquad n \leq \frac{km}{12},$$

where

$$m = [\mathrm{SL}_2(\mathbb{Z}) : \Gamma_0(N)] = \#\mathbb{P}^1(\mathbb{Z}/N\mathbb{Z}) = N \cdot \prod_{p \mid N} \left(1 + \frac{1}{p}\right).$$

Then $f \equiv g \pmod{\mathfrak{m}}$.

Proof. Let $h = f - g$ and let $r = km/12$, so $\mathrm{ord}_{\mathfrak{m}}(h) > r$. Let s be the order of the Dirichlet character ε. Then $h^s \in M_{ks}(\Gamma_0(N))$ and

$$\mathrm{ord}_{\mathfrak{m}}(h^s) > sr = \frac{ksm}{12}.$$

By Theorem 9.18, we have $\mathrm{ord}_{\mathfrak{m}}(h^s) = \infty$, so $\mathrm{ord}_{\mathfrak{m}}(h) = \infty$. It follows that $f \equiv g \pmod{\mathfrak{m}}$. □

9.4.1.3. *Congruence for Newforms.* Sturm's paper [**Stu87**] also applies some results of Asai on q-expansions at various cusps to obtain a more refined result for newforms.

Theorem 9.21 (Sturm). *Let* N *be a positive integer that is square-free, and suppose* f *and* g *are two newforms in* $S_k(N, \varepsilon, \mathcal{O})$, *where* $\mathcal{O}$ *is the ring of integers of a number field, and suppose that* $\mathfrak{m}$ *is a maximal ideal of* $\mathcal{O}$. *Let* I *be an arbitrary subset of the prime divisors of* N. *If* $a_p(f) = a_p(g)$ *for all* $p \in I$ *and if*

$$a_p(f) \equiv a_p(g) \pmod{\mathfrak{m}}$$

for all primes

$$p \leq \frac{k \cdot [\mathrm{SL}_2(\mathbb{Z}) : \Gamma_0(N)]}{12 \cdot 2^{\#I}},$$

then $f \equiv g \pmod{\mathfrak{m}}$.

The paper [**BS02**] contains a similar result about congruences between newforms, which does not require that the level be square-free. Recall from Definition 4.18 that the conductor of a Dirichlet character ε is the largest divisor c of N such that ε factors through $(\mathbb{Z}/c\mathbb{Z})^\times$.

Theorem 9.22. *Let $N > 4$ be any integer, and suppose f and g are two normalized eigenforms in $S_k(N, \varepsilon; \mathcal{O})$, where $\mathcal{O}$ is the ring of integers of a number field, and suppose that $\mathfrak{m}$ is a maximal ideal of $\mathcal{O}$. Let I be the set of prime divisors of N that do not divide $\frac{N}{\mathrm{cond}(\varepsilon)}$. If*

$$a_p(f) \equiv a_p(g) \pmod{\mathfrak{m}}$$

for all primes $p \in I$ and for all primes

$$p \leq \frac{k \cdot [\mathrm{SL}_2(\mathbb{Z}) : \Gamma_0(N)]}{12 \cdot 2^{\#I}},$$

then $f \equiv g \pmod{\mathfrak{m}}$.

For the proof, see Lemma 1.4 and Corollary 1.7 in [**BS02**, §1.3].

9.4.2. Generating the Hecke Algebra. The following theorem appeared in [**LS02**, Appendix], except that we give a better bound here. It is a nice application of the congruence result above, which makes possible explicit computations with Hecke rings $\mathbb{T}$.

Theorem 9.23. *Suppose Γ is a congruence subgroup that contains $\Gamma_1(N)$ and let*

$$r = \frac{km}{12} - \frac{m-1}{N}, \tag{9.4.5}$$

where $m = [\mathrm{SL}_2(\mathbb{Z}) : \Gamma]$. Then the Hecke algebra

$$\mathbb{T} = \mathbb{Z}[\ldots, T_n, \ldots] \subset \mathrm{End}(S_k(\Gamma))$$

is generated as a $\mathbb{Z}$-module by the Hecke operators T_n for $n \leq r$.

Proof. For any ring R, let $S_k(N, R) = S_k(N; \mathbb{Z}) \otimes R$, where $S_k(N; \mathbb{Z}) \subset \mathbb{Z}[[q]]$ is the submodule of cusp forms with integer Fourier expansion at the cusp ∞, and let $\mathbb{T}_R = \mathbb{T} \otimes_{\mathbb{Z}} R$. For any ring R, there is a perfect pairing

$$S_k(N, R) \otimes_R \mathbb{T}_R \to R$$

given by $\langle f, T\rangle \mapsto a_1(T(f))$ (this is true for $R = \mathbb{Z}$, hence for any R).

Let M be the submodule of $\mathbb{T}$ generated by $T_1, T_2, \ldots, T_r$, where r is the largest integer $\leq \frac{kN}{12} \cdot [\mathrm{SL}_2(\mathbb{Z}) : \Gamma]$. Consider the exact sequence of additive abelian groups

$$0 \to M \xrightarrow{i} \mathbb{T} \to \mathbb{T}/M \to 0.$$

Let p be a prime and use the fact that tensor product is right exact to obtain an exact sequence

$$M \otimes \mathbb{F}_p \xrightarrow{\bar{i}} \mathbb{T} \otimes \mathbb{F}_p \to (\mathbb{T}/M) \otimes \mathbb{F}_p \to 0.$$

Suppose that $f \in S_k(N, \mathbb{F}_p)$ pairs to 0 with each of $T_1, \ldots, T_r$. Then

$$a_m(f) = a_1(T_m f) = \langle f, T_m \rangle = 0$$

in $\mathbb{F}_p$ for each $m \leq r$. By Theorem 9.18, it follows that $f = 0$. Thus the pairing restricted to the image of $M \otimes \mathbb{F}_p$ in $\mathbb{T}_{\mathbb{F}_p}$ is nondegenerate, so because (9.4.5) is perfect, it follows that

$$\dim_{\mathbb{F}_p} \bar{i}(M \otimes \mathbb{F}_p) = \dim_{\mathbb{F}_p} S_k(N, \mathbb{F}_p).$$

Thus $(\mathbb{T}/M) \otimes \mathbb{F}_p = 0$. Repeating the argument for all primes p shows that $\mathbb{T}/M = 0$, as claimed. □

Remark 9.24. In general, the conclusion of Theorem 9.23 is not true if one considers only T_n where n runs over the primes less than the bound. Consider, for example, $S_2(11)$, where the bound is 1 and there are no primes ≤ 1. However, the Hecke algebra is generated as an algebra by operators T_p with $p \leq r$.

9.5. Exercises

9.1 Prove that the group $\Gamma_1(N)$ is a normal subgroup of $\Gamma_0(N)$ and that the quotient $\Gamma_0(N)/\Gamma_1(N)$ is isomorphic to $(\mathbb{Z}/N\mathbb{Z})^*$.

9.2 Prove that the operators $\langle d \rangle$ are elements of $\mathbb{Z}[\ldots, T_n, \ldots]$. [Hint: Use Dirichlet's theorem on primes in arithmetic progression.]

9.3 Find an example like Example 9.6 but in which the new subspace is nonzero. More precisely, find an integer N such that the Hecke ring on $S_2(\Gamma_0(N))$ is not equal to the ring generated by Hecke operators T_n with $\gcd(n, N) = 1$ and $S_2(\Gamma_0(N))_{\text{new}} \neq 0$.

9.4 (a) Following Example 9.15, compute a basis for $S_2(\Gamma_0(31))$.
(b) Use Algorithm 9.12 to compute a basis for $S_2(\Gamma_0(31))$.

Chapter 10

Computing Periods

This chapter is about computing period maps associated to newforms. We assume you have read Chapters 8 and 9 and that you are familiar with abelian varieties at the level of [**Ros86**].

In Section 10.1 we introduce the period map and give some examples of situations in which computing it is relevant. Section 10.2 is about how to use the period mapping to attach an abelian variety to any newform. In Section 10.3, we introduce extended modular symbols, which are the key computational tool for quickly computing periods of modular symbols. We turn to numerical computation of period integrals in Section 10.4, and in Section 10.5 we explain how to use Atkin-Lehner operators to speed convergence. In Section 10.6 we explain how to compute the full period map with a minimum amount of work.

Section 10.7 briefly sketches three approaches to computing all elliptic curves of a given conductor.

This chapter was inspired by [**Cre97a**], which contains similar algorithms in the special case of a newform $f = \sum a_n q^n \in S_2(\Gamma_0(N))$ with $a_n \in \mathbb{Z}$.

See also [**Dok04**] for algorithmic methods to compute special values of very general L-functions, which can be used for approximating $L(f, s)$ for arbitrary s.

10.1. The Period Map

Let Γ be a subgroup of $\mathrm{SL}_2(\mathbb{Z})$ that contains $\Gamma_1(N)$ for some N, and suppose

$$f = \sum_{n\geq 1} a_n q^n \in S_k(\Gamma)$$

is a newform (see Definition 9.9). In this chapter we describe how to approximately compute the complex period mapping

$$\Phi_f : \mathbb{M}_k(\Gamma) \to \mathbb{C},$$

given by

$$\Phi_f(P\{\alpha, \beta\}) = \langle f, P\{\alpha, \beta\}\rangle = \int_\alpha^\beta f(z)P(z,1)dz,$$

as in Section 8.5. As an application, we can approximate the special values $L(f, j)$, for $j = 1, 2, \ldots, k-1$ using (8.5.5). We can also compute the period lattice attached to a modular abelian variety, which is an important step, e.g., in enumeration of $\mathbb{Q}$-curves (see, e.g., [**GLQ04**]) or computation of a curve whose Jacobian is a modular abelian variety A_f (see, e.g., [**Wan95**]).

10.2. Abelian Varieties Attached to Newforms

Fix a newform $f \in S_k(\Gamma)$, where $\Gamma_1(N) \subset \Gamma$ for some N. Let $f_1, \ldots, f_d$ be the $\mathrm{Gal}(\overline{\mathbb{Q}}/\mathbb{Q})$-conjugates of f, where $\mathrm{Gal}(\overline{\mathbb{Q}}/\mathbb{Q})$ acts via its action on the Fourier coefficients, which are algebraic integers (since they are the eigenvalues of matrices with integer entries). Let

$$V_f = \mathbb{C}f_1 \oplus \cdots \oplus \mathbb{C}f_d \subset S_k(\Gamma) \qquad (10.2.1)$$

be the subspace of cusp forms spanned by the $\mathrm{Gal}(\overline{\mathbb{Q}}/\mathbb{Q})$-conjugates of f. One can show using the results discussed in Section 9.2 that the above sum is direct, i.e., that V_f has dimension d.

The integration pairing induces a $\mathbb{T}$-equivariant homomorphism

$$\Phi_f : \mathbb{M}_k(\Gamma) \to V_f^* = \mathrm{Hom}_{\mathbb{C}}(V_f, \mathbb{C})$$

from modular symbols to the $\mathbb{C}$-linear dual V_f^* of V_f. Here $\mathbb{T}$ acts on V_f^* via $(\varphi t)(x) = \varphi(tx)$, and this homomorphism is $\mathbb{T}$-stable by Theorem 8.21. The *abelian variety attached to* f is the quotient

$$A_f(\mathbb{C}) = V_f^*/\Phi_f(\mathbb{S}_k(\Gamma;\mathbb{Z})).$$

Here $\mathbb{S}_k(\Gamma;\mathbb{Z}) = \mathbb{S}_k(\Gamma)$, and we include the $\mathbb{Z}$ in the notation to emphasize that these are integral modular symbols. See [**Shi59**] for a proof that $A_f(\mathbb{C})$ is an abelian variety (in particular, $\Phi_f(\mathbb{S}_k(\Gamma;\mathbb{Z}))$ is a lattice, and V_f^* is equipped with a nondegenerate Riemann form).

When $k = 2$, we can also construct A_f as a quotient of the modular Jacobian $\mathrm{Jac}(X_\Gamma)$, so A_f is an abelian variety canonically defined over $\mathbb{Q}$.

In general, we have an exact sequence

$$0 \to \mathrm{Ker}(\Phi_f) \to \mathbb{S}_k(\Gamma) \to V_f^* \to A_f(\mathbb{C}) \to 0.$$

Remark 10.1. When $k = 2$, the abelian variety A_f has a canonical structure of abelian variety over $\mathbb{Q}$. Moreover, there is a conjecture of Ribet and Serre in [**Rib92**] that describes the simple abelian varieties A over $\mathbb{Q}$ that should arise via this construction. In particular, the conjecture is that A is isogenous to some abelian variety A_f if and only if $\mathrm{End}(A/\mathbb{Q}) \otimes \mathbb{Q}$ is a number field of degree $\dim(A)$. The abelian varieties A_f have this property since $\mathbb{Q}(\ldots, a_n(f), \ldots)$ embeds in $\mathrm{End}(A/\mathbb{Q}) \otimes \mathbb{Q}$ and the endomorphism ring over $\mathbb{Q}$ has degree at most $\dim(A)$ (see [**Rib92**] for details). Ribet proves that his conjecture is a consequence of Serre's conjecture [**Ser87**] on modularity of mod p odd irreducible Galois representations (see Section 1.5). Much of Serre's conjecture has been proved by Khare and Wintenberger (not published). In particular, it is a theorem that if A is a simple abelian variety over $\mathbb{Q}$ with $\mathrm{End}(A/\mathbb{Q}) \otimes \mathbb{Q}$ a number field of degree $\dim(A)$ and if A has good reduction at 2, then A is isogenous to some abelian variety A_f.

Remark 10.2. When $k > 2$, there is an object called a *Grothendieck motive* that is attached to f and has a canonical "structure over $\mathbb{Q}$". See [**Sch90**].

10.3. Extended Modular Symbols

In this section, we extend the notion of modular symbols to allows symbols of the form $P\{w, z\}$ where w and z are arbitrary elements of $\mathfrak{h}^* = \mathfrak{h} \cup \mathbb{P}^1(\mathbb{Q})$.

Definition 10.3 (Extended Modular Symbols). The abelian group $\overline{\mathbb{M}}_k$ of *extended modular symbols* of weight k is the $\mathbb{Z}$-span of symbols $P\{w, z\}$, with $P \in V_{k-2}$ a homogeneous polynomial of degree $k-2$ with integer coefficients, modulo the relations

$$P \cdot (\{w, y\} + \{y, z\} + \{z, w\}) = 0$$

and modulo any torsion.

Fix a finite index subgroup $\Gamma \subset \mathrm{SL}_2(\mathbb{Z})$. Just as for usual modular symbols, $\overline{\mathbb{M}}_k$ is equipped with an action of Γ, and we define the space of *extended modular symbols* of weight k for Γ to be the quotient

$$\overline{\mathbb{M}}_k(\Gamma) = (\overline{\mathbb{M}}_k / \langle \gamma x - x : \gamma \in \Gamma, x \in \overline{\mathbb{M}}_k \rangle)/\,\mathrm{tor}\,.$$

The quotient $\overline{\mathbb{M}}_k(\Gamma)$ is torsion-free and fixed by Γ.

The integration pairing extends naturally to a pairing

$$(10.3.1) \qquad \big(S_k(\Gamma) \oplus \overline{S}_k(\Gamma)\big) \times \overline{\mathbb{M}}_k(\Gamma) \to \mathbb{C},$$

where we recall from (8.5.1) that $\overline{S}_k(\Gamma)$ denotes the space of antiholomorphic cusp forms. Moreover, if

$$\iota : \mathbb{M}_k(\Gamma) \to \overline{\mathbb{M}}_k(\Gamma)$$

is the natural map, then ι respects (10.3.1) in the sense that for all $f \in S_k(\Gamma) \oplus \overline{S}_k(\Gamma)$ and $x \in \mathbb{M}_k(\Gamma)$, we have

$$\langle f, x \rangle = \langle f, \iota(x) \rangle.$$

As we will see soon, it is often useful to replace $x \in \mathbb{M}_k(\Gamma)$ first by $\iota(x)$ and then by an equivalent sum $\sum y_i$ of symbols $y_i \in \overline{\mathbb{M}}_k(N, \varepsilon)$ such that $\langle f, \sum y_i \rangle$ is easier to compute numerically than $\langle f, x \rangle$.

Let ε be a Dirichlet character of modulus N. If $\gamma = \left(\begin{smallmatrix} a & b \\ c & d \end{smallmatrix}\right) \in \mathrm{SL}_2(\mathbb{Z})$, let $\varepsilon(\gamma) = \varepsilon(d)$. Let $\overline{\mathbb{M}}_k(N, \varepsilon)$ be the quotient of $\overline{\mathbb{M}}_k(N, \mathbb{Z}[\varepsilon])$ by the relations $\gamma(x) - \varepsilon(\gamma)x$, for all $x \in \mathbb{M}_k(N, \mathbb{Z}[\varepsilon])$, $\gamma \in \Gamma_0(N)$, and modulo any torsion.

10.4. Approximating Period Integrals

In this section we assume Γ is a congruence subgroup of $\mathrm{SL}_2(\mathbb{Z})$ that contains $\Gamma_1(N)$ for some N. Suppose $\alpha \in \mathfrak{h}$, so $\mathrm{Im}(\alpha) > 0$ and m is an integer such that $0 \le m \le k-2$, and consider the extended modular symbol $X^m Y^{k-2-m}\{\alpha, \infty\}$. Let $\langle \cdot, \cdot \rangle$ denote the integration pairing from Section 8.5. Given an arbitrary cusp form $f = \sum_{n=1}^{\infty} a_n q^n \in S_k(\Gamma)$, we have

$$\Phi_f(X^m Y^{k-2-m}\{\alpha, \infty\}) = \left\langle f,\, X^m Y^{k-2-m}\{\alpha, \infty\} \right\rangle \tag{10.4.1}$$

$$= \int_\alpha^\infty f(z) z^m dz \tag{10.4.2}$$

$$= \sum_{n=1}^{\infty} a_n \int_\alpha^\infty e^{2\pi i n z} z^m dz. \tag{10.4.3}$$

The reversal of summation and integration is justified because the imaginary part of α is positive so that the sum converges absolutely. The following lemma is useful for computing the above infinite sum.

Lemma 10.4.

$$\int_\alpha^\infty e^{2\pi i n z} z^m dz = e^{2\pi i n \alpha} \sum_{s=0}^{m} \left(\frac{(-1)^s \alpha^{m-s}}{(2\pi i n)^{s+1}} \prod_{j=(m+1)-s}^{m} j \right). \tag{10.4.4}$$

Proof. See Exercise 10.1 □

In practice we will usually be interested in computing the period map Φ_f when $f \in S_k(\Gamma)$ is a newform. Since f is a newform, there is a Dirichlet character ε such that $f \in S_k(N, \varepsilon)$. The period map $\Phi_f : \mathbb{M}_k(\Gamma) \to \mathbb{C}$ then factors through the quotient $\mathbb{M}_k(N,\varepsilon)$, so it suffices to compute the period map on modular symbols in $\mathbb{M}_k(N,\varepsilon)$.

The following proposition is an analogue of [**Cre97a**, Prop. 2.1.1(5)].

Proposition 10.5. *For any $\gamma \in \Gamma_0(N)$, $P \in V_{k-2}$ and $\alpha \in \mathfrak{h}^*$, we have the following relation in $\overline{\mathbb{M}}_k(N,\varepsilon)$:*

$$\begin{aligned} &(10.4.5) \quad P\{\infty,\gamma(\infty)\} &= P\{\alpha,\gamma(\alpha)\} + (P - \varepsilon(\gamma)\gamma^{-1}P)\{\infty,\alpha\} \\ &(10.4.6) &= \varepsilon(\gamma)(\gamma^{-1}P)\{\alpha,\infty\} - P\{\gamma(\alpha),\infty\}. \end{aligned}$$

Proof. By definition, if $x \in \mathbb{M}_k(N,\varepsilon)$ is a modular symbol and $\gamma \in \Gamma_0(N)$, then $\gamma x = \varepsilon(\gamma)x$. Thus $\varepsilon(\gamma)\gamma^{-1}x = x$, so

$$\begin{aligned} P\{\infty,\gamma(\infty)\} &= P\{\infty,\alpha\} + P\{\alpha,\gamma(\alpha)\} + P\{\gamma(\alpha),\gamma(\infty)\} \\ &= P\{\infty,\alpha\} + P\{\alpha,\gamma(\alpha)\} + \varepsilon(\gamma)\gamma^{-1}(P\{\gamma(\alpha),\gamma(\infty)\}) \\ &= P\{\infty,\alpha\} + P\{\alpha,\gamma(\alpha)\} + \varepsilon(\gamma)(\gamma^{-1}P)\{\alpha,\infty\} \\ &= P\{\alpha,\gamma(\alpha)\} + P\{\infty,\alpha\} - \varepsilon(\gamma)(\gamma^{-1}P)\{\infty,\alpha\} \\ &= P\{\alpha,\gamma(\alpha)\} + (P - \varepsilon(\gamma)\gamma^{-1}P)\{\infty,\alpha\}. \end{aligned}$$

The second equality in the statement of the proposition now follows easily. □

In the case of weight 2 and trivial character, the "error term"

$$(10.4.7) \qquad (P - \varepsilon(\gamma)\gamma^{-1}P)\{\infty,\alpha\}$$

vanishes since P is constant and $\varepsilon(\gamma) = 1$. In general this term does not vanish. However, we can suitably modify the formulas found in [**Cre97a**, 2.10] and still obtain an algorithm for computing period integrals.

Algorithm 10.6 (Period Integrals). *Given $\gamma \in \Gamma_0(N)$, $P \in V_{k-2}$ and $f \in S_k(N,\varepsilon)$ presented as a q-expansion to some precision, this algorithm outputs an approximation to the period integral $\langle f,\, P\{\infty,\gamma(\infty)\}\rangle$.*

(1) Write $\gamma = \left(\begin{smallmatrix} a & b \\ cN & d \end{smallmatrix}\right) \in \Gamma_0(N)$, with $a,b,c,d \in \mathbb{Z}$, and set $\alpha = \frac{-d+i}{cN}$ in Proposition 10.5.
(2) Replacing γ by $-\gamma$ if necessary, we find that the imaginary parts of α and $\gamma(\alpha) = \frac{a+i}{cN}$ are both equal to the positive number $\frac{1}{cN}$.
(3) Use (10.4.3) and Lemma 10.4 to compute the integrals that appear in Proposition 10.5.

It would be nice if the modular symbols of the form $P\{\infty,\gamma(\infty)\}$ for $P \in V_{k-2}$ and $\gamma \in \Gamma_0(N)$ were to generate a large subspace of $\mathbb{M}_k(N,\varepsilon)\otimes\mathbb{Q}$. When $k = 2$ and $\varepsilon = 1$, Manin proved in [**Man72**] that the map $\Gamma_0(N) \to H_1(X_0(N),\mathbb{Z})$ sending γ to $\{0,\gamma(0)\}$ is a surjective group homomorphism. When $k > 2$, the author does not know a similar group-theoretic statement. However, we have the following theorem.

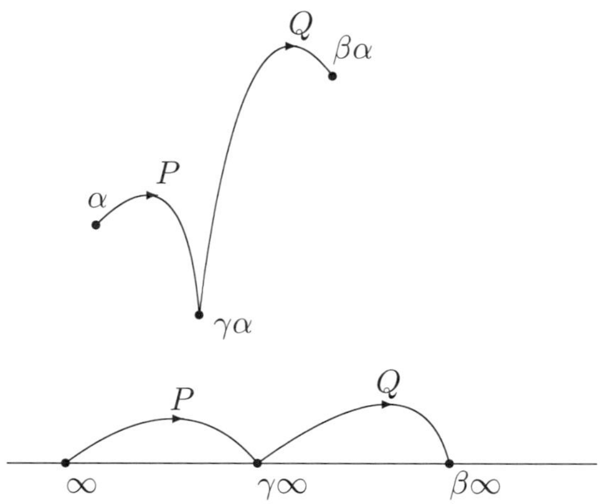

Figure 10.4.1. "Transporting" a transportable modular symbol.

Theorem 10.7. *Any element of* $\mathbb{S}_k(N,\varepsilon)$ *can be written in the form*

$$\sum_{i=1}^{n} P_i\{\infty, \gamma_i(\infty)\}$$

for some $P_i \in V_{k-2}$ *and* $\gamma_i \in \Gamma_0(N)$. *Moreover,* P_i *and* γ_i *can be chosen so that* $\sum P_i = \sum \varepsilon(\gamma_i)\gamma_i^{-1}(P_i)$, *so the error term* (10.4.7) *vanishes.*

The author and Helena Verrill prove this theorem in [**SV01**]. The condition that the error term vanishes means that one can replace ∞ by any α in the expression for the modular symbol and obtain an equivalent modular symbol. For this reason, we call such modular symbols *transportable*, as illustrated in Figure 10.4.1.

Note that in general not every element of the form $P\{\infty, \gamma(\infty)\}$ must lie in $\mathbb{S}_k(N,\varepsilon)$. However, if $\gamma P = P$, then $P\{\infty, \gamma(\infty)\}$ does lie in $\mathbb{S}_k(N,\varepsilon)$. It would be interesting to know under what circumstances $\mathbb{S}_k(N,\varepsilon)$ is generated by symbols of the form $P\{\infty, \gamma(\infty)\}$ with $\gamma P = P$. This sometimes fails for k odd; for example, when $k = 3$, the condition $\gamma P = P$ implies that $\gamma \in \Gamma_0(N)$ has an eigenvector with eigenvalue 1, and hence is of finite order. When k is even, the author can see no obstruction to generating $\mathbb{S}_k(N,\varepsilon)$ using such symbols.

10.5. Speeding Convergence Using Atkin-Lehner

Let $w_N = \left(\begin{smallmatrix} 0 & -1 \\ N & 0 \end{smallmatrix}\right) \in \mathrm{Mat}_2(\mathbb{Z})$. Consider the Atkin-Lehner involution W_N on $M_k(\Gamma_1(N))$, which is defined by

$$\begin{aligned} W_N(f) &= N^{(2-k)/2} \cdot f|_{[w_N]_k} \\ &= N^{(2-k)/2} \cdot f\left(-\frac{1}{Nz}\right) \cdot N^{k-1} \cdot (Nz)^{-k} \\ &= N^{-k/2} \cdot z^{-k} \cdot f\left(-\frac{1}{Nz}\right). \end{aligned}$$

Here we take the positive square root if k is odd. Then $W_N^2 = (-1)^k$ is an involution when k is even.

There is an operator on modular symbols, which we also denote W_N, which is given by

$$\begin{aligned} W_N(P\{\alpha,\beta\}) &= N^{(2-k)/2} \cdot w_N(P)\{w_N(\alpha), w_N(\beta)\} \\ &= N^{(2-k)/2} \cdot P(-Y, NX)\left\{-\frac{1}{\alpha N}, -\frac{1}{\beta N}\right\}, \end{aligned}$$

and one has that if $f \in S_k(\Gamma_1(N))$ and $x \in \mathbb{M}_k(\Gamma_1(N))$, then

$$\langle W_N(f), x\rangle = \langle f, W_N(x)\rangle.$$

If ε is a Dirichlet character of modulus N, then the operator W_N sends $S_k(N,\varepsilon)$ to $S_k(\Gamma_1(N),\overline{\varepsilon})$. Thus if $\varepsilon^2 = 1$, then W_N preserves $S_k(N,\varepsilon)$. In particular, W_N acts on $S_k(\Gamma_0(N))$.

The next proposition shows how to compute the pairing $\langle f, P\{\infty,\gamma(\infty)\}\rangle$ under certain restrictive assumptions. It generalizes a result of [**Cre97b**] to higher weight.

Proposition 10.8. *Let $f \in S_k(N,\varepsilon)$ be a cusp form which is an eigenform for the Atkin-Lehner operator W_N having eigenvalue $w \in \{\pm 1\}$ (thus $\varepsilon^2 = 1$ and k is even). Then for any $\gamma \in \Gamma_0(N)$ and any $P \in V_{k-2}$, with the property that $\gamma P = \varepsilon(\gamma)P$, we have the following formula, valid for any $\alpha \in \mathfrak{h}$:*

$$\begin{aligned} \langle f, P\{\infty,\gamma(\infty)\}\rangle = \Big\langle f, \quad & w\frac{P(Y,-NX)}{N^{k/2-1}}\{w_N(\alpha),\infty\} \\ &+ \left(P - w\frac{P(Y,-NX)}{N^{k/2-1}}\right)\left\{i/\sqrt{N},\infty\right\} - P\left\{\gamma(\alpha),\infty\right\}\Big\rangle. \end{aligned}$$

Here $w_N(\alpha) = -\dfrac{1}{N\alpha}$.

Proof. By Proposition 10.5 our condition on P implies that $P\{\infty,\gamma(\infty)\} = P\{\alpha,\gamma(\alpha)\}$. We describe the steps of the following computation below.

$$\begin{aligned}
&\left\langle f,\quad P\{\alpha,\gamma(\alpha)\}\right\rangle\\
&=\left\langle f,\quad P\{\alpha,i/\sqrt{N}\}+P\{i/\sqrt{N},W(\alpha)\}+P\{W(\alpha),\gamma(\alpha)\}\right\rangle\\
&=\left\langle f,\quad w\frac{W(P)}{N^{k/2-1}}\{W(\alpha),i/\sqrt{N}\}+P\{i/\sqrt{N},W(\alpha)\}+P\{W(\alpha),\gamma(\alpha)\}\right\rangle.
\end{aligned}$$

For the first equality, we break the path into three paths, and in the second, we apply the W-involution to the first term and use that the action of W is compatible with the pairing $\langle\,,\,\rangle$ and that f is an eigenvector with eigenvalue w. In the following sequence of equalities we combine the first two terms and break up the third; then we replace $\{W(\alpha),i/\sqrt{N}\}$ by $\{W(\alpha),\infty\}+\{\infty,i/\sqrt{N}\}$ and regroup:

$$\begin{aligned}
&w\frac{W(P)}{N^{k/2-1}}\{W(\alpha),i/\sqrt{N}\}+P\{i/\sqrt{N},W(\alpha)\}+P\{W(\alpha),\gamma(\alpha)\}\\
&=\left(w\frac{W(P)}{N^{k/2-1}}-P\right)\{W(\alpha),i/\sqrt{N}\}+P\{W(\alpha),\infty\}-P\{\gamma(\alpha),\infty\}\\
&=w\frac{W(P)}{N^{k/2-1}}\{W(\alpha),\infty\}+\left(P-w\frac{W(P)}{N^{k/2-1}}\right)\{i/\sqrt{N},\infty\}-P\{\gamma(\alpha),\infty\}.
\end{aligned}$$

□

A good choice for α is $\alpha=\gamma^{-1}\left(\frac{b}{d}+\frac{i}{d\sqrt{N}}\right)$, so that $W(\alpha)=\frac{c}{d}+\frac{i}{d\sqrt{N}}$. This maximizes the minimum of the imaginary parts of α and $W(\alpha)$, which results in series that converge more quickly.

Let $\gamma=\left(\begin{smallmatrix}a&b\\c&d\end{smallmatrix}\right)\in\Gamma_0(N)$. The polynomial

$$P(X,Y)=(cX^2+(d-a)XY-bY^2)^{\frac{k-2}{2}}$$

satisfies $\gamma(P)=P$. We obtained this formula by viewing V_{k-2} as the $(k-2)$th symmetric product of the 2-dimensional space on which $\Gamma_0(N)$ acts naturally. For example, observe that since $\det(\gamma)=1$, the symmetric product of two eigenvectors for γ is an eigenvector in V_2 having eigenvalue 1. For the same reason, if $\varepsilon(\gamma)\neq 1$, there need not be a polynomial $P(X,Y)$ such that $\gamma(P)=\varepsilon(\gamma)P$. One remedy is to choose another γ so that $\varepsilon(\gamma)=1$.

Since the imaginary parts of the terms $i/\sqrt{N}$, α and $W(\alpha)$ in the proposition are all relatively large, the sums appearing at the beginning of Section 10.4 converge quickly if d is small. It is *important* to choose γ in Proposition 10.8 with d small; otherwise the series will converge very slowly.

Remark 10.9. Is there a generalization of Proposition 10.8 without the restrictions that $\varepsilon^2=1$ and k is even?

10.5.1. Another Atkin-Lehner Trick. Suppose E is an elliptic curve and let $L(E,s)$ be the corresponding L-function. Let $\varepsilon \in \{\pm 1\}$ be the root number of E, i.e., the sign of the functional equation for $L(E,s)$, so $\Lambda(E,s) = \varepsilon\Lambda(E,2-s)$, where $\Lambda(E,s) = N^{s/2}(2\pi)^{-s}\Gamma(s)L(E,s)$. Let $f = f_E$ be the modular form associated to E (which exists by [**Wil95, BCDT01**]). If $W_N(f) = wf$, then $\varepsilon = -w$ (see Exercise 10.2). We have

$$\begin{aligned} L(E,1) &= -2\pi \int_0^{\infty} f(z)\,dz \\ &= -2\pi i\,\langle f,\,\{0,\infty\}\rangle \\ &= -2\pi i\,\left\langle f, \{0, i/\sqrt{N}\} + \{i/\sqrt{N},\infty\}\right\rangle \\ &= -2\pi i\,\left\langle wf, \{w_N(0), w_N(i/\sqrt{N})\} + \{i/\sqrt{N},\infty\}\right\rangle \\ &= -2\pi i\,\left\langle wf, \{\infty, i/\sqrt{N}\} + \{i/\sqrt{N},\infty\}\right\rangle \\ &= -2\pi i\,(w-1)\left\langle f, \{\infty, i/\sqrt{N}\}\right\rangle. \end{aligned}$$

If $w = 1$, then $L(E,1) = 0$. If $w = -1$, then

$$L(E,1) = 4\pi i\left\langle f, \{\infty, i/\sqrt{N}\}\right\rangle = 2\sum_{n=1}^{\infty} \frac{a_n}{n} e^{-2\pi n/\sqrt{N}}. \tag{10.5.1}$$

For more about computing with L-functions of elliptic curves, including a trick for computing ε quickly without directly computing W_N, see [**Coh93**, §7.5] and [**Cre97a**, §2.11]. One can also find higher derivatives $L^{(r)}(E,1)$ by a formula similar to (10.5.1) (see [**Cre97a**, §2.13]). The methods in this chapter for obtaining rapidly converging series are not just of computational interest; see, e.g., [**Gre83**] for a nontrivial theoretical application to the Birch and Swinnerton-Dyer conjecture.

10.6. Computing the Period Mapping

Fix a newform $f = \sum a_n q^n \in S_k(\Gamma)$, where $\Gamma_1(N) \subset \Gamma$ for some N. Let V_f be as in (10.2.1).

Let $\Theta_f : M_k(\Gamma;\mathbf{Q}) \to V$ be *any* $\mathbf{Q}$-linear map with the same kernel as Φ_f; we call any such map a *rational period mapping* associated to f. Let Φ_f be the period mapping associated to the $\mathrm{Gal}(\overline{\mathbf{Q}}/\mathbf{Q})$-conjugates of f. We

have a commutative diagram

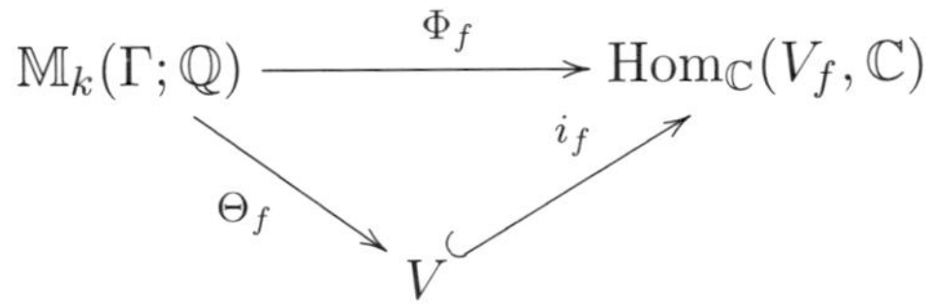

Recall from Section 10.2 that the cokernel of Φ_f is the abelian variety $A_f(\mathbb{C})$.

The Hecke algebra $\mathbb{T}$ acts on the linear dual

$$\mathbb{M}_k(\Gamma;\mathbb{Q})^* = \mathrm{Hom}(\mathbb{M}_k(\Gamma),\mathbb{Q})$$

by $(t\varphi)(x) = \varphi(tx)$. Let $I = I_f \subset \mathbb{T}$ be the kernel of the ring homomorphism $\mathbb{T} \to \mathbb{Z}[a_2, a_3, \ldots]$ that sends T_n to a_n. Let

$$\mathbb{M}_k(\Gamma;\mathbb{Q})^*[I] = \{\varphi \in \mathbb{M}_k(\Gamma;\mathbb{Q})^* : t\varphi = 0 \text{ all } t \in I\}.$$

Since f is a newform, one can show that $\mathbb{M}_k(\Gamma;\mathbb{Q})^*[I]$ has dimension d. Let $\theta_1, \ldots, \theta_d$ be a basis for $\mathbb{M}_k(\Gamma;\mathbb{Q})^*[I]$, so

$$\mathrm{Ker}(\Phi_f) = \mathrm{Ker}(\theta_1) \oplus \cdots \oplus \mathrm{Ker}(\theta_d).$$

We can thus compute $\mathrm{Ker}(\Phi_f)$, hence a choice of Θ_f. To compute Φ_f, it remains to compute i_f.

Let $S_k(\Gamma;\mathbb{Q})$ denote the space of cusp forms with q-expansion in $\mathbb{Q}[[q]]$. By Exercise 10.3

$$S_k(\Gamma;\mathbb{Q})[I] = S_k(\Gamma)[I] \cap \mathbb{Q}[[q]]$$

is a $\mathbb{Q}$-vector space of dimension d. Let $g_1, \ldots, g_d$ be a basis for this $\mathbb{Q}$-vector space. We will compute Φ_f with respect to the basis of $\mathrm{Hom}_{\mathbb{Q}}(S_k(\Gamma;\mathbb{Q})[I];\mathbb{C})$ dual to this basis. Choose elements $x_1, \ldots, x_d \in \mathbb{M}_k(\Gamma)$ with the following properties:

(1) Using Proposition 10.5 or Proposition 10.8, it is possible to compute the period integrals $\langle g_i, x_j\rangle$, $i, j \in \{1, \ldots, d\}$, efficiently.

(2) The $2d$ elements $v + \eta(v)$ and $v - \eta(v)$ for $v = \Theta_f(x_1), \ldots, \Theta_f(x_d)$ span a space of dimension $2d$ (i.e., they span $\mathbb{M}_k(\Gamma)/\mathrm{Ker}(\Phi_f)$).

Given this data, we can compute

$$i_f(v + \eta(v)) = 2\mathrm{Re}(\langle g_1, x_i\rangle, \ldots, \langle g_d, x_i\rangle)$$

and

$$i_f(v - \eta(v)) = 2i\mathrm{Im}(\langle g_1, x_i\rangle, \ldots, \langle g_d, x_i\rangle).$$

We break the integrals into real and imaginary parts because this increases the precision of our answers. Since the vectors $v_n + \eta(v_n)$ and $v_n - \eta(v_n)$, $n = 1, \ldots, d$, span $\mathbb{M}_k(N,\varepsilon;\mathbb{Q})/\mathrm{Ker}(\Phi_f)$, we have computed i_f.

Remark 10.10. We want to find symbols x_i satisfying the conditions of Proposition 10.8. This is usually possible when d is very small, but in practice it is difficult when d is large.

Remark 10.11. The above strategy was motivated by [**Cre97a**, §2.10].

10.7. All Elliptic Curves of Given Conductor

Using modular symbols and the period map, we can compute all elliptic curves over $\mathbb{Q}$ of conductor N, up to isogeny. The algorithm in this section gives all *modular elliptic curves* (up to isogeny), i.e., elliptic curves attached to modular forms, of conductor N. Fortunately, it is now known by [**Wil95, BCDT01, TW95**] that every elliptic curve over $\mathbb{Q}$ is modular, so the procedure of this section gives all elliptic curves (up to isogeny) of given conductor. See [**Cre06**] for a nice historical discussion of this problem.

Algorithm 10.12 (Elliptic Curves of Conductor N). *Given $N > 0$, this algorithm outputs equations for all elliptic curves of conductor N, up to isogeny.*

(1) [Modular Symbols] Compute $\mathbb{M}_2(\Gamma_0(N))$ using Section 8.7.
(2) [Find Rational Eigenspaces] Find the 2-dimensional eigenspaces V in $\mathbb{M}_2(\Gamma_0(N))_{\text{new}}$ that correspond to elliptic curves. Do *not* use the algorithm for decomposition from Section 7.5, which is too complicated and gives more information than we need. Instead, for the first few primes $p \nmid N$, compute all eigenspaces $\text{Ker}(T_p - a)$, where a runs through integers with $-2\sqrt{p} < a < 2\sqrt{p}$. Intersect these eigenspaces to find the eigenspaces that correspond to elliptic curves. To find just the new ones, either compute the degeneracy maps to lower level or find all the rational eigenspaces of all levels that strictly divide N and exclude them.
(3) [Find Newforms] Use Algorithm 9.14 to compute to some precision each newform $f = \sum_{n=1}^{\infty} a_n q^n \in \mathbb{Z}[[q]]$ associated to each eigenspace V found in step (2).
(4) [Find Each Curve] For each newform f found in step (3), do the following:
    (a) [Period Lattice] Compute the corresponding period lattice $\Lambda = \mathbb{Z}\omega_1 + \mathbb{Z}\omega_2$ by computing the image of Φ_f, as described in Section 10.6.
    (b) [Compute τ] Let $\tau = \omega_1/\omega_2$. If $\text{Im}(\tau) < 0$, swap ω_1 and ω_2, so $\text{Im}(\tau) > 0$. By successively applying generators of $\text{SL}_2(\mathbb{Z})$, we find an $\text{SL}_2(\mathbb{Z})$ equivalent element τ' in $\mathcal{F}$, i.e., $|\text{Re}(\tau')| \leq 1/2$ and $|\tau| \geq 1$.

(c) [c-invariants] Compute the invariants c_4 and c_6 of the lattice Λ using the following rapidly convergent series:

$$c_4 = \left(\frac{2\pi}{\omega_2}\right)^4 \cdot \left(1 + 240 \sum_{n=1}^{\infty} \frac{n^3 q^n}{1-q^n}\right),$$

$$c_6 = \left(\frac{2\pi}{\omega_2}\right)^6 \cdot \left(1 - 504 \sum_{n=1}^{\infty} \frac{n^5 q^n}{1-q^n}\right),$$

where $q = e^{2\pi i \tau'}$, where τ' is as in step (4b). A theorem of Edixhoven (that the Manin constant is an integer) implies that the invariants c_4 and c_6 of Λ are integers, so it is only necessary to compute Λ to large precision to completely determine them.

(d) [Elliptic Curve] An elliptic curve with invariants c_4 and c_6 is

$$E: \quad y^2 = x^3 - \frac{c_4}{48}x - \frac{c_6}{864}.$$

(e) [Prove Correctness] Using Tate's algorithm, find the conductor of E. If the conductor is not N, then recompute c_4 and c_6 using more terms of f and real numbers to larger precision, etc. If the conductor is N, compute the coefficients b_p of the modular form $g = g_E$ attached to the elliptic curve E, for $p \leq \#\mathbb{P}^1(\mathbb{Z}/N\mathbb{Z})/6$. Verify that $a_p = b_p$, where a_p are the coefficients of f. If this equality holds, then E must be isogenous to the elliptic curve attached to f, by the Sturm bound (Theorem 9.18) and Faltings's isogeny theorem. If the equality fails for some p, recompute c_4 and c_6 to larger precision.

There are numerous tricks to optimize the above algorithm. For example, often one can work separately with $\mathbb{M}_k(\Gamma_0(N))^+_{\text{new}}$ and $\mathbb{M}_k(\Gamma_0(N))^-_{\text{new}}$ and get enough information to find E, up to isogeny (see [**Cre97b**]).

Once we have one curve from each isogeny class of curves of conductor N, we find each curve in each isogeny class (which is another interesting problem discussed in [**Cre97a**]), hence all curves of conductor N. If $E/\mathbb{Q}$ is an elliptic curve, then any curve isogenous to E is isogenous via a chain of isogenies of prime degree. There is an *a priori* bound on the degrees of these isogenies due to Mazur. Also, there are various methods for finding all isogenies of a given degree with domain E. See [**Cre97a**, §3.8] for more details.

10.7.1. Finding Curves: S-Integral Points. In this section we briefly survey an alternative approach to finding curves of a given conductor by finding integral points on other elliptic curves.

Cremona and others have developed a complementary approach to the problem of computing all elliptic curves of given conductor (see [**CL04**]).

Instead of computing all curves of given conductor, we instead consider the seemingly more difficult problem of finding all curves with good reduction outside a finite set S of primes. Since one can compute the conductor of a curve using Tate's algorithm [**Tat75, Cre97a**, §3.2], if we know all curves with good reduction outside S, we can find all curves of conductor N by letting S be the set of prime divisors of N.

There is a strategy for finding all curves with good reduction outside S. It is not an algorithm, in the sense that it is always guaranteed to terminate (the modular symbols method above *is* an algorithm), but in practice it often works. Also, this strategy makes sense over any number field, whereas the modular symbols method does not (there are generalizations of modular symbols to other number fields).

Fix a finite set S of primes of a number field K. It is a theorem of Shafarevich that there are only finitely many elliptic curves with good reduction outside S (see [**Sil92**, Section IX.6]). His proof uses that the group of S-units in K is finite and Siegel's theorem that there are only finitely many S-integral points on an elliptic curve. One can make all this explicit, and sometimes in practice one can compute all these S-integral points.

The problem of finding all elliptic curves with good reduction outside of S can be broken into several subproblems, the main ones being

(1) determine the following finite subgroup of $K^*/(K^*)^m$:

$$K(S,m) = \{x \in K^*/(K^*)^m \ : \ m \mid \mathrm{ord}_{\mathfrak{p}}(x) \text{ all } \mathfrak{p} \notin S\};$$

(2) find all S-integral points on certain elliptic curves $y^2 = x^3 + k$.

In [**CL04**], there is one example, where they find all curves of conductor $N = 2^8 \cdot 17^2 = 73984$ by finding all curves with good reduction outside $\{2, 17\}$. They finds 32 curves of conductor 73984 that divide into 16 isogeny classes. (Note that $\dim S_2(\Gamma_0(N)) = 9577$.)

10.7.2. Finding Curves: Enumeration. One can also find curves by simply enumerating Weierstrass equations. For example, the paper [**SW02**] discusses a database that the author and Watkins created that contains hundreds of millions of elliptic curves. It was constructed by enumerating Weierstrass equations of a certain form. This database does not contain *every* curve of each conductor included in the database. It is, however, fairly complete in some cases. For example, using the Mestre method of graphs [**Mes86**], we verified in [**JBS03**] that the database contains all elliptic curve of prime conductor < 234446, which implies that the smallest conductor rank 4 curve is composite.

10.8. Exercises

10.1 Prove Lemma 10.4.

10.2 Suppose $f \in S_2(\Gamma_0(N))$ is a newform and that $W_N(f) = wf$. Let $\Lambda(E, s) = N^{s/2}(2\pi)^{-s}\Gamma(s)L(E, s)$. Prove that

$$\Lambda(E, s) = -w\Lambda(E, 2 - s).$$

[Hint: Show that $\Lambda(f, s) = \int_{0,\infty} f(iy/\sqrt{N})y^{s-1}\, dy$. Then substitute $1/y$ for y.]

10.3 Let $f = \sum a_n q^n \in \mathbb{C}[[q]]$ be a power series whose coefficients a_n together generate a number field K of degree d over $\mathbb{Q}$. Let V_f be the complex vector space spanned by the $\mathrm{Gal}(\overline{\mathbb{Q}}/\mathbb{Q})$-conjugates of f.

(a) Give an example to show that V_f need not have dimension d.

(b) Suppose V_f has dimension d. Prove that $V_f \cap \mathbb{Q}[[q]]$ is a $\mathbb{Q}$-vector space of dimension d.

10.4 Find an elliptic curve of conductor 11 using Section 10.7.

Chapter 11

Solutions to Selected Exercises

11.1. Chapter 1

(1) Exercise 1.1. Suppose $\gamma = \left(\begin{smallmatrix} a & nb \\ c & d \end{smallmatrix}\right) \in \mathrm{GL}_2(\mathbb{R})$ is a matrix with positive determinant. Then γ is a linear fractional transformation that fixes the real line, so it must either fix or swap the upper and lower half planes. Now

$$\gamma(i) = \frac{ai+b}{ci+d} = \frac{ac+bd+(ad-bc)i}{d^2+c^2},$$

so since $\det\gamma = ad-bc > 0$, the imaginary part of $\gamma(i)$ is positive; hence γ sends the upper half plane to itself.

(2) Exercise 1.2. Avoiding poles, the quotient rule for differentiation goes through exactly as in the real case, so any rational function $f(z) = p(z)/q(z)$ ($p, q \in \mathbb{C}[z]$) is holomorphic on $\mathbb{C} - \{\alpha : q(\alpha) = 0\}$. By the fundamental theorem of algebra, this set of poles is finite, and hence it is discrete. Write $q(z) = a_n(z-\alpha_1)^{r_1}\cdots(z-\alpha_k)^{r_k}$ for each α_i and let $q_i(z) = q(z)/(z-\alpha_i)^{r_i}$ which is a polynomial nonzero at α_i. Thus for each i we have $(z-\alpha_i)^{r_i}f(z) = p(z)/q'(z)$ is holomorphic at α_i and hence $f(z)$ is meromorphic on $\mathbb{C}$.

(3) Exercise 1.3.

(a) The product fg of two meromorphic functions on the upper half plane is itself meromorphic. Also, for all $\gamma \in \mathrm{SL}_2(\mathbb{Z})$ we have

$$\begin{aligned}(fg)^{[\gamma]_{k+j}} &= \frac{1}{(cz+d)^{k+j}}((fg)\circ\gamma)\\ &= \frac{1}{(cz+d)^k}(f\circ\gamma)\frac{1}{(cz+d)^j}(g\circ\gamma) = fg,\end{aligned}$$

so fg is weakly modular.

(b) If f is meromorphic on the upper half plane, then so is $1/f$. Now

$$\frac{1}{f} = \frac{1}{(cz+d)^{-k} f\circ\gamma} = (cz+d)^k((1/f)\circ\gamma) = \frac{1}{f}^{[\gamma]_{-k}},$$

so $1/f$ is a weakly modular form of weight $-k$.

(c) Let f and g be modular functions. Then, as above, fg is a weakly modular function. Let $\sum_{n=m}^{\infty} a_n q^n$ and $\sum_{n=m'}^{\infty} b_n q^n$ be their q-expansions around any $\alpha \in \mathbb{P}^1(\mathbb{Q})$; then their formal product is the q-expansion of fg. But the formal product of two Laurant series about the same point is itself a Laurant series with convergence in the intersection of the convergent domains of the original series, so fg has a meromorphic q-expansion at each $\alpha \in \mathbb{P}^1(\mathbb{Q})$ and hence at each cusp.

(d) We are in exactly the same case as in part (c), but because f and g are modular functions, $m, m' \geq 0$ and hence the function is holomorphic at each of its cusps.

(4) Exercise 1.4. Let f be a weakly modular function of odd weight k. Since $\gamma = \left(\begin{smallmatrix}-1 & 0\\ 0 & -1\end{smallmatrix}\right) \in \mathrm{SL}_2(\mathbb{Z})$, we have $f(z) = (-1)^{-k} f(\gamma(z)) = -f(z)$ so $f = 0$.

(5) Exercise 1.5. Because $\mathrm{SL}_2(\mathbb{Z}/1\mathbb{Z})$ is the trivial group, $\Gamma(1) = \ker(\mathrm{SL}_2(\mathbb{Z}) \to \mathrm{SL}_2(\mathbb{Z}/1\mathbb{Z}))$ must be all of $\mathrm{SL}_2(\mathbb{Z})$. As $\mathrm{SL}_2(\mathbb{Z}) = \Gamma(1) \subset \Gamma_1(1) \subset \Gamma_0(1) \subset \mathrm{SL}_2(\mathbb{Z})$, we must have $\Gamma(1) = \Gamma_1(1) = \Gamma_0(1) = \mathrm{SL}_2(\mathbb{Z})$.

(6) Exercise 1.6.

(a) The group $\Gamma_1(N)$ is the inverse image of the subgroup of $\mathrm{SL}_2(\mathbb{Z}/N\mathbb{Z})$ generated by $\left(\begin{smallmatrix}1 & 1\\ 0 & 1\end{smallmatrix}\right)$, and the inverse image of a group (under a group homomorphism) is a group.

(b) The group contains the kernel of the homomorphism $\mathrm{SL}_2(\mathbb{Z}) \to \mathrm{SL}_2(\mathbb{Z}/N\mathbb{Z})$, and that kernel has finite index since the quotient is contained in $\mathrm{SL}_2(\mathbb{Z}/N\mathbb{Z})$, which is finite.

(c) Same argument as previous part.

(d) The level is at most N since both groups contain $\Gamma(N)$. It can be no greater than N since $\left(\begin{smallmatrix}1 & N\\ 0 & 1\end{smallmatrix}\right)$ is in both groups.

(7) Exercise 1.7. See [**DS05**, Lemma 1.2.2].

(8) Exercise 1.8. Let $\alpha = p/q \in \mathbb{Q}$, where p and q are relatively prime. By the Euclidian algorithm, we can find $x, y \in \mathbb{Z}$ such that $px+qy = 1$. Let $\gamma_\alpha = \left(\begin{smallmatrix} p & -y \\ q & x \end{smallmatrix}\right)$. Note that $\gamma_\alpha \in \mathrm{SL}_2(\mathbb{Z})$ and $\gamma_\alpha(\infty) = \alpha$. Also let γ_∞ be the identity map on $\mathbb{P}^1(\mathbb{Q})$. Now γ_β^{-1} sends β to ∞ so we have $\gamma_\alpha \circ \gamma_\beta^{-1}$ which sends α to β.

11.2. Chapter 2

(1) Exercise 2.1. We have

$$\zeta(26) = \frac{1315862 \cdot \pi^{26}}{11094481976030578125}.$$

Variation: Compute $\zeta(28)$.

(2) Exercise 2.2. Omitted.

(3) Exercise 2.3.

$$\begin{aligned} E_8 &= -\frac{B_8}{16} + q + \sum_{n=2}^{\infty} \sigma_7(n) q^n \\ &= \frac{1}{480} + q + 129q^2 + 2188q^3 + \cdots . \end{aligned}$$

Variation: Compute E_{10}.

(4) Exercise 2.4. Omitted.

(5) Exercise 2.5. We have $d = \dim S_{28} = 2$. A choice of a, b with $4a + 6b \leq 14$ and $4a + 6b \equiv 4 \pmod{12}$ is $a = 1, b = 0$. A basis for S_{28} is then

$$\begin{aligned} g_1 &= \Delta F_6^{2(2-1)+0} F_4 = q - 792q^2 - 324q^3 + 67590208q^4 + \cdots , \\ g_2 &= \Delta^2 F_6^{2(2-2)+0} F_4 = q^2 + 192q^3 - 8280q^4 + \cdots . \end{aligned}$$

The Victor Miller basis is then

$$\begin{aligned} f_1 &= g_1 + 729g_2 = q + 151740q^3 + 61032448q^4 + \cdots , \\ f_2 &= g_2 = q^2 + 192q^3 - 8280q^4 + \cdots . \end{aligned}$$

Variation: Compute the Victor Miller basis for S_{30}.

(6) Exercise 2.6. From the previous exercise we have $f = \Delta^2 F_4$. Then

$$\begin{aligned} f = \Delta^2 F_4 &= \left(\frac{F_4^3 - F_6^2}{-1728}\right)^2 \cdot F_4 \\ &= \left(\frac{\left(-\frac{8}{B_4}E_4\right)^3 - \left(-\frac{12}{B_6}E_6\right)^2}{-1728}\right)^2 \cdot \left(-\frac{8}{B_4}E_4\right) \\ &= 5186160E_4E_6^4 - 564480000E_4^4E_6^2 + 15360000000E_4^7. \end{aligned}$$

(7) Exercise 2.7. No, it is not always integral. For example, for $k = 12$, the coefficient of q is $-2 \cdot 12/B_{12} = 65520/691 \notin \mathbb{Z}$. Variation: Find, with proof, the set of all k such that the normalized series F_k *is* integral (use that B_k is eventually very large compared to $2k$).

(8) Exercise 2.8. We compute the Victor Miller basis to precision great enough to determine T_2. This means we need up to $O(q^5)$.

$$\begin{aligned} f_0 &= 1 + 2611200q^3 + 19524758400q^4 + \cdots, \\ f_1 &= q + 50220q^3 + 87866368q^4 + \cdots, \\ f_2 &= q^2 + 432q^3 + 39960q^4 + \cdots. \end{aligned}$$

Then the matrix of T_2 on this basis is

$$\begin{pmatrix} 2147483649 & 0 & 19524758400 \\ 0 & 0 & 2235350016 \\ 0 & 1 & 39960 \end{pmatrix}.$$

(The rows of this matrix are the linear combinations that give the images of the f_i under T_2.) This matrix has characteristic polynomial

$$(x - 2147483649) \cdot (x^2 - 39960x - 2235350016).$$

11.3. Chapter 3

(1) Exercise 3.1. Write $g = \left(\begin{smallmatrix} a & b \\ c & d \end{smallmatrix}\right)$, so $\lambda' = \frac{a\lambda+b}{c\lambda+d}$. Let f be the isomorphism $\mathbb{C}/\Lambda \to \mathbb{C}/\Lambda'$ given by $f(z) = z/(c\lambda + d)$. We have

$$f\left(\frac{1}{N}\right) = \frac{1}{N(c\lambda+d)} = \frac{a}{N} - \frac{c}{N} \cdot \frac{a\lambda+b}{c\lambda+d} \cong \frac{a}{N} \pmod{\mathbb{Z} + \mathbb{Z}\lambda'},$$

where the second equality can be verified easily by expanding out each side, and for the congruence we use that $N \mid c$. Thus the subgroup of $\mathbb{C}/\Lambda$ generated by $\frac{1}{N}$ is taken isomorphically to the subgroup of $\mathbb{C}/\Lambda'$ generated by $\frac{1}{N}$.

(2) Exercise 3.2. For any integer r, we have $\left(\begin{smallmatrix}1 & r\\ 0 & 1\end{smallmatrix}\right) \in \Gamma_0(N)$, so $\{0,\infty\} = \{r,\infty\}$. Thus

$$0 = \{0,\infty\} - \{0,\infty\} = \{n,\infty\} - \{m,\infty\} = \{n,\infty\} + \{\infty,m\} = \{n,m\}.$$

(3) Exercise 3.3.
   (a) $(0:1),(1:0),(1:1),\ldots,(1,p-1)$.
   (b) $p+1$.
   (c) See [**Cre97a**, Prop. 2.2.1].

(4) Exercise 3.4. We start with $b=4$, $a=7$. Then $4\cdot 2 \equiv 1 \pmod 7$. Let $\delta_1 = \left(\begin{smallmatrix}4 & 1\\ 7 & 2\end{smallmatrix}\right) \in \mathrm{SL}_2(\mathbb{Z})$. Since $\delta_1 \in \Gamma_0(7)$, we use the right coset representative $\left(\begin{smallmatrix}1 & 0\\ 0 & 1\end{smallmatrix}\right)$ and see that

$$\{0,4/7\} = \{0,1/2\} + \left(\begin{smallmatrix}1 & 0\\ 0 & 1\end{smallmatrix}\right)\{0,\infty\}.$$

Repeating the process, we have $\delta_2 = \left(\begin{smallmatrix}1 & 1\\ 2 & 0\end{smallmatrix}\right)$, which is in the same coset at $\left(\begin{smallmatrix}0 & -1\\ 1 & 0\end{smallmatrix}\right)$. Thus

$$\{0,1/2\} = \left(\begin{smallmatrix}0 & 6\\ 1 & 0\end{smallmatrix}\right)\{0,\infty\} + \{0,0\}.$$

Putting it together gives

$$\{0,4/7\} = \left(\begin{smallmatrix}1 & 0\\ 0 & 1\end{smallmatrix}\right)\{0,\infty\} + \left(\begin{smallmatrix}0 & 6\\ 1 & 0\end{smallmatrix}\right)\{0,\infty\} = [(0,1)] + [(1,0)].$$

(5) Exercise 3.5.
   (a) Coset representatives for $\Gamma_0(3)$ in $\mathrm{SL}_2(\mathbb{Z})$ are

$$\begin{pmatrix}1 & 0\\ 0 & 1\end{pmatrix},\quad \begin{pmatrix}1 & 0\\ 1 & 1\end{pmatrix},\quad \begin{pmatrix}1 & 0\\ 2 & 1\end{pmatrix},\quad \begin{pmatrix}0 & -1\\ 1 & 0\end{pmatrix},$$

   which we refer to below as $[r_0],[r_1],[r_2]$, and $[r_3]$, respectively.
   (b) In terms of representatives we have

$$\begin{aligned}
&[r_0]+[r_3]=0, && [r_0]+[r_3]+[r_2]=0,\\
&[r_1]+[r_2]=0, && [r_1]+[r_1]+[r_1]=0,\\
&[r_2]+[r_1]=0, && [r_2]+[r_0]+[r_3]=0,\\
&[r_3]+[r_0]=0, && [r_3]+[r_2]+[r_0]=0.
\end{aligned}$$

   (c) By the first three relations we have $[r_2]=[r_1]=0=0[r_0]$ and $[r_3]=-1[r_0]$.
   (d)

$$\begin{aligned}
T_2([r_0]) &= [r_0]\left(\begin{smallmatrix}1 & 0\\ 0 & 2\end{smallmatrix}\right) + [r_0]\left(\begin{smallmatrix}2 & 0\\ 0 & 1\end{smallmatrix}\right) + [r_0]\left(\begin{smallmatrix}2 & 1\\ 0 & 1\end{smallmatrix}\right) + [r_0]\left(\begin{smallmatrix}1 & 0\\ 1 & 2\end{smallmatrix}\right)\\
&= [\left(\begin{smallmatrix}1 & 0\\ 0 & 2\end{smallmatrix}\right)] + [\left(\begin{smallmatrix}2 & 0\\ 0 & 1\end{smallmatrix}\right)] + [\left(\begin{smallmatrix}2 & 1\\ 0 & 1\end{smallmatrix}\right)] + [\left(\begin{smallmatrix}1 & 0\\ 1 & 2\end{smallmatrix}\right)]\\
&= [r_0]+[r_0]+[r_0]+[r_2]\\
&= 3[r_0].
\end{aligned}$$

11.4. Chapter 4

(1) Exercise 4.1. Suppose f is a Dirichlet character with modulus N. Then $-1 = f(-1) = f(-1+N) = 1$, a contradiction.

(2) Exercise 4.2.

(a) Any finite subgroup of the multiplicative group of a field is cyclic (since the number of roots of a polynomial over a field is at most its degree), so $(\mathbb{Z}/p\mathbb{Z})^*$ is cyclic. Let g be an integer that reduces to a generator of $(\mathbb{Z}/p\mathbb{Z})^*$. Let $x = 1 + p \in (\mathbb{Z}/p^n\mathbb{Z})^*$; by the binomial theorem

$$x^{p^{n-2}} = 1 + p^{n-2} \cdot p + \cdots \equiv 1 + p^{n-1} \not\equiv 0 \pmod{p^n},$$

so x has order p^{n-1}. Since p is odd, $\gcd(p^{n-1}, p-1) = 1$, so xg has order $p^{n-1} \cdot (p-1) = \varphi(p^n)$; hence $(\mathbb{Z}/p^n\mathbb{Z})^*$ is cyclic.

(b) By the binomial theorem $(1+2^2)^{2^{n-3}} \not\equiv 1 \pmod{2^n}$, so 5 has order 2^{n-2} in $(\mathbb{Z}/2^n\mathbb{Z})^*$, and clearly -1 has order 2. Since $5 \equiv 1 \pmod 4$, -1 is not a power of 5 in $(\mathbb{Z}/2^n\mathbb{Z})^*$. Thus the subgroups $\langle -1 \rangle$ and $\langle 5 \rangle$ have trivial intersection. The product of their orders is $2^{n-1} = \varphi(2^n) = \#(\mathbb{Z}/2^n\mathbb{Z})^*$, so the claim follows.

(3) Exercise 4.3. Write $n = \prod p_i^{e_i}$. The order of g divides n, so the condition implies that $p_i^{e_i}$ divides the order of g for each i. Thus the order of g is divisible by the least common multiple of the $p_i^{e_i}$, i.e., by n.

(4) Exercise 4.4.

(a) The bijection given by $1 + p^{n-1}a \pmod{p^n} \mapsto a \pmod p$ is a homomorphism since

$$(1+p^{n-1}a)(1+p^{n-1}b) \equiv 1 + p^{n-1}(a+b) \pmod{p^n}.$$

(b) We have an exact sequence

$$1 \to 1 + p\mathbb{Z}/p^n\mathbb{Z} \to (\mathbb{Z}/p^n\mathbb{Z})^* \to (\mathbb{Z}/p\mathbb{Z})^* \to 1,$$

so it suffices to solve the discrete log problem in the kernel and cokernel. We prove by induction on n that we can solve the discrete log problem in the kernel easily (compared to known methods for solving the discrete log problem in $(\mathbb{Z}/p\mathbb{Z})^*$). We have an exact sequence

$$1 \to 1 + p^{n-1}\mathbb{Z}/p^n\mathbb{Z} \to (\mathbb{Z}/p^n\mathbb{Z})^* \to (\mathbb{Z}/p^{n-1}\mathbb{Z})^* \to 1.$$

The first part of this problem shows that we can solve the discrete log problem in the kernel, and by induction we can solve it in the cokernel. This completes the proof.

(5) Exercise 4.5. If $\varepsilon(5) = 1$, then since ε is nontrivial, Exercise 4.2 implies that ε factors through $(\mathbb{Z}/4\mathbb{Z})^*$, hence has conductor $4 = 2^{1+1}$, as claimed. If $\varepsilon(5) \neq 1$, then again from Exercise 4.2 we see that if ε has order r, then ε factors through $(\mathbb{Z}/2^{r+2}\mathbb{Z})^*$ but nothing smaller.

(6) Exercise 4.6.
   (a) Take $f = x^2 + 2$.
   (b) The element 2 has order 4.
   (c) A minimal generator for $(\mathbb{Z}/25\mathbb{Z})^*$ is 2, and the characters are [1], [2], [3], [4].
   (d) Each of the four Galois orbits has size 1.

11.5. Chapter 5

(1) Exercise 5.1. The eigenspace E_λ of A with eigenvalue λ is preserved by B, since if $v \in E_\lambda$, then

$$ABv = BAv = B(\lambda v) = \lambda Bv.$$

Because B is diagonalizable, its minimal polynomial equals its characteristic polynomial; hence the same is true for the restriction of B to E_λ, i.e., the restriction of B is diagonalizable. Choose basis for all E_λ so that the restrictions of B to these eigenspaces is diagonal with respect to these bases. Then the concatentation of these bases is a basis that simultaneously diagonalizes A and B.

(2) Exercise 5.2. When ε is the trivial character, the $B_{k,\varepsilon}$ are defined by

$$\sum_{a=1}^{1} \frac{\varepsilon(a) x e^{ax}}{e^x - 1} = \frac{xe^x}{e^x - 1} = x + \frac{x}{e^x - 1} = \sum_{k=0}^{\infty} B_{k,\varepsilon} \frac{x^k}{k!}.$$

Thus $B_{1,\varepsilon} = 1 + B_1 = \frac{1}{2}$, and for $k > 1$, we have $B_{k,\varepsilon} = B_k$.

(3) Exercise 5.3. Omitted.

(4) Exercise 5.4. The Eisenstein series in our basis for $E_3(\Gamma_1(13))$ are of the form $E_{3,1,\varepsilon}$ or $E_{3,\varepsilon,1}$ with $\varepsilon(-1) = (-1)^3 = -1$. There are six characters ε with modulus 13 such that $\varepsilon(-1) = -1$, and we have the two series $E_{3,1,\varepsilon}$ and $E_{3,\varepsilon,1}$ associated to each of these. This gives a dimension of 12.

11.6. Chapter 6

(1) Exercise 6.1.

(a) By Proposition 3.10, we have $[\mathrm{SL}_2(\mathbb{Z}) : \Gamma_0(N)] = \#\mathbb{P}^1(\mathbb{Z}/N\mathbb{Z})$. By the Chinese Remainder Theorem,

$$\#\mathbb{P}^1(\mathbb{Z}/N\mathbb{Z}) = \prod_{p|N} \#\mathbb{P}^1(\mathbb{Z}/p^{\mathrm{ord}_p(N)}\mathbb{Z}).$$

So we are reduced to computing $\#\mathbb{P}^1(\mathbb{Z}/p^{\mathrm{ord}_p(N)}\mathbb{Z})$. We have $(a,b) \in (\mathbb{Z}/p^n\mathbb{Z})^2$ with $\gcd(a,b,p) = 1$ if and only if $(a,b) \notin (p\mathbb{Z}/p^n\mathbb{Z})^2$, so there are $p^{2n} - p^{2(n-1)}$ such pairs. The unit group $(\mathbb{Z}/p^n\mathbb{Z})^*$ has order $\varphi(p^n) = p^n - p^{n-1}$. It follows that

$$\#\mathbb{P}^1(\mathbb{Z}/N\mathbb{Z}) = \frac{p^{2n} - p^{2(n-1)}}{p^n - p^{n-1}} = p^n + p^{n-1}.$$

(b) Omitted.

(2) Exercise 6.2. Omitted.

(3) Exercise 6.3. Omitted.

(4) Exercise 6.4. Omitted.

(5) Exercise 6.5. See the source code to SAGE.

11.7. Chapter 7

(1) Exercise 7.1. Take a basis of W and let G be the matrix whose rows are these basis elements. Let B be the row echelon form of G. After a permutation p of columns, we may write $B = p_i(I|C)$, where I is the identity matrix. The matrix $A = p^{-1}(-C^t|I)$, where I is a different sized identity matrix, has the property that $W = \mathrm{Ker}(A)$.

(2) Exercise 7.2. The answer is no. For example if $A = nI$ is n times the identity matrix and if $p \mid n$, then $\mathrm{rref}(A \pmod{p}) = 0$ but $\mathrm{rref}(A) \pmod{p} = I$.

(3) Exercise 7.3. Let $T = \prod E_i$ be an invertible matrix such that $TA = E$ is in (reduced) echelon form and the E_i are elementary matrices, i.e., the result of applying an elementary row operation to the identity matrix. If p is a prime that does not divide any of the nonzero numerators or denominators of the entries of A and any E_i, then $\mathrm{rref}(A \pmod{p}) = \mathrm{rref}(A) \pmod{p}$. This is because $E \pmod{p}$ is in echelon form and $A \pmod{p}$ can be transformed to $E \pmod{p}$ via a series of elementary row operations modulo p.

(4) Exercise 7.4.

(a) The echelon form (over $\mathbb{Q}$) is

$$\begin{pmatrix} 1 & 0 & -1 \\ 0 & 1 & 2 \\ 0 & 0 & 0 \end{pmatrix}.$$

   (b) The kernel is the 1-dimensional span of $(1, -2, 1)$.
   (c) The characteristic polynomial is $x \cdot (x^2 - 15x - 18)$.

(5) Exercise 7.5.
   (a) The answer is given in the problem.
   (b) See [**Coh93**, §2.4].

11.8. Chapter 8

(1) Exercise 8.1. Using the Chinese Remainder Theorem we immediately reduce to proving the statement when both $M = p^r$ and $N = p^s$ are powers of a prime p. Then $[a] \in (\mathbb{Z}/p^s\mathbb{Z})^*$ is represented by an integer a with $\gcd(a, p) = 1$. That same integer a defines an element of $(\mathbb{Z}/p^r\mathbb{Z})^*$ that reduces modulo p^s to $[a]$.

(2) Exercise 8.2. See [**Shi94**, Lemma 1.38].

(3) Exercise 8.3. Coset representitives for $\Gamma_1(3)$ are in bijection with (c, d) where $c, d \in \mathbb{Z}/3\mathbb{Z}$ and $\gcd(c, d, N) = 1$, so the following are representatives:

$$\left(\begin{smallmatrix}1&0\\0&1\end{smallmatrix}\right), \left(\begin{smallmatrix}2&0\\0&2\end{smallmatrix}\right), \left(\begin{smallmatrix}0&2\\1&0\end{smallmatrix}\right), \left(\begin{smallmatrix}1&0\\1&1\end{smallmatrix}\right), \left(\begin{smallmatrix}2&0\\1&2\end{smallmatrix}\right), \left(\begin{smallmatrix}1&1\\2&0\end{smallmatrix}\right), \left(\begin{smallmatrix}1&0\\2&1\end{smallmatrix}\right), \left(\begin{smallmatrix}2&0\\2&2\end{smallmatrix}\right),$$

which we call $r_1, \ldots, r_8$, respectively. Now our Manin symbols are of the form $[X, r_i]$ and $[Y, r_i]$ for $1 \le i \le 8$ modulo the relations

$$x + x\sigma = 0, \quad x + x\tau + x\tau^2 = 0, \quad \text{and } x - xJ = 0.$$

First, note that J acts trivially on Manin symbols of odd weight because it sends X to $-X$, Y to $-Y$ and r_i to $-r_i$, so

$$[z, g]J = [-z, -g] = [z, g].$$

Thus the last relation is trivially true.

Now $\sigma^{-1}X = -Y$ and $\sigma^{-1}Y = X$. Also $\tau^{-1}X = -Y, \tau^{-1}Y = X - Y, \tau^{-2}X = -X + Y$ and $\tau^{-2}Y = -X$.

The first relation on the first symbol says that

$$[X, r_1] = -[-Y, r_3] = [Y, r_3]$$

and the second relation tells us that

$$[X, r_1] + [-Y, r_5] + [-X + Y, r_6] = 0.$$

(4) Exercise 8.4. Let $f \in S_k(\Gamma)$ and $g \in \Gamma$. All that remains to be shown is that this pairing respects the relation $x = xg$ for all modular symbols x. By linearity it suffices to show the invariance

of $\langle f, X^{k-i-2}Y^i\{\alpha,\beta\}\rangle$. We have

$$\begin{aligned}
&\left\langle f, (X^{k-2-i}Y^i\{\alpha,\beta\})g^{-1}\right\rangle \\
&= \left\langle f, (aX+bY)^{k-i-2}(cX+dY)^i\{g^{-1}(\alpha), g^{-1}(\beta)\}\right\rangle \\
&= \int_{g^{-1}(\alpha)}^{g^{-1}(\beta)} f(z)(az+b)^{k-i-2}(cz+d)^i\, dz \\
&= \int_{g^{-1}(\alpha)}^{g^{-1}(\beta)} f(z)\frac{(az+b)^{k-i-2}}{(cz+d)^{k-i-2}}(cz+d)^{k-2}\, dz \\
&= \int_{g^{-1}(\alpha)}^{g^{-1}(\beta)} f(z)\, g(z)^{k-i-2}(cz+d)^{k-2}\, dz \\
&= \int_{\alpha}^{\beta} f(g^{-1}(z))\, g(g^{-1}(z))^{k-i-2}(cg^{-1}(z)+d)^{k-2}\, d(g^{-1}(z)) \\
&= \int_{\alpha}^{\beta} f(g^{-1}(z))\, z^{k-i-2}(cg^{-1}(z)+d)^{k-2}\, (cg^{-1}(z)+d)^{2}\, dz \\
&= \int_{\alpha}^{\beta} f(z)\, z^{k-i-2}\, dz \\
&= \left\langle f, X^{k-i-2}Y^i\{\alpha,\beta\}\right\rangle,
\end{aligned}$$

where the second to last simplification is due to invariance under $[g]_k$, i.e.,

$$f(g^{-1}(z)) = f^{[g]_k}(g^{-1}(z)) = (cg^{-1}(z)+d)^{-k} f(g(g^{-1}(z))).$$

(The proof for $f \in \overline{S}_k(\Gamma)$ works in exactly the same way.)

(5) Exercise 8.5.

(a) Let $\eta = \left(\begin{smallmatrix} -1 & 0 \\ 0 & 1 \end{smallmatrix}\right)$. For any $\gamma = \left(\begin{smallmatrix} a & b \\ c & d \end{smallmatrix}\right)$ we have

$$\gamma\eta = \left(\begin{smallmatrix} -a & b \\ -c & d \end{smallmatrix}\right),\ \eta\gamma = \left(\begin{smallmatrix} -a & -b \\ c & d \end{smallmatrix}\right),\ \text{and } \eta\gamma\eta = \left(\begin{smallmatrix} a & -b \\ -c & d \end{smallmatrix}\right).$$

First, if $\gamma \in SL_2(\mathbb{Z})$, then $\eta\gamma\eta \in GL_2(\mathbb{Z})$ and

$$\det(\eta\gamma\eta) = \det\eta \det\gamma \det\eta = (-1)(1)(-1) = 1$$

so $\eta\gamma\eta \in SL_2(\mathbb{Z})$. As $\eta^2 = 1$, conjugation by η is self-inverse, so it must be a bijection.
Now if $\gamma \in \Gamma_0(N)$, then $c \equiv 0 \pmod N$, so $-c \equiv 0 \pmod N$, and so $\eta\gamma\eta \in \Gamma_0(N)$. Thus $\eta\Gamma_0(N)\eta = \Gamma_0(N)$.
If $\gamma \in \Gamma_1(N)$, then $-c \equiv 0 \pmod N$ as before and also $a \equiv d \equiv 1 \pmod N$, so $\eta\gamma\eta \in \Gamma_1(N)$. Thus $\eta\Gamma_1(N)\eta = \Gamma_1(N)$.

(b) Omitted.

11.9. Chapter 9

(1) Exercise 9.1. Consider the surjective homomorphism

$$r : \mathrm{SL}_2(\mathbb{Z}) \to \mathrm{SL}_2(\mathbb{Z}/N\mathbb{Z}).$$

Notice that $\Gamma_1(N)$ is the exact inverse image of the subgroup H of matrices of the form $\left(\begin{smallmatrix} 1 & * \\ 0 & 1 \end{smallmatrix}\right)$ and $\Gamma_0(N)$ is the inverse image of the subroup T of upper triangular matrices. It thus suffices to observe that H is normal in T, which is clear. Finally, the quotient T/H is isomorphic to the group of diagonal matrices in $\mathrm{SL}_2(\mathbb{Z}/N\mathbb{Z})^*$, which is isomorphic to $(\mathbb{Z}/N\mathbb{Z})^*$.

(2) Exercise 9.2. It is enough to show $\langle p\rangle \in \mathbb{Z}[\ldots, T_n, \ldots]$ for primes p, since each $\langle d\rangle$ can be written in terms of the $\langle p\rangle$. Since $p \nmid N$, we have that

$$T_{p^2} = T_p^2 - \langle p\rangle p^{k-1},$$

so

$$\langle p\rangle p^{k-1} = T_p^2 - T_{p^2}.$$

By Dirichlet's theorem on primes in arithmetic progression, there is a prime $q \neq p$ congruent to p mod N. Since p^{k-1} and q^{k-1} are relatively prime, there exist integers a and b such that $ap^{k-1} + bq^{k-1} = 1$. Then

$$\langle p\rangle = \langle p\rangle(ap^{k-1} + bq^{k-1}) = a({T_p}^2 - T_{p^2}) + b({T_q}^2 - T_{q^2}) \in \mathbb{Z}[\ldots, T_n, \ldots].$$

(3) Exercise 9.3. Take $N = 33$. The space $S_2(\Gamma_0(33))$ is a direct sum of the two old subspaces coming from $S_2(\Gamma_0(11))$ and the new subspace, which has dimension 1. If f is a basis for $S_2(\Gamma_0(11))$ and g is a basis for $S_2(\Gamma_0(33))_{\mathrm{new}}$, then $\alpha_1(f), \alpha_3(f), g$ is a basis for $S_2(\Gamma_0(33))$ on which all Hecke operators T_n, with $\gcd(n, 33) = 1$, have diagonal matrix. However, the operator T_3 on $S_2(\Gamma_0(33))$ does not act as a scalar on $\alpha_1(f)$, so it cannot be in the ring generated by all operators T_n with $\gcd(n, 33) = 1$.

(4) Exercise 9.4. Omitted.

11.10. Chapter 10

(1) Exercise 10.1. Hint: Use either repeated integration by parts or a change of variables that relates the integral to the Γ function.

(2) Exercise 10.2. See [**Cre97a**, §2.8].

(3) Exercise 10.3.
   (a) Let $f = \sqrt{-1}\sum q^n$. Then $d = 2$, but the nontrivial conjugate of f is $-f$, so V_f has dimension 1.

(b) Choose $\alpha \in K$ such that $K = \mathbb{Q}(\alpha)$. Write

$$f = \sum_{i=0}^{d-1} \alpha^i g_i \tag{11.10.1}$$

with $g_i \in \mathbb{Q}[[q]]$. Let W_g be the $\mathbb{Q}$-span of the g_i, and let $W_f = V_f \cap \mathbb{Q}[[q]]$. By considering the $\mathrm{Gal}(\overline{\mathbb{Q}}/\mathbb{Q})$ conjugates of (11.10.1), we see that the Galois conjugates of f are in the $\mathbb{C}$-span of the g_i, so

$$d = \dim_{\mathbb{C}} V_f \leq \dim_{\mathbb{Q}} W_g. \tag{11.10.2}$$

Likewise, taking the above modulo $O(q^n)$ for any n, we obtain a matrix equation

$$F = AG,$$

where the columns of F are the $\mathrm{Gal}(\overline{\mathbb{Q}}/\mathbb{Q})$-conjugates of f, the matrix A is the Vandermonde matrix corresponding to α (and its $\mathrm{Gal}(\overline{\mathbb{Q}}/\mathbb{Q})$ conjugates), and G has columns g_i. Since A is a Vandermonde matrix, it is invertible, so $A^{-1}F = G$. Taking the limit as n goes to infinity, we see that each g_i is a linear combination of the f_i, hence an element of V_f. Thus $W_g \subset W_f$, so (11.10.2) implies that $\dim_{\mathbb{Q}} W_f \geq d$. But $W_f \otimes_{\mathbb{Q}} \mathbb{C} \subset V_f$ so finally

$$d \leq \dim_{\mathbb{Q}} W_f = \dim_{\mathbb{C}}(W_f \otimes_{\mathbb{Q}} \mathbb{C}) \leq \dim_{\mathbb{C}} V_f = d.$$

(4) Exercise 10.4. See the appendix to Chapter II in [**Cre97a**], where this example is worked out in complete detail.

Appendix A

Computing in Higher Rank

by Paul E. Gunnells

A.1. Introduction

This book has addressed the theoretical and practical problems of performing computations with modular forms. Modular forms are the simplest examples of the general theory of automorphic forms attached to a reductive algebraic group G with an arithmetic subgroup Γ; they are the case $G = \mathrm{SL}_2(\mathbb{R})$ with Γ a congruence subgroup of $\mathrm{SL}_2(\mathbb{Z})$. For such pairs (G, Γ) the Langlands philosophy asserts that there should be deep connections between automorphic forms and arithmetic, connections that are revealed through the action of the Hecke operators on spaces of automorphic forms. There have been many profound advances in recent years in our understanding of these phenomena, for example:

- the establishment of the modularity of elliptic curves defined over $\mathbb{Q}$ [**Wil95, TW95, Dia96, CDT99, BCDT01**],
- the proof by Harris–Taylor of the local Langlands correspondence [**HT01**], and
- Lafforgue's proof of the global Langlands correspondence for function fields [**Laf02**].

Nevertheless, we are still far from seeing that the links between automorphic forms and arithmetic hold in the broad scope in which they are generally believed. Hence one has the natural problem of studying spaces of automorphic forms computationally.

The goal of this appendix is to describe some computational techniques for automorphic forms. We focus on the case $G = \mathrm{SL}_n(\mathbb{R})$ and $\Gamma \subset \mathrm{SL}_n(\mathbb{Z})$, since the automorphic forms that arise are one natural generalization of modular forms, and since this is the setting for which we have the most tools available. In fact, we do not work directly with automorphic forms, but rather with the cohomology of the arithmetic group Γ with certain coefficient modules. This is the most natural generalization of the tools developed in previous chapters.

Here is a brief overview of the contents. Section A.2 gives background on automorphic forms and the cohomology of arithmetic groups and explains why the two are related. In Section A.3 we describe the basic topological tools used to compute the cohomology of Γ explicitly. Section A.4 defines the Hecke operators, describes the generalization of the modular symbols from Chapter 8 to higher rank, and explains how to compute the action of the Hecke operators on the top degree cohomology group. Section A.5 discusses computation of the Hecke action on cohomology groups below the top degree. Finally, Section A.6 briefly discusses some related material and presents some open problems.

A.1.1. The theory of automorphic forms is notorious for the difficulty of its prerequisites. Even if one is only interested in the cohomology of arithmetic groups—a small part of the full theory—one needs considerable background in algebraic groups, algebraic topology, and representation theory. This is somewhat reflected in our presentation, which falls far short of being self-contained. Indeed, a complete account would require a long book of its own. We have chosen to sketch the foundational material and to provide many pointers to the literature; good general references are [**BW00, Harb, LS90, Vog97**]. We hope that the energetic reader will follow the references and fill many gaps on his/her own.

The choice of topics presented here is heavily influenced (as usual) by the author's interests and expertise. There are many computational topics in the cohomology of arithmetic groups we have completely omitted, including the trace formula in its many incarnations [**GP05**], the explicit Jacquet–Langlands correspondence [**Dem04, SW05**], and moduli space techniques [**FvdG, vdG**]. We encourage the reader to investigate these extremely interesting and useful techniques.

A.1.2. Acknowledgements. I thank Avner Ash, John Cremona, Mark McConnell, and Dan Yasaki for helpful comments. I also thank the NSF for support.

A.2. Automorphic Forms and Arithmetic Groups

A.2.1. Let $\Gamma = \Gamma_0(N) \subset \mathrm{SL}_2(\mathbb{Z})$ be the usual Hecke congruence subgroup of matrices upper-triangular mod N. Let $Y_0(N)$ be the modular curve $\Gamma\backslash\mathfrak{h}$, and let $X_0(N)$ be its canonical compactification obtained by adjoining cusps. For any integer $k \geq 2$, let $S_k(N)$ be the space of weight k holomorphic cuspidal modular forms on Γ. According to Eichler–Shimura [**Shi94**, Chapter 8], we have the isomorphism

$$H^1(X_0(N); \mathbb{C}) \xrightarrow{\sim} S_2(N) \oplus \overline{S_2(N)}, \tag{A.2.1}$$

where the bar denotes complex conjugation and where the isomorphism is one of Hecke modules.

More generally, for any integer $n \geq 0$, let $M_n \subset \mathbb{C}[x, y]$ be the subspace of degree n homogeneous polynomials. The space M_n admits a representation of Γ by the "change of variables" map

$$\begin{pmatrix} a & b \\ c & d \end{pmatrix} \cdot p(x, y) = p(dx - by, -cx + ay). \tag{A.2.2}$$

This induces a local system $\widetilde{M}_n$ on the curve $X_0(N)$.[1] Then the analogue of (A.2.1) for higher-weight modular forms is the isomorphism

$$H^1(X_0(N); \widetilde{M}_{k-2}) \xrightarrow{\sim} S_k(N) \oplus \overline{S_k(N)}. \tag{A.2.3}$$

Note that (A.2.3) reduces to (A.2.1) when $k = 2$.

Similar considerations apply if we work with the open curve $Y_0(N)$ instead, except that Eisenstein series also contribute to the cohomology. More precisely, let $E_k(N)$ be the space of weight k Eisenstein series on $\Gamma_0(N)$. Then (A.2.3) becomes

$$H^1(Y_0(N); \widetilde{M}_{k-2}) \xrightarrow{\sim} S_k(N) \oplus \overline{S_k(N)} \oplus E_k(N). \tag{A.2.4}$$

These isomorphisms lie at the heart of the modular symbols method.

A.2.2. The first step on the path to general automorphic forms is a reinterpretation of modular forms in terms of functions on $\mathrm{SL}_2(\mathbb{R})$. Let $\Gamma \subset \mathrm{SL}_2(\mathbb{Z})$ be a congruence subgroup. A weight k modular form on Γ is a holomorphic function $f\colon \mathfrak{h} \to \mathbb{C}$ satisfying the transformation property

$$f((az + b)/(cz + d)) = j(\gamma, z)^k f(z), \quad \gamma = \begin{pmatrix} a & b \\ c & d \end{pmatrix} \in \Gamma, \quad z \in \mathfrak{h}.$$

[1]The classic references for cohomology with local systems are [**Ste99a**, Section 31] and [**Eil47**, Ch. V]. A more recent exposition (in the language of Čech cohomology and locally constant sheaves) can be found in [**BT82**, II.13]. For an exposition tailored to our needs, see [**Harb**, Section 2.9].

Here $j(\gamma, z)$ is the *automorphy factor* $cz + d$. There are some additional conditions f must satisfy at the cusps of $\mathfrak{h}$, but these are not so important for our discussion.

The group $G = \mathrm{SL}_2(\mathbb{R})$ acts transitively on $\mathfrak{h}$, with the subgroup $K = \mathrm{SO}(2)$ fixing i. Thus $\mathfrak{h}$ can be written as the quotient G/K. From this, we see that f can be viewed as a function $G \to \mathbb{C}$ that is K*-invariant on the right* and that satisfies a certain symmetry condition with respect to the Γ*-action on the left.* Of course not every f with these properties is a modular form: some extra data is needed to take the role of holomorphicity and to handle the behavior at the cusps. Again, this can be ignored right now.

We can turn this interpretation around as follows. Suppose φ is a function $G \to \mathbb{C}$ that is Γ*-invariant on the left*, that is, $\varphi(\gamma g) = \varphi(g)$ for all $\gamma \in \Gamma$. Hence φ can be thought of as a function $\varphi\colon \Gamma\backslash G \to \mathbb{C}$. We further suppose that φ satisfies a certain symmetry condition with respect to the K*-action on the right.* In particular, any matrix $m \in K$ can be written

$$m = \begin{pmatrix} \cos\theta & -\sin\theta \\ \sin\theta & \cos\theta \end{pmatrix}, \quad \theta \in \mathbb{R}, \tag{A.2.5}$$

with θ uniquely determined modulo 2π. Let ζ_m be the complex number $e^{i\theta}$. Then the K-symmetry we require is

$$\varphi(gm) = \zeta_m^{-k}\varphi(z), \quad m \in K,$$

where k is some fixed nonnegative integer.

It turns out that such functions φ are very closely related to modular forms: *any* $f \in S_k(\Gamma)$ uniquely determines such a function $\varphi_f\colon \Gamma\backslash G \to \mathbb{C}$. The correspondence is very simple. Given a weight k modular form f, define

$$\varphi_f(g) := f(g \cdot i) j(g, i)^{-k}. \tag{A.2.6}$$

We claim φ_f is left Γ-invariant and satisfies the desired K-symmetry on the right. Indeed, since j satisfies the cocycle property

$$j(gh, z) = j(g, h \cdot z) j(h, z),$$

we have

$$\varphi_f(\gamma g) = f((\gamma g)\cdot i) j(\gamma g, i)^{-k} = j(\gamma, g\cdot i)^k f(g\cdot i) j(\gamma, g\cdot i)^{-k} j(g, i)^{-k} = \varphi_f(g).$$

Moreover, any $m \in K$ stabilizes i. Hence

$$\varphi_f(gm) = f((gm) \cdot i) j(gm, i)^{-k} = f(g \cdot i) j(m, i)^{-k} j(g, m \cdot i)^{-k}.$$

From (A.2.5) we have $j(m, i)^{-k} = (\cos\theta + i \sin\theta)^{-k} = \zeta_m^{-k}$, and thus $\varphi_f(gm) = \zeta_m^{-k}\varphi_f(g)$.

Hence in (A.2.6) the weight and the automorphy factor "untwist" the Γ-action to make φ_f left Γ-invariant. The upshot is that we can study modular forms by studying the spaces of functions that arise through the construction (A.2.6).

Of course, not every $\varphi\colon \Gamma\backslash G \to \mathbb{C}$ will arise as φ_f for some $f \in S_K(\Gamma)$: after all, f is holomorphic and satisfies rather stringent growth conditions. Pinning down all the requirements is somewhat technical and is (mostly) done in the sequel.

A.2.3. Before we define automorphic forms, we need to find the correct generalizations of our groups $\mathrm{SL}_2(\mathbb{R})$ and $\Gamma_0(N)$. The correct setup is rather technical, but this really reflects the power of the general theory, which handles so many different situations (e.g., Maass forms, Hilbert modular forms, Siegel modular forms, etc.).

Let G be a connected Lie group, and let $K \subset G$ be a maximal compact subgroup. We assume that G is the set of real points of a connected semisimple algebraic group $\mathbf{G}$ defined over $\mathbb{Q}$. These conditions mean the following [**PR94**, §2.1.1]:

(1) The group $\mathbf{G}$ has the structure of an affine algebraic variety given by an ideal I in the ring $R = \mathbb{C}[x_{ij}, D^{-1}]$, where the variables $\{x_{ij} \mid 1 \leq i, j \leq n\}$ should be interpreted as the entries of an "indeterminate matrix," and D is the polynomial $\det(x_{ij})$. Both the group multiplication $\mathbf{G} \times \mathbf{G} \to \mathbf{G}$ and inversion $\mathbf{G} \to \mathbf{G}$ are required to be morphisms of algebraic varieties.

    The ring R is the coordinate ring of the algebraic group GL_n. Hence this condition means that $\mathbf{G}$ can be essentially viewed as a subgroup of $\mathrm{GL}_n(\mathbb{C})$ defined by polynomial equations in the matrix entries of the latter.

(2) *Defined over* $\mathbb{Q}$ means that I is generated by polynomials with rational coefficients.

(3) *Connected* means that $\mathbf{G}$ is connected as an algebraic variety.

(4) *Set of real points* means that G is the set of real solutions to the equations determined by I. We write $G = \mathbf{G}(\mathbb{R})$.

(5) *Semisimple* means that the maximal connected solvable normal subgroup of $\mathbf{G}$ is trivial.

Example A.1. The most important example for our purposes is the *split form of* SL_n. For this choice we have

$$G = \mathrm{SL}_n(\mathbb{R}) \text{ and } K = \mathrm{SO}(n).$$

Example A.2. Let $F/\mathbb{Q}$ be a number field. Then there is a $\mathbb{Q}$-group $\mathbf{G}$ such that $\mathbf{G}(\mathbb{Q}) = \mathrm{SL}_n(F)$. The group $\mathbf{G}$ is constructed as $\mathbf{R}_{F/\mathbb{Q}}(\mathrm{SL}_n)$, where $\mathbf{R}_{F/\mathbb{Q}}$ denotes the *restriction of scalars* from F to $\mathbb{Q}$ [**PR94**, §2.1.2]. For example, if F is totally real, the group $\mathbf{R}_{F/\mathbb{Q}}(\mathrm{SL}_2)$ appears when one studies Hilbert modular forms.

Let (r, s) be the signature of the field F, so that $F \otimes \mathbb{R} \simeq \mathbb{R}^r \times \mathbb{C}^s$. Then $G = \mathrm{SL}_n(\mathbb{R})^r \times \mathrm{SL}_n(\mathbb{C})^s$ and $K = \mathrm{SO}(n)^r \times \mathrm{SU}(n)^s$.

Example A.3. Another important example is the *split symplectic group* Sp_{2n}. This is the group that arises when one studies Siegel modular forms. The group of real points $\mathrm{Sp}_{2n}(\mathbb{R})$ is the subgroup of $\mathrm{SL}_{2n}(\mathbb{R})$ preserving a fixed nondegenerate alternating bilinear form on $\mathbb{R}^{2n}$. We have $K = \mathrm{U}(n)$.

A.2.4. To generalize $\Gamma_0(N)$, we need the notion of an *arithmetic group*. This is a discrete subgroup Γ of the group of rational points $\mathbf{G}(\mathbb{Q})$ that is commensurable with the set of integral points $\mathbf{G}(\mathbb{Z})$. Here commensurable simply means that $\Gamma \cap \mathbf{G}(\mathbb{Z})$ is a finite index subgroup of both Γ and $\mathbf{G}(\mathbb{Z})$; in particular $\mathbf{G}(\mathbb{Z})$ itself is an arithmetic group.

Example A.4. For the split form of SL_n we have $\mathbf{G}(\mathbb{Z}) = \mathrm{SL}_n(\mathbb{Z}) \subset \mathbf{G}(\mathbb{Q}) = \mathrm{SL}_n(\mathbb{Q})$. A trivial way to obtain other arithmetic groups is by conjugation: if $g \in \mathrm{SL}_n(\mathbb{Q})$, then $g \cdot \mathrm{SL}_n(\mathbb{Z}) \cdot g^{-1}$ is also arithmetic.

A more interesting collection of examples is given by the congruence subgroups. The *principal congruence subgroup* $\Gamma(N)$ is the group of matrices congruent to the identity modulo N for some fixed integer $N \geq 1$. A *congruence subgroup* is a group containing $\Gamma(N)$ for some N.

In higher dimensions there are many candidates to generalize the Hecke subgroup $\Gamma_0(N)$. For example, one can take the subgroup of $\mathrm{SL}_n(\mathbb{Z})$ that is upper-triangular mod N. From a computational perspective, this choice is not so good since its index in $\mathrm{SL}_n(\mathbb{Z})$ is large. A better choice, and the one that usually appears in the literature, is to define $\Gamma_0(N)$ to be the subgroup of $\mathrm{SL}_n(\mathbb{Z})$ with bottom row congruent to $(0, \ldots, 0, *) \mod N$.

A.2.5. We are almost ready to define automorphic forms. Let $\mathfrak{g}$ be the Lie algebra of G, and let $U(\mathfrak{g})$ be its universal enveloping algebra over $\mathbb{C}$. Geometrically, $\mathfrak{g}$ is just the tangent space at the identity of the smooth manifold G. The algebra $U(\mathfrak{g})$ is a certain complex associative algebra canonically built from $\mathfrak{g}$. The usual definition would lead us a bit far afield, so we will settle for an equivalent characterization: $U(\mathfrak{g})$ can be realized as a certain subalgebra of the ring of differential operators on $C^\infty(G)$, the space of smooth functions on G.

In particular, G acts on $C^\infty(G)$ by *left translations*: given $g \in G$ and $f \in C^\infty(G)$, we define

$$L_g(f)(x) := f(g^{-1}x).$$

Then $U(\mathfrak{g})$ can be identified with the ring of all differential operators on $C^\infty(G)$ that are invariant under left translation. For our purposes the most important part of $U(\mathfrak{g})$ is its center $Z(\mathfrak{g})$. In terms of differential operators, $Z(\mathfrak{g})$ consists of those operators that are also invariant under *right translation*:

$$R_g(f)(x) := f(xg).$$

Definition A.5. An *automorphic form* on G with respect to Γ is a function $\varphi\colon G \to \mathbb{C}$ satisfying

(1) $\varphi(\gamma g) = \varphi(g)$ for all $\gamma \in \Gamma$,

(2) the right translates $\{\varphi(gk) \mid k \in K\}$ span a finite-dimensional space ξ of functions,

(3) there exists an ideal $J \subset Z(\mathfrak{g})$ of finite codimension such that J annihilates φ, and

(4) φ satisfies a certain growth condition that we do not wish to make precise. (In the literature, φ is said to be *slowly increasing.*)

For fixed ξ and J, we denote by $\mathscr{A}(\Gamma, \xi, J, K)$ the space of all functions satisfying the above four conditions. It is a basic theorem, due to Harish-Chandra [**HC68**], that $\mathscr{A}(\Gamma, \xi, J, K)$ is finite-dimensional.

Example A.6. We can identify the cuspidal modular forms $S_k(N)$ in the language of Definition A.5. Given a modular form f, let $\varphi_f \in C^\infty(\mathrm{SL}_2(\mathbb{R}))$ be the function from (A.2.6). Then the map $f \mapsto \varphi_f$ identifies $S_k(N)$ with the subspace $\mathscr{A}_k(N)$ of functions φ satisfying

(1) $\varphi(\gamma g) = \varphi(g)$ for all $\gamma \in \Gamma_0(N)$,

(2) $\varphi(gm) = \zeta_m^{-k}\varphi(g)$ for all $m \in \mathrm{SO}(2)$,

(3) $(\Delta + \lambda_k)\varphi = 0$, where $\Delta \in Z(\mathfrak{g})$ is the *Laplace–Beltrami–Casimir operator* and

$$\lambda_k = \frac{k}{2}\left(\frac{k}{2} - 1\right),$$

(4) φ is slowly increasing, and

(5) φ is *cuspidal.*

The first four conditions parallel Definition A.5. Item (1) is the Γ-invariance. Item (2) implies that the right translates of φ by $\mathrm{SO}(2)$ lie in a fixed finite-dimensional representation of $\mathrm{SO}(2)$. Item (3) is how holomorphicity appears, namely that φ is killed by a certain differential operator. Finally, item (4) is the usual growth condition.

The only condition missing from the general definition is (5), which is an extra constraint placed on φ to ensure that it comes from a cusp form. This condition can be expressed by the vanishing of certain integrals ("constant terms"); for details we refer to [**Bum97, Gel75**].

Example A.7. Another important example appears when we set $k = 0$ in (2) in Example A.6 and relax (3) by requiring only that $(\Delta - \lambda)\varphi = 0$ for *some* nonzero $\lambda \in \mathbb{R}$. Such automorphic forms cannot possibly arise from modular forms, since there are no nontrivial cusp forms of weight 0. However, there are plenty of solutions to these conditions: they correspond to *real-analytic* cuspidal modular forms of weight 0 and are known as *Maass forms*. Traditionally one writes $\lambda = (1 - s^2)/4$. The positivity of Δ implies that $s \in (-1, 1)$ or is purely imaginary.

Maass forms are highly elusive objects. Selberg proved that there are infinitely many linearly independent Maass forms of full level (i.e., on $\mathrm{SL}_2(\mathbb{Z})$), but to this date no explicit construction of a single one is known. (Selberg's argument is indirect and relies on the trace formula; for an exposition see [**Sar03**].) For higher levels some explicit examples can be constructed using theta series attached to indefinite quadratic forms [**Vig77**]. Numerically Maass forms have been well studied; see for example [**FL**].

In general the arithmetic nature of the eigenvalues λ that correspond to Maass forms is unknown, although a famous conjecture of Selberg states that for congruence subgroups they satisfy the inequality $\lambda \geq 1/4$ (in other words, only purely imaginary s appear above). The truth of this conjecture would have far-reaching consequences, from analytic number theory to graph theory [**Lub94**].

A.2.6. As Example A.6 indicates, there is a notion of *cuspidal automorphic form*. The exact definition is too technical to state here, but it involves an appropriate generalization of the notion of constant term familiar from modular forms.

There are also *Eisenstein series* [**Lan66, Art79**]. Again the complete definition is technical; we only mention that there are different types of Eisenstein series corresponding to certain subgroups of G. The Eisenstein series that are easiest to understand are those built from cusp forms on lower rank groups. Very explicit formulas for Eisenstein series on GL_3 can be seen in [**Bum84**]. For a down-to-earth exposition of some of the Eisenstein series on GL_n, we refer to [**Gol05**].

The decomposition of $M_k(\Gamma_0(N))$ into cusp forms and Eisenstein series also generalizes to a general group G, although the statement is much more complicated. The result is a theorem of Langlands [**Lan76**] known as the *spectral decomposition of* $L^2(\Gamma\backslash G)$. A thorough recent presentation of this can be found in [**MW94**].

A.2.7. Let $\mathscr{A} = \mathscr{A}(\Gamma, K)$ be the space of all automorphic forms, where ξ and J range over all possibilities. The space $\mathscr{A}$ is huge, and the arithmetic significance of much of it is unknown. This is already apparent for $G =$

$\mathrm{SL}_2(\mathbb{R})$. The automorphic forms directly connected with arithmetic are the holomorphic modular forms, not the Maass forms[2]. Thus the question arises: which automorphic forms in $\mathscr{A}$ are the most natural generalization of the modular forms?

One answer is provided by the isomorphisms (A.2.1), (A.2.3), (A.2.4). These show that modular forms appear naturally in the cohomology of modular curves. Hence a reasonable approach is to generalize the *left* of (A.2.1), (A.2.3), (A.2.4), and to study the resulting cohomology groups. This is the approach we will take. One drawback is that it is not obvious that our generalization has anything to do with automorphic forms, but we will see eventually that it certainly does. So we begin by looking for an appropriate generalization of the modular curve $Y_0(N)$.

Let G and K be as in Section A.2.3. and let X be the quotient G/K. This is a global Riemannian symmetric space [**Hel01**]. One can prove that X is contractible. Any arithmetic group $\Gamma \subset G$ acts on X properly discontinuously. In particular, if Γ is torsion-free, then the quotient $\Gamma\backslash X$ is a smooth manifold.

Unlike the modular curves, $\Gamma\backslash X$ will not have a complex structure in general[3]; nevertheless, $\Gamma\backslash X$ is a very nice space. In particular, if Γ is torsion-free, it is an *Eilenberg–Mac Lane* space for Γ, otherwise known as a $K(\Gamma, 1)$. This means that the only nontrivial homotopy group of $\Gamma\backslash X$ is its fundamental group, which is isomorphic to Γ, and that the universal cover of $\Gamma\backslash X$ is contractible. Hence $\Gamma\backslash X$ is in some sense a "topological incarnation"[4] of Γ.

This leads us to the notion of the *group cohomology* $H^*(\Gamma;\mathbb{C})$ of Γ with trivial complex coefficients. In the early days of algebraic topology, this was defined to be the complex cohomology of an Eilenberg–Mac Lane space for Γ [**Bro94**, Introduction, I.4]:

$$H^*(\Gamma;\mathbb{C}) = H^*(\Gamma\backslash X;\mathbb{C}). \tag{A.2.7}$$

Today there are purely algebraic approaches to $H^*(\Gamma;\mathbb{C})$ [**Bro94**, III.1], but for our purposes (A.2.7) is exactly what we need. In fact, the group cohomology $H^*(\Gamma;\mathbb{C})$ can be identified with the cohomology of the quotient $\Gamma\backslash X$ even if Γ has torsion, since we are working with complex coefficients. The cohomology groups $H^*(\Gamma;\mathbb{C})$, where Γ is an arithmetic group, are our proposed generalization for the weight 2 modular forms.

What about higher weights? For this we must replace the trivial coefficient module $\mathbb{C}$ with local systems, just as we did in (A.2.3). For our

[2]However, Maass forms play a very important *indirect* role in arithmetic.

[3]The symmetric spaces that have a complex structure are known as *bounded domains*, or *Hermitian symmetric spaces* [**Hel01**].

[4]This apt phrase is due to Vogan [**Vog97**].

purposes it is enough to let $\mathscr{M}$ be a rational finite-dimensional representation of G over the complex numbers. Any such $\mathscr{M}$ gives a representation of $\Gamma \subset G$ and thus induces a local system $\widetilde{\mathscr{M}}$ on $\Gamma\backslash X$. As before, the group cohomology $H^*(\Gamma; \mathscr{M})$ is the cohomology $H^*(\Gamma\backslash X; \widetilde{\mathscr{M}})$. In (A.2.3) we took $\mathscr{M} = M_n$, the nth symmetric power of the standard representation. For a general group G there are many kinds of representations to consider. In any case, we contend that the cohomology spaces

$$H^*(\Gamma; \mathscr{M}) = H^*(\Gamma\backslash X; \widetilde{\mathscr{M}})$$

are a good generalization of the spaces of modular forms.

A.2.8. It is certainly not obvious that the cohomology groups $H^*(\Gamma; \mathscr{M})$ have *anything* to do with automorphic forms, although the isomorphisms (A.2.1), (A.2.3), (A.2.4) look promising.

The connection is provided by a deep theorem of Franke [**Fra98**], which asserts that

(1) the cohomology groups $H^*(\Gamma; \mathscr{M})$ can be directly computed in terms of certain automorphic forms (the automorphic forms of "cohomological type," also known as those with "nonvanishing $(\mathfrak{g}, K)$ cohomology" [**VZ84**]); and
(2) there is a direct sum decomposition

$$H^*(\Gamma; \mathscr{M}) = H^*_{\mathrm{cusp}}(\Gamma; \mathscr{M}) \oplus \bigoplus_{\{P\}} H^*_{\{P\}}(\Gamma; \mathscr{M}), \qquad \text{(A.2.8)}$$

where the sum is taken over the set of classes of *associate proper $\mathbb{Q}$-parabolic subgroups of G*.

The precise version of statement (1) is known in the literature as the *Borel conjecture*. Statement (2) parallels Langlands's spectral decomposition of $L^2(\Gamma\backslash G)$.

Example A.8. For $\Gamma = \Gamma_0(N) \subset \mathrm{SL}_2(\mathbb{Z})$, the decomposition (A.2.8) is exactly (A.2.4). The cuspforms $S_k(N) \oplus \overline{S_k(N)}$ correspond to the summand $H^1_{\mathrm{cusp}}(\Gamma; \mathscr{M})$. There is one class of proper $\mathbb{Q}$-parabolic subgroups in $\mathrm{SL}_2(\mathbb{R})$, represented by the Borel subgroup of upper-triangular matrices. Hence only one term appears in big direct sum on the right of (A.2.8), which is the Eisenstein term E_k.

The summand $H^*_{\mathrm{cusp}}(\Gamma; \mathscr{M})$ of (A.2.8) is called the *cuspidal cohomology*; this is the subspace of classes represented by cuspidal automorphic forms. The remaining summands constitute the *Eisenstein cohomology* of Γ [**Har91**]. In particular the summand indexed by $\{P\}$ is constructed using Eisenstein series attached to certain cuspidal automorphic forms on lower rank groups. Hence $H^*_{\mathrm{cusp}}(\Gamma; \mathscr{M})$ is in some sense the most important part of the cohomology: all the rest can be built systematically from cuspidal cohomology on lower rank groups[5]. This leads us to our basic computational problem:

Problem A.9. Develop tools to compute explicitly the cohomology spaces $H^*(\Gamma; \mathscr{M})$ and to identify the cuspidal subspace $H^*_{\mathrm{cusp}}(\Gamma; \mathscr{M})$.

A.3. Combinatorial Models for Group Cohomology

A.3.1. In this section, we restrict attention to $G = \mathrm{SL}_n(\mathbb{R})$ and Γ, a congruence subgroup of $\mathrm{SL}_n(\mathbb{Z})$. By the previous section, we can study the group cohomology $H^*(\Gamma; \mathscr{M})$ by studying the cohomology $H^*(\Gamma\backslash X; \widetilde{\mathscr{M}})$. The latter spaces can be studied using standard topological techniques, such as taking the cohomology of complexes associated to cellular decompositions of $\Gamma\backslash X$. For $\mathrm{SL}_n(\mathbb{R})$, one can construct such decompositions using a version of explicit reduction theory of real positive-definite quadratic forms due to Voronoĭ [**Vor08**]. The goal of this section is to explain how this is done. We also discuss how the cohomology can be explicitly studied for congruence subgroups of $\mathrm{SL}_3(\mathbb{Z})$.

A.3.2. Let V be the $\mathbb{R}$-vector space of all symmetric $n \times n$ matrices, and let $C \subset V$ be the subset of positive-definite matrices. The space C can be identified with the space of all real positive-definite quadratic forms in n variables: in coordinates, if $x = (x_1, \ldots, x_n)^t \in \mathbb{R}^n$ (column vector), then the matrix $A \in C$ induces the quadratic form

$$x \longmapsto x^t A x,$$

and it is well known that any positive-definite quadratic form arises in this way. The space C is a cone, in that it is preserved by homotheties: if $x \in C$, then $\lambda x \in C$ for all $\lambda \in \mathbb{R}_{>0}$. It is also convex: if $x_1, x_2 \in C$, then $tx_1 + (1-t)x_2 \in C$ for $t \in [0,1]$. Let D be the quotient of C by homotheties.

Example A.10. The case $n = 2$ is illustrative. We can take coordinates on $V \simeq \mathbb{R}^3$ by representing any matrix in V as

$$\begin{pmatrix} x & y \\ y & z \end{pmatrix}, \quad x, y, z \in \mathbb{R}.$$

The subset of singular matrices $Q = \{xz - y^2 = 0\}$ is a quadric cone in V dividing the complement $V \smallsetminus Q$ into three connected components. The component containing the identity matrix is the cone C of positive-definite matrices. The quotient D can be identified with an open 2-disk.

[5]This is a bit of an oversimplification, since it is a highly nontrivial problem to decide when cusp cohomology from lower rank groups appears in Γ. However, many results are known; as a selection we mention [**Har91, Har87, LS04**]

The group G acts on C on the left by

$$(g, c) \longmapsto gcg^t.$$

This action commutes with that of the homotheties and thus descends to a G-action on D. One can show that G acts transitively on D and that the stabilizer of the image of the identity matrix is $K = \mathrm{SO}(n)$. Hence we may identify D with our symmetric space $X = \mathrm{SL}_n(\mathbb{R})/\mathrm{SO}(n)$. We will do this in the sequel, using the notation D when we want to emphasize the coordinates coming from the linear structure of $C \subset V$ and using the notation X for the quotient G/K.

We can make the identification $D \simeq X$ more explicit. If $g \in \mathrm{SL}_n(\mathbb{R})$, then the map

$$g \longmapsto gg^t \tag{A.3.1}$$

takes g to a symmetric positive-definite matrix. Any coset gK is taken to the same matrix since $KK^t = \mathrm{Id}$. Thus (A.3.1) identifies G/K with a subset C_1 of C, namely those positive-definite symmetric matrices with determinant 1. It is easy to see that C_1 maps diffeomorphically onto D.

The inverse map $C_1 \to G/K$ is more complicated. Given a determinant 1 positive-definite symmetric matrix A, one must find $g \in \mathrm{SL}_n(\mathbb{R})$ such that $gg^t = A$. Such a representation always exists, with g determined uniquely up to right multiplication by an element of K. In computational linear algebra, such a g can be constructed through *Cholesky decomposition* of A.

The group $\mathrm{SL}_n(\mathbb{Z})$ acts on C via the G-action and does so properly discontinuously. This is the "unimodular change of variables" action on quadratic forms [**Ser73**, V.1.1]. Under our identification of D with X, this is the usual action of $\mathrm{SL}_n(\mathbb{Z})$ by left translation from Section A.2.7.

A.3.3. Now consider the group cohomology $H^*(\Gamma; \mathscr{M}) = H^*(\Gamma\backslash X; \widetilde{\mathscr{M}})$. The identification $D \simeq X$ shows that the dimension of X is $n(n+1)/2 - 1$. Hence $H^i(\Gamma; \mathscr{M})$ vanishes if $i > n(n+1)/2 - 1$. Since $\dim X$ grows quadratically in n, there are many potentially interesting cohomology groups to study.

However, it turns out that there is some additional vanishing of the cohomology for deeper (topological) reasons. For $n = 2$, this is easy to see. The quotient $\Gamma\backslash\mathfrak{h}$ is homeomorphic to a topological surface with punctures, corresponding to the cusps of Γ. Any such surface S can be retracted onto a finite graph simply by "stretching" S along its punctures. Thus $H^2(\Gamma; \mathscr{M}) = 0$, even though $\dim \Gamma\backslash\mathfrak{h} = 2$.

For $\Gamma \subset \mathrm{SL}_n(\mathbb{Z})$, a theorem of Borel–Serre implies that $H^i(\Gamma; \mathscr{M})$ vanishes if $i > \dim X - n + 1 = n(n-1)/2$ [**BS73**, Theorem 11.4.4]. The number $\nu = n(n-1)/2$ is called the *virtual cohomological dimension* of Γ

and is denoted $\operatorname{vcd}\Gamma$. Thus we only need to consider cohomology in degrees $i \leq \nu$.

Moreover we know from Section A.2.8 that the most interesting part of the cohomology is the cuspidal cohomology. In what degrees can it live? For $n = 2$, there is only one interesting cohomology group $H^1(\Gamma; \mathscr{M})$, and it contains the cuspidal cohomology. For higher dimensions, the situation is quite different: for most i, the subspace $H^i_{\text{cusp}}(\Gamma; \mathscr{M})$ vanishes! In fact in the late 1970's Borel, Wallach, and Zuckerman observed that the cuspidal cohomology can only live in the cohomological degrees lying in an interval around $(\dim X)/2$ of size linear in n. An explicit description of this interval is given in [**Sch86**, Proposition 3.5]; one can also look at Table A.3.1, from which the precise statement is easy to determine.

Another feature of Table A.3.1 deserves to be mentioned. There are exactly two values of n, namely $n = 2, 3$, such that virtual cohomological dimension equals the upper limit of the cuspidal range. This will have implications later, when we study the action of the Hecke operators on the cohomology.

n	2	3	4	5	6	7	8	9
$\dim X$	2	5	9	14	20	27	35	44
$\operatorname{vcd}\Gamma$	1	3	6	10	15	21	28	36
top degree of H^*_{cusp}	1	3	5	8	11	15	19	24
bottom degree of H^*_{cusp}	1	2	4	6	9	12	16	20

Table A.3.1. The virtual cohomological dimension and the cuspidal range for subgroups of $\mathrm{SL}_n(\mathbb{Z})$.

A.3.4. Recall that a point in $\mathbb{Z}^n$ is said to be *primitive* if the greatest common divisor of its coordinates is 1. In particular, a primitive point is nonzero. Let $\mathscr{P} \subset \mathbb{Z}^n$ be the set of primitive points. Any $v \in \mathscr{P}$, written as a column vector, determines a rank-1 symmetric matrix $q(v)$ in the closure $\bar{C}$ via $q(v) = vv^t$. The *Voronoĭ polyhedron* Π is defined to be the closed convex hull in $\bar{C}$ of the points $q(v)$, as v ranges over $\mathscr{P}$. Note that by construction, $\mathrm{SL}_n(\mathbb{Z})$ acts on Π, since $\mathrm{SL}_n(\mathbb{Z})$ preserves the set $\{q(v)\}$ and acts linearly on V.

Example A.11. Figure A.3.1 represents a crude attempt to show what Π looks like for $n = 2$. These images were constructed by computing a large subset of the points $q(v)$ and taking the convex hull (we took all points $v \in \mathscr{P}$ such that $\operatorname{Trace} q(v) < N$ for some large integer N). From a distance, the polyhedron Π looks almost indistinguishable from the cone C; this is somewhat conveyed by the right of Figure A.3.1. Unfortunately Π is not locally finite, so we really cannot produce an accurate picture. To get a more accurate image, the reader should imagine that each vertex meets infinitely many edges. On the other hand, Π is not hopelessly complex: each maximal face is a triangle, as the pictures suggest.

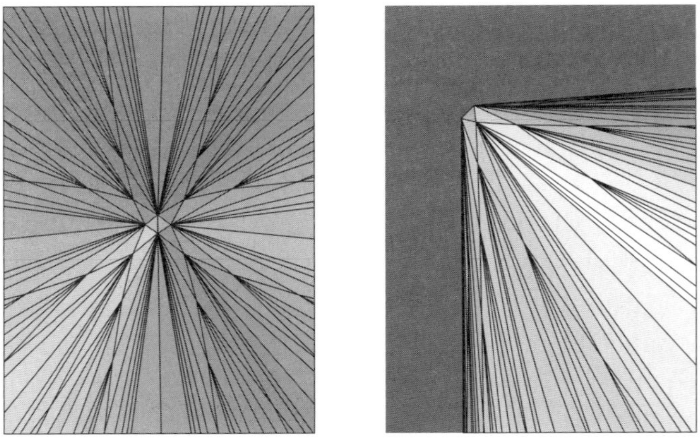

(a) (b)

Figure A.3.1. The polyhedron Π for $\mathrm{SL}_2(\mathbb{Z})$. In (a) we see Π from the origin, in (b) from the side. The small triangle at the right center of (a) is the facet with vertices $\{q(e_1), q(e_2), q(e_1+e_2)\}$, where $\{e_1, e_2\}$ is the standard basis of $\mathbb{Z}^2$. In (b) the x-axis runs along the top from left to right, and the z-axis runs down the left side. The facet from (a) is the little triangle at the top left corner of (b).

A.3.5. The polyhedron Π is quite complicated: it has infinitely many faces and is not locally finite. However, one of Voronoĭ's great insights is that Π is actually not as complicated as it seems.

For any $A \in C$, let $\mu(A)$ be the minimum value attained by A on $\mathscr{P}$ and let $M(A) \subset \mathscr{P}$ be the set on which A attains $\mu(A)$. Note that $\mu(A) > 0$ and $M(A)$ is finite since A is positive-definite. Then A is called *perfect* if it is recoverable from the knowledge of the pair $(\mu(A), M(A))$. In other words, given $(\mu(A), M(A))$, we can write a system of linear equations

$$mZm^t = \mu(A), \quad m \in M(A), \tag{A.3.2}$$

where $Z = (z_{ij})$ is a symmetric matrix of variables. Then A is perfect if and only if A is the unique solution to the system (A.3.2).

Example A.12. The quadratic form $Q(x, y) = x^2 - xy + y^2$ is perfect. The smallest nontrivial value it attains on $\mathbb{Z}^2$ is $\mu(Q) = 1$, and it does so on the columns of

$$M(Q) = \begin{pmatrix} 1 & 0 & 1 \\ 0 & 1 & 1 \end{pmatrix}$$

and their negatives. Letting $\alpha x^2+\beta xy+\gamma y^2$ be an undetermined quadratic form and applying the data $(\mu(Q), M(Q))$, we are led to the system of linear equations

$$\alpha = 1, \quad \gamma = 1, \quad \alpha+\beta+\gamma = 1.$$

From this we recover $Q(x, y)$.

Example A.13. The quadratic form $Q'(x, y) = x^2 + y^2$ is not perfect. Again the smallest nontrivial value of Q' on $\mathbb{Z}^2$ is $m(Q') = 1$, attained on the columns of

$$M(Q') = \begin{pmatrix} 1 & 0 \\ 0 & 1 \end{pmatrix}$$

and their negatives. But every member of the one-parameter family of quadratic forms

$$x^2 + \alpha xy + y^2, \quad \alpha \in (-1, 1) \tag{A.3.3}$$

has the same set of minimal vectors, and so Q' cannot be recovered from the knowledge of $m(Q'), M(Q')$.

Example A.14. Example A.12 generalizes to all n. Define

$$A_n(x) := \sum_{i=1}^{n} x_i^2 - \sum_{1\leq i<j\leq n} x_i x_j. \tag{A.3.4}$$

Then A_n is perfect for all n. We have $\mu(A_n) = 1$, and $M(A_n)$ consists of all points of the form

$$\pm(e_i + e_{i+1} + \cdots + e_{i+k}), \quad 1 \leq i \leq n, \quad i \leq i+k \leq n,$$

where $\{e_i\}$ is the standard basis of $\mathbb{Z}^n$. This quadratic form is closely related to the A_n root lattice [**FH91**], which explains its name. It is one of two infinite families of perfect forms studied by Voronoĭ (the other is related to the D_n root lattice).

We can now summarize Voronoĭ's main results:

(1) There are finitely many equivalence classes of perfect forms modulo the action of $\mathrm{SL}_n(\mathbb{Z})$. Voronoĭ even gave an explicit algorithm to determine all the perfect forms of a given dimension.

(2) The facets of Π, in other words the codimension 1 faces, are in bijection with the rank n perfect quadratic forms. Under this correspondence the minimal vectors $M(A)$ determine a facet F_A by taking the convex hull in $\bar{C}$ of the finite point set $\{q(m) \mid m \in M(A)\}$. Hence there are finitely many faces of Π modulo $\mathrm{SL}_n(\mathbb{Z})$ and thus finitely many modulo any finite index subgroup Γ.

(3) Let $\mathscr{V}$ be the set of cones over the faces of Π. Then $\mathscr{V}$ is a *fan*, which means (i) if $\sigma \in \mathscr{V}$, then any face of σ is also in $\mathscr{V}$; and (ii) if $\sigma, \sigma' \in \mathscr{V}$, then $\sigma \cap \sigma'$ is a common face of each[6]. The fan $\mathscr{V}$ provides a reduction theory for C in the following sense: any point $x \in C$ is contained in a unique $\sigma(x) \in \mathscr{V}$, and the set $\{\gamma \in \mathrm{SL}_n(\mathbb{Z}) \mid \gamma \cdot \sigma(x) = \sigma(x)\}$ is finite. Voronoĭ also gave an explicit algorithm to determine $\sigma(x)$ given x, the *Voronoĭ reduction algorithm.*

The number N_{perf} of equivalence classes of perfect forms modulo the action of $\mathrm{GL}_n(\mathbb{Z})$ grows rapidly with n (Table A.3.2); the complete classification is known only for $n \leq 8$. For a list of perfect forms up to $n = 7$, see [**CS88**]. For a recent comprehensive treatment of perfect forms, with many historical remarks, see [**Mar03**].

Dimension	N_{perf}	Authors
2	1	Voronoĭ [**Vor08**]
3	1	*ibid.*
4	2	*ibid.*
5	3	*ibid.*
6	7	Barnes [**Bar57**]
7	33	Jaquet-Chiffelle [**Jaq91, JC93**]
8	10916	Dutour–Schürmann–Vallentin [**DVS05**]

Table A.3.2. The number N_{perf} of equivalence classes of perfect forms.

A.3.6. Our goal now is to describe how the Voronoĭ fan $\mathscr{V}$ can be used to compute the cohomology $H^*(\Gamma; \mathscr{M})$. The idea is to use the cones in $\mathscr{V}$ to chop the quotient D into pieces.

For any $\sigma \in \mathscr{V}$, let σ° be the open cone obtained by taking the complement in σ of its proper faces. Then after taking the quotient by homotheties, the cones $\{\sigma^\circ \cap C \mid \sigma \in \mathscr{V}\}$ pass to locally closed subsets of D. Let $\mathscr{C}$ be the set of these images.

Any $c \in \mathscr{C}$ is a *topological cell*, i.e., it is homeomorphic to an open ball, since c is homeomorphic to a face of Π. Because $\mathscr{C}$ comes from the fan $\mathscr{V}$, the cells in $\mathscr{C}$ have good incidence properties: the closure in D of any $c \in \mathscr{C}$ can be written as a finite disjoint union of elements of $\mathscr{C}$. Moreover, $\mathscr{C}$ is locally finite: by taking quotients of all the σ° meeting C, we have eliminated the open cones lying in $\bar{C}$, and it is these cones that are responsible for the failure of local finiteness of $\mathscr{V}$. We summarize these properties by saying

[6]Strictly speaking, Voronoĭ actually showed that every codimension 1 cone is contained in two top-dimensional cones.

that $\mathscr{C}$ gives a *cellular decomposition* of D. Clearly $\mathrm{SL}_n(\mathbb{Z})$ acts on $\mathscr{C}$, since $\mathscr{C}$ is constructed using the fan $\mathscr{V}$. Thus we obtain a cellular decomposition[7] of $\Gamma\backslash D$ for any torsion-free Γ. We call $\mathscr{C}$ the *Voronoĭ decomposition* of D.

Some care must be taken in using these cells to perform topological computations. The problem is that even though the individual pieces are homeomorphic to balls and are glued together nicely, the boundaries of the closures of the pieces are not homeomorphic to spheres in general. (If they were, then the Voronoĭ decomposition would give rise to a *regular* cell complex [**CF67**], which can be used as a substitute for a simplicial or CW complex in homology computations.) Nevertheless, there is a way to remedy this.

Recall that a subspace A of a topological space B is a *strong deformation retract* if there is a continuous map $f\colon B \times [0,1] \to B$ such that $f(b,0) = b$, $f(b,1) \in A$, and $f(a,t) = a$ for all $a \in A$. For such pairs $A \subset B$ we have $H^*(A) = H^*(B)$. One can show that there is a strong deformation retraction from C to itself equivariant under the actions of both $\mathrm{SL}_n(\mathbb{Z})$ and the homotheties and that the image of the retraction modulo homotheties, denoted W, is naturally a locally finite regular cell complex of dimension ν. Moreover, the cells in W are in bijective, inclusion-reversing correspondence with the cells in $\mathscr{C}$. In particular, if a cell in $\mathscr{C}$ has *codimension* d, the corresponding cell in W has *dimension* d. Thus, for example, the vertices of W modulo $\mathrm{SL}_n(\mathbb{Z})$ are in bijection with the top-dimensional cells in $\mathscr{C}$, which are in bijection with equivalence classes of perfect forms.

In the literature W is called the *well-rounded retract*. The subspace $W \subset D \simeq X$ has a beautiful geometric interpretation. The quotient

$$\mathrm{SL}_n(\mathbb{Z})\backslash X = \mathrm{SL}_n(\mathbb{Z})\backslash \mathrm{SL}_n(\mathbb{R})/\mathrm{SO}(n)$$

can be interpreted as the moduli space of lattices in $\mathbb{R}^n$ modulo the equivalence relation of rotation and positive scaling (cf. [**AG00**]; for $n = 2$ one can also see [**Ser73**, VII, Proposition 3]). Then W corresponds to those lattices whose shortest nonzero vectors span $\mathbb{R}^n$. This is the origin of the name: the shortest vectors of such a lattice are "more round" than those of a generic lattice.

The space W was known classically for $n = 2$ and was constructed for $n \geq 3$ by Lannes and Soulé, although Soulé only published the case $n = 3$ [**Sou75**]. The construction for all n appears in work of Ash [**Ash80, Ash84**], who also generalized W to a much larger class of groups. Explicit computations of the cell structure of W have only been performed up to

[7] If Γ has torsion, then cells in $\mathscr{C}$ can have nontrivial stabilizers in Γ, and thus $\Gamma\backslash\mathscr{C}$ should be considered as an "orbifold" cellular decomposition.

$n = 6$ [**EVGS02**]. Certainly computing W explicitly for $n = 8$ seems very difficult, as Table A.3.2 indicates.

Example A.15. Figure A.3.2 illustrates $\mathscr{C}$ and W for $\mathrm{SL}_2(\mathbb{Z})$. As in Example A.11, the polyhedron Π is 3-dimensional, and so the Voronoĭ fan $\mathscr{V}$ has cones of dimensions $0, 1, 2, 3$. The 1-cones of $\mathscr{V}$, which correspond to the vertices of Π, pass to infinitely many points on the boundary $\partial\bar{D} = \bar{D} \smallsetminus D$. The 3-cones become triangles in $\bar{D}$ with vertices on $\partial\bar{D}$. In fact, the identifications $D \simeq \mathrm{SL}_2(\mathbb{R})/\mathrm{SO}(2) \simeq \mathfrak{h}$ realize D as the Klein model for the hyperbolic plane, in which geodesics are represented by Euclidean line segments. Hence, the images of the 1-cones of $\mathscr{V}$ are none other than the usual cusps of $\mathfrak{h}$, and the triangles are the $\mathrm{SL}_2(\mathbb{Z})$-translates of the ideal triangle with vertices $\{0, 1, \infty\}$. These triangles form a tessellation of $\mathfrak{h}$ sometimes known as the *Farey tessellation.* The edges of the Voronoĭ are the $\mathrm{SL}_2(\mathbb{Z})$-translates of the ideal geodesic between 0 and ∞. After adjoining cusps and passing to the quotient $X_0(N)$, these edges become the supports of the Manin symbols from Section 8.2 (cf. Figure 3.2.1). This example also shows how the Voronoĭ decomposition fails to be a regular cell complex: the boundaries of the closures of the triangles in D do not contain the vertices and thus are not homeomorphic to circles.

The virtual cohomological dimension of $\mathrm{SL}_2(\mathbb{Z})$ is 1. Hence the well-rounded retract W is a graph (Figures A.3.2 and A.3.3). Note that W is not a manifold. The vertices of W are in bijection with the Farey triangles—each vertex lies at the center of the corresponding triangle—and the edges are in bijection with the Manin symbols. Under the map $D \to \mathfrak{h}$, the graph W becomes the familiar "PSL_2-tree" embedded in $\mathfrak{h}$, with vertices at the order 3 elliptic points (Figure A.3.3).

A.3.7. We now discuss the example $\mathrm{SL}_3(\mathbb{Z})$ in some detail. This example gives a good feeling for how the general situation compares to the case $n = 2$.

We begin with the Voronoĭ fan $\mathscr{V}$. The cone C is 6-dimensional, and the quotient D is 5-dimensional. There is one equivalence class of perfect forms modulo the action of $\mathrm{SL}_3(\mathbb{Z})$, represented by the form (A.3.4). Hence there are 12 minimal vectors; six are the columns of the matrix

$$\text{(A.3.5)} \qquad \begin{pmatrix} 1 & 0 & 0 & 1 & 0 & 1 \\ 0 & 1 & 0 & 1 & 1 & 1 \\ 0 & 0 & 1 & 0 & 1 & 1 \end{pmatrix},$$

and the remaining six are the negatives of these. This implies that the cone σ corresponding to this form is 6-dimensional and simplicial. The latter implies that the faces of σ are the cones generated by $\{q(v) \mid v \in S\}$, where S ranges over all subsets of (A.3.5). To get the full structure of the fan, one

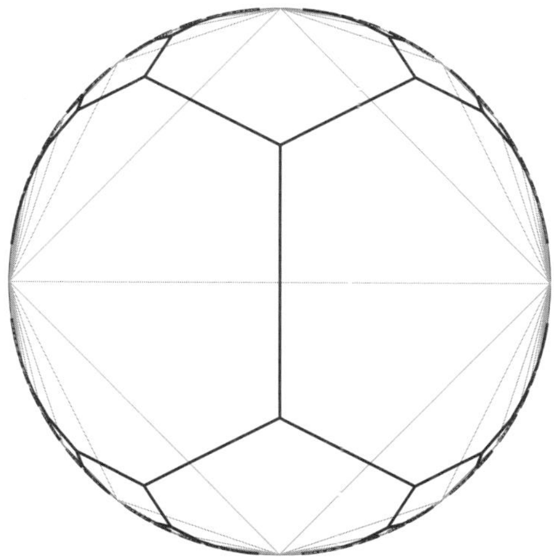

Figure A.3.2. The Voronoĭ decomposition and the retract in D.

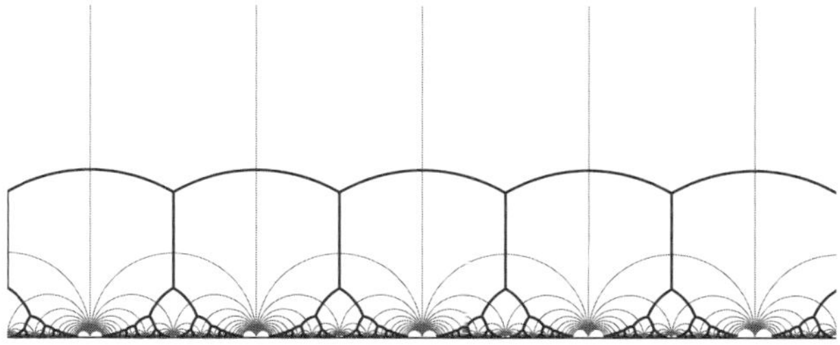

Figure A.3.3. The Voronoĭ decomposition and the retract in $\mathfrak{h}$.

must determine the $SL_3(\mathbb{Z})$ orbits of faces, as well as which faces lie in the boundary $\partial\bar{C} = \bar{C} \smallsetminus C$. After some pleasant computation, one finds:

(1) There is one equivalence class modulo $SL_3(\mathbb{Z})$ for each of the 6-, 5-, 2-, and 1-dimensional cones.

(2) There are two equivalence classes of the 4-dimensional cones, represented by the sets of minimal vectors

$$\begin{pmatrix} 1 & 0 & 0 & 1 \\ 0 & 1 & 0 & 1 \\ 0 & 0 & 1 & 1 \end{pmatrix} \quad \text{and} \quad \begin{pmatrix} 1 & 0 & 0 & 1 \\ 0 & 1 & 0 & 1 \\ 0 & 0 & 1 & 0 \end{pmatrix}.$$

(3) There are two equivalence classes of the 3-dimensional cones, represented by the sets of minimal vectors

$$\begin{pmatrix} 1 & 0 & 0 \\ 0 & 1 & 0 \\ 0 & 0 & 1 \end{pmatrix} \quad \text{and} \quad \begin{pmatrix} 1 & 0 & 1 \\ 0 & 1 & 1 \\ 0 & 0 & 0 \end{pmatrix}.$$

The second type of 3-cone lies in $\partial\bar{C}$ and thus does not determine a cell in $\mathscr{C}$.

(4) The 2- and 1-dimensional cones lie entirely in $\partial\bar{C}$ and do not determine cells in $\mathscr{C}$.

After passing from C to D, the cones of dimension k determine cells of dimension $k-1$. Therefore, modulo the action of $\mathrm{SL}_3(\mathbb{Z})$ there are five types of cells in the Voronoï decomposition $\mathscr{C}$, with dimensions from 5 to 2. We denote these cell types by c_5, c_4, c_{3a}, c_{3b}, and c_2. Here c_{3a} corresponds to the first type of 4-cone in item (2) above, and c_{3b} to the second. For a beautiful way to index the cells of $\mathscr{C}$ using configurations in projective spaces, see [**McC91**].

The virtual cohomological dimension of $\mathrm{SL}_3(\mathbb{Z})$ is 3, which means that the retract W is a 3-dimensional cell complex. The closures of the top-dimensional cells in W, which are in bijection with the Voronoï cells of type c_2, are homeomorphic to solid cubes truncated along two pairs of opposite corners (Figure A.3.4). To compute this, one must see how many Voronoï cells of a given type contain a fixed cell of type c_2 (since the inclusions of cells in W are the *opposite* of those in $\mathscr{C}$).

A table of the incidence relations between the cells of $\mathscr{C}$ and W is given in Table A.3.3. To interpret the table, let $m = m(X, Y)$ be the integer in row X and column Y.

- If m is below the diagonal, then the boundary of a cell of type Y contains m cells of type X.
- If m is above the diagonal, then a cell of type Y appears in the boundary of m cells of type X.

For instance, the entry 16 in row c_5 and column c_2 means that a Voronoï cell of type c_2 meets the boundaries of 16 cells of type c_5. This is the same as the number of vertices in the Soulé cube (Figure A.3.4). Investigation of the table shows that the triangular (respectively, hexagonal) faces of the Soulé cube correspond to the Voronoï cells of type c_{3a} (resp., c_{3b}).

Figure A.3.5 shows a Schlegel diagram for the Soulé cube. One vertex is at infinity; this is indicated by the arrows on three of the edges. This Soulé cube is dual to the Voronoï cell C of type c_2 with minimal vectors given by the columns of the identity matrix. The labels on the 2-faces are additional minimal vectors that show which Voronoï cells contain C. For example, the central triangle labelled with $(1,1,1)^t$ is dual to the Voronoï cell of type c_{3a} with minimal vectors given by those of C together with $(1,1,1)^t$. Cells of type c_4 containing C in their closure correspond to the edges of the figure; the minimal vectors for a given edge are those of C together with the two vectors on the 2-faces containing the edge. Similarly, one can read off the minimial vectors of the top-dimensional Voronoĭ cells containing C, which correspond to the vertices of Figure A.3.5.

	c_5	c_4	c_{3a}	c_{3b}	c_2
c_5	•	2	3	6	16
c_4	6	•	3	6	24
c_{3a}	3	1	•	•	4
c_{3b}	12	4	•	•	6
c_2	12	8	4	3	•

Table A.3.3. Incidence relations in the Voronoĭ decomposition and the retract for $\mathrm{SL}_3(\mathbb{Z})$.

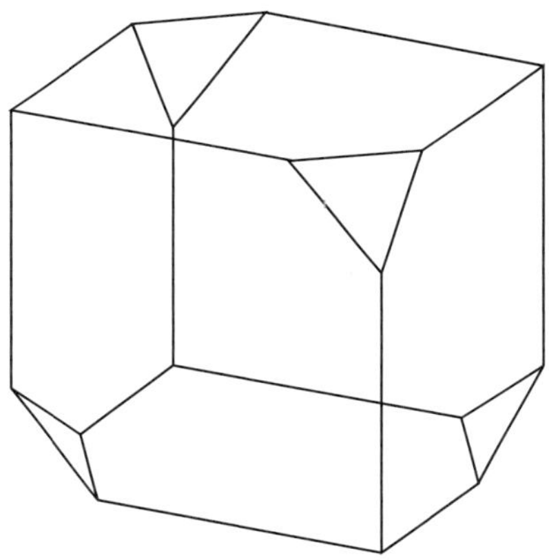

Figure A.3.4. The Soulé cube.

A.3.8. Now let p be a prime, and let $\Gamma = \Gamma_0(p) \subset \mathrm{SL}_3(\mathbb{Z})$ be the Hecke subgroup of matrices with bottom row congruent to $(0, 0, *) \mod p$ (Example A.4). The virtual cohomological dimension of Γ is 3, and the cusp cohomology with constant coefficients can appear in degrees 2 and 3. One can show that the cusp cohomology in degree 2 is dual to that in degree 3, so for computational purposes it suffices to focus on degree 3.

In terms of W, these will be cochains supported on the 3-cells. Unfortunately we cannot work directly with the quotient $\Gamma\backslash W$ since Γ has torsion: there will be cells taken to themselves by the Γ-action, and thus the cells of W need to be subdivided to induce the structure of a cell complex on $\Gamma\backslash W$. Thus when Γ has torsion, the "set of 3-cells modulo Γ" unfortunately makes no sense.

To circumvent this problem, one can mimic the idea of Manin symbols. The quotient $\Gamma\backslash \mathrm{SL}_3(\mathbb{Z})$ is in bijection with the finite projective plane $\mathbb{P}^2(\mathbb{F}_p)$,

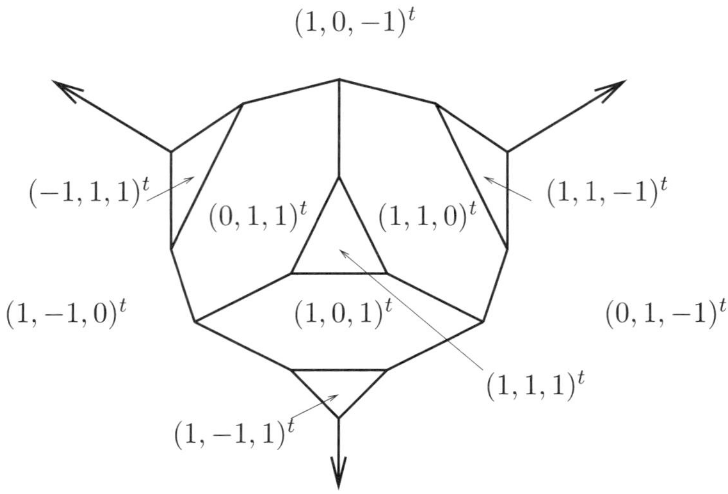

Figure A.3.5. A Schlegel diagram of a Soulé cube, showing the minimal vectors that correspond to the 2-faces.

where $\mathbb{F}_p$ is the field with p elements (cf. Proposition 3.10). The group $\mathrm{SL}_3(\mathbb{Z})$ acts transitively on the set of all 3-cells of W; if we fix one such cell w, its stabilizer $\mathrm{Stab}(w) = \{\gamma \in \mathrm{SL}_3(\mathbb{Z}) \mid \gamma w = w\}$ is a finite subgroup of $\mathrm{SL}_3(\mathbb{Z})$. Hence the set of 3-cells modulo Γ should be interpreted as the set of orbits in $\mathbb{P}^2(\mathbb{F}_p)$ of the finite group $\mathrm{Stab}(w)$. This suggests describing $H^3(\Gamma; \mathbb{C})$ in terms of the space $\mathscr{S}$ of complex-valued functions $f\colon \mathbb{P}^2(\mathbb{F}_p) \to \mathbb{C}$. To carry this out, there are two problems:

(1) How do we explicitly describe $H^3(\Gamma; \mathbb{C})$ in terms of $\mathscr{S}$?

(2) How can we isolate the cuspidal subspace $H^3_{\mathrm{cusp}}(\Gamma; \mathbb{C}) \subset H^3(\Gamma; \mathbb{C})$ in terms of our description?

Fully describing the solutions to these problems is rather complicated. We content ourselves with presenting the following theorem, which collects together several statements in [**AGG84**]. This result should be compared to Theorems 3.13 and 8.4.

Theorem A.16 (Theorem 3.19 and Summary 3.23 of [**AGG84**]). *We have*

$$\dim H^3(\Gamma_0(p); \mathbb{C}) = \dim H^3_{cusp}(\Gamma_0(p); \mathbb{C}) + 2S_p,$$

where S_p is the dimension of the space of weight 2 holomorphic cusp forms on $\Gamma_0(p) \subset \mathrm{SL}_2(\mathbb{Z})$. Moreover, the cuspidal cohomology $H^3_{cusp}(\Gamma_0(p); \mathbb{C})$ is isomorphic to the vector space of functions $f\colon \mathbb{P}^2(\mathbb{F}_p) \to \mathbb{C}$ satisfying

(1) $f(x,y,z) = f(z,x,y) = f(-x,y,z) = -f(y,x,z)$,

(2) $f(x,y,z) + f(-y,x-y,z) + f(y-x,-x,z) = 0$,

(3) $f(x,y,0) = 0$, *and*

(4) $\sum_{z=1}^{p-1} f(x, y, z) = 0$.

Unlike subgroups of $\mathrm{SL}_2(\mathbb{Z})$, cuspidal cohomology is apparently much rarer for $\Gamma_0(p) \subset \mathrm{SL}_3(\mathbb{Z})$. The computations of [**AGG84, vGvdKTV97**] show that the only prime levels $p \leq 337$ with nonvanishing cusp cohomology are 53, 61, 79, 89, and 223. In all these examples, the cuspidal subspace is 2-dimensional.

For more details of how to implement such computations, we refer to [**AGG84, vGvdKTV97**]. For further details about the additional complications arising for higher rank groups, in particular subgroups of $\mathrm{SL}_4(\mathbb{Z})$, see [**AGM02**, Section 3].

A.4. Hecke Operators and Modular Symbols

A.4.1. There is one ingredient missing so far in our discussion of the cohomology of arithmetic groups, namely the Hecke operators. These are an essential tool in the study of modular forms. Indeed, the forms with the most arithmetic significance are the Hecke eigenforms, and the connection with arithmetic is revealed by the Hecke eigenvalues.

In higher rank the situation is similar. There is an algebra of Hecke operators acting on the cohomology spaces $H^*(\Gamma; \mathscr{M})$. The eigenvalues of these operators are conjecturally related to certain representations of the Galois group. Just as in the case $G = \mathrm{SL}_2(\mathbb{R})$, we need tools to compute the Hecke action.

In this section we discuss this problem. We begin with a general description of the Hecke operators and how they act on cohomology. Then we focus on one particular cohomology group, namely the top degree $H^\nu(\Gamma; \mathbb{C})$, where $\nu = \mathrm{vcd}(\Gamma)$ and Γ has finite index in $\mathrm{SL}_n(\mathbb{Z})$. This is the setting that generalizes the modular symbols method from Chapter 8. We conclude by giving examples of Hecke eigenclasses in the cuspidal cohomology of $\Gamma_0(p) \subset \mathrm{SL}_3(\mathbb{Z})$.

A.4.2. Let $g \in \mathrm{SL}_n(\mathbb{Q})$. The group $\Gamma' = \Gamma \cap g^{-1}\Gamma g$ has finite index in both Γ and $g^{-1}\Gamma g$. The element g determines a diagram $C(g)$

$$\begin{array}{ccc} & \Gamma'\backslash X & \\ {}^{s}\swarrow & & \searrow^{t} \\ \Gamma\backslash X & & \Gamma\backslash X \end{array}$$

called a *Hecke correspondence.* The map s is induced by the inclusion $\Gamma' \subset \Gamma$, while t is induced by the inclusion $\Gamma' \subset g^{-1}\Gamma g$ followed by the diffeomorphism $g^{-1}\Gamma g\backslash X \to \Gamma\backslash X$ given by left multiplication by g. Specifically,

$$s(\Gamma' x) = \Gamma x, \quad t(\Gamma' x) = \Gamma g x, \quad x \in X.$$

The maps s and t are finite-to-one, since the indices $[\Gamma' : \Gamma]$ and $[\Gamma' : g^{-1}\Gamma g]$ are finite. This implies that we obtain maps on cohomology

$$s^*\colon H^*(\Gamma\backslash X) \to H^*(\Gamma'\backslash X), \quad t_*\colon H^*(\Gamma'\backslash X) \to H^*(\Gamma\backslash X).$$

Here the map s^* is the usual induced map on cohomology, while the "wrong-way" map[8] t_* is given by summing a class over the finite fibers of t. These maps can be composed to give a map

$$T_g := t_* s^*\colon H^*(\Gamma\backslash X; \widetilde{\mathscr{M}}) \longrightarrow H^*(\Gamma\backslash X; \widetilde{\mathscr{M}}).$$

This is called the *Hecke operator* associated to g. There is an obvious notion of isomorphism of Hecke correspondences. One can show that up to isomorphism, the correspondence $C(g)$ and thus the Hecke operator T_g depend only on the double coset $\Gamma g \Gamma$. One can compose Hecke correspondences, and thus we obtain an algebra of operators acting on the cohomology, just as in the classical case.

Example A.17. Let $n = 2$, and let $\Gamma = \mathrm{SL}_2(\mathbb{Z})$. If we take $g = \mathrm{diag}(1, p)$, where p is a prime, then the action of T_g on $H^1(\Gamma; M_{k-2})$ is the same as the action of the classical Hecke operator T_p on the weight k holomorphic modular forms. If we take $\Gamma = \Gamma_0(N)$, we obtain an operator $T(p)$ for all p prime to N, and the algebra of Hecke operators coincides with the (semisimple) Hecke algebra generated by the T_p, $(p, N) = 1$. For $p|N$, one can also describe the U_p operators in this language.

Example A.18. Now let $n > 2$ and let $\Gamma = \mathrm{SL}_n(\mathbb{Z})$. The picture is very similar, except that now there are several Hecke operators attached to any prime p. In fact there are $n-1$ operators $T(p,k)$, $k = 1, \ldots, n-1$. The operator $T(p,k)$ is associated to the correspondence $C(g)$, where $g = \mathrm{diag}(1, \ldots, 1, p, \ldots, p)$ and where p occurs k times. If we consider the congruence subgroups $\Gamma_0(N)$, we have operators $T(p,k)$ for $(p, N) = 1$ and analogues of the U_p operators for $p|N$.

Just as in the classical case, any double coset $\Gamma g \Gamma$ can be written as a disjoint union of left cosets

$$\Gamma g \Gamma = \coprod_{h \in \Omega} \Gamma h$$

[8]Under the identification $H^*(\Gamma\backslash X; \widetilde{\mathscr{M}}) \simeq H^*(\Gamma; \mathscr{M})$, the map t_* becomes the transfer map in group cohomology [**Bro94**, III.9].

for a certain finite set of $n \times n$ integral matrices Ω. For the operator $T(p,k)$, the set Ω can be taken to be all upper-triangular matrices of the form [**Kri90**, Proposition 7.2]

$$\begin{pmatrix} p^{e_1} & & a_{ij} \\ & \ddots & \\ & & p^{e_n} \end{pmatrix},$$

where

- $e_i \in \{0,1\}$ and exactly k of the e_i are equal to 1 and
- $a_{ij} = 0$ unless $e_i = 0$ and $e_j = 1$, in which case a_{ij} satisfies $0 \leq a_{ij} < p$.

Remark A.19. The number of coset representatives for the operator $T(p,k)$ is the same as the number of points in the finite Grassmannian $G(k,n)(\mathbb{F}_p)$. A similar phenomenon is true for the Hecke operators for any group G, although there are some subtleties [**Gro98**].

A.4.3. Recall that in Section A.3.6 we constructed the Voronoĭ decomposition $\mathscr{C}$ and the well-rounded retract W and that we can use them to compute the cohomology $H^*(\Gamma;\mathscr{M})$. Unfortunately, we cannot directly use them to compute the action of the Hecke operators on cohomology, since the Hecke operators do not act cellularly on $\mathscr{C}$ or W. The problem is that the Hecke image of a cell in $\mathscr{C}$ (or W) is usually not a union of cells in $\mathscr{C}$ (or W). This is already apparent for $n = 2$. The edges of $\mathscr{C}$ are the $\mathrm{SL}_2(\mathbb{Z})$-translates of the ideal geodesic τ from 0 to ∞ (Example A.15). Applying a Hecke operator takes such an edge to a union of ideal geodesics, each with vertices at a pair of cusps. In general such geodesics are not an $\mathrm{SL}_2(\mathbb{Z})$-translate of τ.

For $n = 2$, one solution is to work with all possible ideal geodesics with vertices at the cusps, in other words the space of modular symbols $\mathbb{M}_2$ from Section 3.2. Manin's trick (Proposition 3.11) shows how to write any modular symbol as a linear combination of unimodular symbols, by which we mean modular symbols supported on the edges of $\mathscr{C}$. These are the ideas we now generalize to all n.

Definition A.20. Let S_0 be the $\mathbb{Q}$-vector space spanned by the symbols $\mathbf{v} = [v_1, \ldots, v_n]$, where $v_i \in \mathbb{Q}^n \smallsetminus \{0\}$, modulo the following relations:

(1) If τ is a permutation on n letters, then

$$[v_1, \ldots, v_n] = \mathrm{sign}(\tau)[\tau(v_1), \ldots, \tau(v_n)],$$

where $\mathrm{sign}(\tau)$ is the sign of τ.

(2) If $q \in \mathbb{Q}^\times$, then

$$[qv_1, v_2 \ldots, v_n] = [v_1, \ldots, v_n].$$

(3) If the points $v_1, \ldots, v_n$ are linearly dependent, then $\mathbf{v} = 0$.

Let $B \subset S_0$ be the subspace generated by linear combinations of the form

$$\sum_{i=0}^{n}(-1)^i[v_0, \ldots, \hat{v}_i, \ldots, v_n], \tag{A.4.1}$$

where $v_0, \ldots, v_n \in \mathbb{Q}^n \smallsetminus \{0\}$ and where $\hat{v}_i$ means to omit v_i.

We call S_0 the space of *modular symbols*. We caution the reader that there are some differences in what we call modular symbols and those found in Section 3.2 and Definition 8.2; we compare them in Section A.4.4. The group $\mathrm{SL}_n(\mathbb{Q})$ acts on S_0 by left multiplication: $g \cdot \mathbf{v} = [gv_1, \ldots, gv_n]$. This action preserves the subspace B and thus induces an action on the quotient $M = S_0/B$. For $\Gamma \subset \mathrm{SL}_n(\mathbb{Z})$ a finite index subgroup, let M_Γ be the space of Γ-coinvariants in M. In other words, M_Γ is the quotient of M by the subspace generated by $\{m - \gamma \cdot m \mid \gamma \in \Gamma\}$.

The relationship between modular symbols and the cohomology of Γ is given by the following theorem, first proved for SL_n by Ash and Rudolph [**AR79**] and by Ash for general G [**Ash86**]:

Theorem A.21 ([**Ash86, AR79**]). *Let $\Gamma \subset \mathrm{SL}_n(\mathbb{Z})$ be a finite index subgroup. There is an isomorphism*

$$M_\Gamma \xrightarrow{\sim} H^\nu(\Gamma; \mathbb{Q}), \tag{A.4.2}$$

where Γ acts trivially on $\mathbb{Q}$ and where $\nu = \mathrm{vcd}(\Gamma)$.

We remark that Theorem A.21 remains true if $\mathbb{Q}$ is replaced with nontrivial coefficients as in Section A.2.7. Moreover, if Γ is assumed to be torsion-free then we can replace $\mathbb{Q}$ with $\mathbb{Z}$.

The great virtue of M_Γ is that it admits an action of the Hecke operators. Given a Hecke operator T_g, write the double coset $\Gamma g \Gamma$ as a disjoint union of left cosets

$$\Gamma g \Gamma = \coprod_{h \in \Omega} \Gamma h \tag{A.4.3}$$

as in Example A.18. Any class in M_Γ can be lifted to a representative $\eta = \sum q(\mathbf{v})\mathbf{v} \in S_0$, where $q(\mathbf{v}) \in \mathbb{Q}$ and almost all $q(\mathbf{v})$ vanish. Then we define

$$T_g(\mathbf{v}) = \sum_{h \in \Omega} h \cdot \mathbf{v} \tag{A.4.4}$$

and extend to η by linearity. The right side of (A.4.4) depends on the choices of η and Ω, but after taking quotients and coinvariants, we obtain a well-defined action on cohomology via (A.4.2).

A.4.4. The space S_0 is closely related to the space $\mathbb{M}_2$ from Section 3.2 and Section 8.1. Indeed, $\mathbb{M}_2$ was defined to be the quotient $(F/R)/(F/R)_{\mathrm{tor}}$, where F is the free abelian group generated by ordered pairs

$$\{\alpha, \beta\}, \quad \alpha, \beta \in \mathbb{P}^1(\mathbb{Q}), \tag{A.4.5}$$

and R is the subgroup generated by elements of the form

$$\{\alpha, \beta\} + \{\beta, \gamma\} + \{\gamma, \alpha\}, \quad \alpha, \beta, \gamma \in \mathbb{P}^1(\mathbb{Q}). \tag{A.4.6}$$

The only new feature in Definition A.20 is item (3). For $n = 2$ this corresponds to the condition $\{\alpha, \alpha\} = 0$, which follows from (A.4.6). We have

$$S_0/B \simeq \mathbb{M}_2 \otimes \mathbb{Q}.$$

Hence there are two differences between S_0 and $\mathbb{M}_2$: our notion of modular symbols uses rational coefficients instead of integral coefficients and is the space of symbols *before* dividing out by the subspace of relations B; we further caution the reader that this is somewhat at odds with the literature.

We also remark that the general arbitrary weight definition of modular symbols for a subgroup $\Gamma \subset \mathrm{SL}_2(\mathbb{Z})$ given in Section 8.1 also includes taking Γ-coinvariants, as well as extra data for a coefficient system. We have not included the latter data since our emphasis is trivial coefficients, although it would be easy to do so in the spirit of Section 8.1.

Elements of $\mathbb{M}_2$ also have a geometric interpretation: the symbol $\{\alpha, \beta\}$ corresponds to the ideal geodesic in $\mathfrak{h}$ with endpoints at the cusps α and β. We have a similar picture for the symbols $\mathbf{v} = [v_1, \dots, v_n]$. We can assume that each v_i is primitive, which means that each v_i determines a vertex of the Voronoĭ polyhedron Π. The rational cone generated by these vertices determines a subset $\Delta(\mathbf{v}) \subset D$, where D is the linear model of the symmetric space $X = \mathrm{SL}_n(\mathbb{R})/\mathrm{SO}(n)$ from Section A.3.2. This subset $\Delta(\mathbf{v})$ is then an "ideal simplex" in X. There is also a connection between $\Delta(\mathbf{v})$ and torus orbits in X; we refer to [**Ash86**] for a related discussion.

A.4.5. Now we need a generalization of the Manin trick (Section 3.3.1). This is known in the literature as the *modular symbols algorithm.*

We can define a kind of norm function on S_0 as follows. Let $\mathbf{v} = [v_1, \dots, v_n]$ be a modular symbol. For each v_i, choose $\lambda_i \in \mathbb{Q}^\times$ such that $\lambda_i v_i$ is primitive. Then we define

$$\|\mathbf{v}\| := |\det(\lambda_1 v_1, \dots, \lambda_n v_n)| \in \mathbb{Z}.$$

Note that $\|\mathbf{v}\|$ is well defined, since the λ_i are unique up to sign, and permuting the v_i only changes the determinant by a sign. We extend $\|\ \|$ to all of S_0 by taking the maximum of $\|\ \|$ over the support of any $\eta \in S_0$: if

$\eta = \sum q(\mathbf{v})\mathbf{v}$, where $q(\mathbf{v}) \in \mathbb{Q}$ and almost all $q(\mathbf{v})$ vanish, then we put

$$\|\eta\| = \operatorname*{Max}_{q(\mathbf{v})\neq 0} \|\mathbf{v}\|.$$

We say a modular symbol η is *unimodular* if $\|\eta\| = 1$. It is clear that the images of the unimodular symbols generate a finite-dimensional subspace of M_Γ. The next theorem shows that this subspace is actually *all* of M_Γ.

Theorem A.22 ([**AR79, Bar94**]). *The space M_Γ is spanned by the images of the unimodular symbols. More precisely, given any symbol $\mathbf{v} \in S_0$ with $\|\mathbf{v}\| > 1$,*

(1) *in S_0/B we may write*

$$\mathbf{v} = \sum q(\mathbf{w})\mathbf{w}, \quad q(\mathbf{w}) \in \mathbb{Z}, \tag{A.4.7}$$

where if $q(\mathbf{w}) \neq 0$, then $\|\mathbf{w}\| = 1$, and

(2) *the number of terms on the right side of* (A.4.7) *is bounded by a polynomial in* $\log \|\mathbf{v}\|$ *that depends only on the dimension n.*

Proof. (Sketch) Given a modular symbol $\mathbf{v} = [v_1, \ldots, v_n]$, we may assume that the points v_i are primitive. We will show that if $\|\mathbf{v}\| > 1$, we can find a point u such that when we apply the relation (A.4.1) using the points $u, v_1, \ldots, v_n$, all terms other than $\mathbf{v}$ have norm less than $\|\mathbf{v}\|$. We call such a point a *reducing point* for $\mathbf{v}$.

Let $P \subset \mathbb{R}^n$ be the open parallelotope

$$P := \left\{ \sum \lambda_i v_i \;\middle|\; |\lambda_i| < \|\mathbf{v}\|^{-1/n} \right\}.$$

Then P is an n-dimensional centrally symmetric convex body with volume 2^n. By Minkowski's theorem from the geometry of numbers (cf. [**FT93**, IV.2.6]), $P \cap \mathbb{Z}^n$ contains a nonzero point u. Using (A.4.1), we find

$$\mathbf{v} = \sum_{i=1}^{n} (-1)^{i-1} \mathbf{v}_i(u), \tag{A.4.8}$$

where $\mathbf{v}_i(u)$ is the symbol

$$\mathbf{v}_i(u) = [v_1, \ldots, v_{i-1}, u, v_{i+1}, \ldots, v_n].$$

Moreover, it is easy to see that the new symbols satisfy

$$0 \leq \|\mathbf{v}_i(u)\| < \|\mathbf{v}\|^{(n-1)/n}, \quad i = 1, \ldots, n. \tag{A.4.9}$$

This completes the proof of the first statement.

To prove the second statement, we must estimate how many times relations of the form (A.4.8) need to be applied to obtain (A.4.7). A nonunimodular symbol produces at most n new modular symbols after (A.4.8) is performed; we potentially have to apply (A.4.8) again to each of the symbols that result, which in turn could produce as many as n new symbols for each. Hence we can visualize the process of constructing (A.4.7) as building a rooted tree, where the root is $\mathbf{v}$, the leaves are the symbols $\mathbf{w}$, and where each node has at most n children. It is not hard to see that the bound (A.4.9) implies that the depth of this tree (i.e., the longest length of a path from the root to a leaf) is $O(\log\log\|\mathbf{v}\|)$. From this the second statement follows easily. □

Statement (1) of Theorem A.22 is due to Ash and Rudolph [**AR79**]. Instead of P, they used the larger parallelotope P' defined by

$$P' := \Big\{\sum \lambda_i v_i \;\Big|\; |\lambda_i| < 1\Big\},$$

which has volume $2^n\|\mathbf{v}\|$. The observation that P' can be replaced by P and the proof of (2) are both due to Barvinok [**Bar94**].

A.4.6. The relationship between Theorem A.22 and Manin's trick should be clear. For $\Gamma \subset \mathrm{SL}_2(\mathbb{Z})$, the Manin symbols correspond exactly to the unimodular symbols mod Γ. So Theorem A.22 implies that every modular symbol (in the language of Section 8.1) is a linear combination of Manin symbols. This is exactly the conclusion of Proposition 8.3.

In higher rank the relationship between Manin symbols and unimodular symbols is more subtle. In fact there are two possible notions of "Manin symbol," which agree for $\mathrm{SL}_2(\mathbb{Z})$ but not in general. One possibility is the obvious one: a Manin symbol is a unimodular symbol.

The other possibility is to define a Manin symbol to be a modular symbol corresponding to a top-dimensional cell of the retract W. But for $n \geq 5$, such modular symbols need not be unimodular. In particular, for $n = 5$ there are two equivalence classes of top-dimensional cells. One class corresponds to the unimodular symbols, the other to a set of modular symbols of norm 2. However, Theorems A.21 and A.22 show that $H^\nu(\Gamma;\mathbb{Q})$ is spanned by unimodular symbols. Thus as far as this cohomology group is concerned, the second class of symbols is in some sense unnecessary.

A.4.7. We return to the setting of Section A.3.8 and give examples of Hecke eigenclasses in the cusp cohomology of $\Gamma = \Gamma_0(p) \subset \mathrm{SL}_3(\mathbb{Z})$. We closely follow [**AGG84, vGvdKTV97**]. Note that since the top of the cuspidal range for SL_3 is the same as the virtual cohomological dimension ν, we can use modular symbols to compute the Hecke action on cuspidal classes.

Given a prime l coprime to p, there are two Hecke operators of interest $T(l,1)$ and $T(l,2)$. We can compute the action of these operators on

$H^3_{\text{cusp}}(\Gamma;\mathbb{C})$ as follows. Recall that $H^3_{\text{cusp}}(\Gamma;\mathbb{C})$ can be identified with a certain space of functions $f\colon \mathbb{P}^2(\mathbb{F}_p)\to\mathbb{C}$ (Theorem A.16). Given $x\in\mathbb{P}^2(\mathbb{F}_p)$, let $Q_x\in\mathrm{SL}_3(\mathbb{Z})$ be a matrix such that $Q_x\mapsto x$ under the identification $\mathbb{P}^2(\mathbb{F}_p)\xrightarrow{\sim}\Gamma\backslash\mathrm{SL}_3(\mathbb{Z})$. Then Q_x determines a unimodular symbol $[Q_x]$ by taking the v_i to be the columns of Q_x. Given any Hecke operator T_g, we can find coset representatives h_i such that $\Gamma g\Gamma=\coprod\Gamma h_i$ (explicit representatives for $\Gamma=\Gamma_0(p)$ and $T_g=T(l,k)$ are given in [**AGG84, vGvdKTV97**]). The modular symbols $[h_iQ_x]$ are no longer unimodular in general, but we can apply Theorem A.22 to write

$$[h_iQ_x]=\sum_j[R_{ij}],\quad R_{ij}\in\mathrm{SL}_3(\mathbb{Z}).$$

Then for $f\colon\mathbb{P}^2(\mathbb{F}_p)\to\mathbb{C}$ as in Theorem A.16, we have

$$(T_gf)(x)=\sum_{i,j}f(\overline{R_{ij}}),$$

where $\overline{R_{ij}}$ is the class of R_{ij} in $\mathbb{P}^2(\mathbb{F}_p)$.

Now let $\xi\in H^3_{\text{cusp}}(\Gamma;\mathbb{C})$ be a simultaneous eigenclass for all the Hecke operators $T(l,1)$, $T(l,2)$, as l ranges over all primes coprime with p. General considerations from the theory of automorphic forms imply that the eigenvalues $a(l,1)$, $a(l,2)$ are complex conjugates of one other. Hence it suffices to compute $a(l,1)$. We give two examples of cuspidal eigenclasses for two different prime levels.

Example A.23. Let $p=53$. Then $H^3_{\text{cusp}}(\Gamma_0(53);\mathbb{C})$ is 2-dimensional. Let $\eta=(1+\sqrt{-11})/2$. One eigenclass is given by the data

l	2	3	5	7	11	13
$a(l,1)$	$-1-2\eta$	$-2+2\eta$	1	-3	1	$-2-12\eta$

and the other is obtained by complex conjugation.

Example A.24. Let $p=61$. Then $H^3_{\text{cusp}}(\Gamma_0(61);\mathbb{C})$ is 2-dimensional. Let $\omega=(1+\sqrt{-3})/2$. One eigenclass is given by the data

l	2	3	5	7	11	13
$a(l,1)$	$1-2\omega$	$-5+4\omega$	$-2+4\omega$	-6ω	$-2+2\omega$	$-2-4\omega$

and the other is obtained by complex conjugation.

A.5. Other Cohomology Groups

A.5.1. In Section A.4 we saw how to compute the Hecke action on the top cohomology group $H^\nu(\Gamma;\mathbb{C})$. Unfortunately for $n\geq 4$, this cohomology group does not contain any cuspidal cohomology. The first case is $\Gamma\subset\mathrm{SL}_4(\mathbb{Z})$; we have $\mathrm{vcd}(\Gamma)=6$, and the cusp cohomology lives in degrees 4 and 5. One can show that the cusp cohomology in degree 4 is dual to that in degree 5, so for computational purposes it suffices to be able to compute the Hecke action on $H^5(\Gamma; \mathbb{C})$. But modular symbols do not help us here.

In this section we describe a technique to compute the Hecke action on $H^{\nu-1}(\Gamma; \mathbb{C})$, following [**Gun00a**]. The technique is an extension of the modular symbol algorithm to these cohomology groups. In principle the ideas in this section can be modified to compute the Hecke action on other cohomology groups $H^{\nu-k}(\Gamma; \mathbb{C})$, $k > 1$, although this has not been investigated[9]. For $n = 4$, we have applied the algorithm in joint work with Ash and McConnell to investigate computationally the cohomology $H^5(\Gamma; \mathbb{C})$, where $\Gamma_0(N) \subset \mathrm{SL}_4(\mathbb{Z})$ [**AGM02**].

A.5.2. To begin, we need an analogue of Theorem A.21 for lower degree cohomology groups. In other words, we need a generalization of the modular symbols for other cohomology groups. This is achieved by the *sharbly complex* S_*:

Definition A.25 ([**Ash94**])**.** Let $\{S_*, \partial\}$ be the chain complex given by the following data:

(1) For $k \geq 0$, S_k is the $\mathbb{Q}$-vector space generated by the symbols $\mathbf{u} = [v_1, \ldots, v_{n+k}]$, where $v_i \in \mathbb{Q}^n \smallsetminus \{0\}$, modulo the relations:
   (a) If τ is a permutation on $(n + k)$ letters, then
   $$[v_1, \ldots, v_{n+k}] = \mathrm{sign}(\tau)[\tau(v_1), \ldots, \tau(v_{n+k})],$$
   where $\mathrm{sign}(\tau)$ is the sign of τ.
   (b) If $q \in \mathbb{Q}^\times$, then
   $$[qv_1, v_2 \ldots, v_{n+k}] = [v_1, \ldots, v_{n+k}].$$
   (c) If the rank of the matrix $(v_1, \ldots, v_{n+k})$ is less than n, then $\mathbf{u} = 0$.

(2) For $k > 0$, the boundary map $\partial\colon S_k \to S_{k-1}$ is
   $$[v_1, \ldots, v_{n+k}] \longmapsto \sum_{i=1}^{n+k} (-1)^i [v_1, \ldots, \hat{v}_i, \ldots, v_{n+k}].$$
   We define ∂ to be identically zero on S_0.

The elements
$$\mathbf{u} = [v_1, \ldots, v_{n+k}]$$

[9] The first interesting case is $n = 5$, for which the cuspidal cohomology lives in $H^{\nu-2}$.

are called *k-sharblies*[10]. The 0-sharblies are exactly the modular symbols from Definition A.20, and the subspace $B \subset S_0$ is the image of the boundary map $\partial\colon S_1 \to S_0$.

There is an obvious left action of Γ on S_* commuting with ∂. For any $k \geq 0$, let $S_{k,\Gamma}$ be the space of Γ-coinvariants. Since the boundary map ∂ commutes with the Γ-action, we obtain a complex $(S_{*,\Gamma}, \partial_\Gamma)$. The following theorem shows that this complex computes the cohomology of Γ:

Theorem A.26 ([**Ash94**]). *There is a natural isomorphism*

$$H^{\nu-k}(\Gamma;\mathbb{C}) \xrightarrow{\sim} H_k(S_{*,\Gamma} \otimes \mathbb{C}).$$

A.5.3. We can extend our norm function $\| \; \|$ from modular symbols to all of S_k as follows. Let $\mathbf{u} = [v_1, \dots, v_{n+k}]$ be a k-sharbly, and let $Z(\mathbf{u})$ be the set of all submodular symbols determined by $\mathbf{u}$. In other words, $Z(\mathbf{u})$ consists of the modular symbols of the form $[v_{i_1}, \dots, v_{i_n}]$, where $\{i_1, \dots, i_n\}$ ranges over all n-fold subsets of $\{1, \dots, n+k\}$. Define $\|\mathbf{u}\|$ by

$$\|\mathbf{u}\| = \operatorname*{Max}_{\mathbf{v}\in Z(\mathbf{u})} \|\mathbf{v}\|.$$

Note that $\|\mathbf{u}\|$ is well defined modulo the relations in Definition A.25. As for modular symbols, we extend the norm to sharbly chains $\xi = \sum q(\mathbf{u})\mathbf{u}$ taking the maximum norm over the support. Formally, we let $\operatorname{supp}(\xi) = \{\mathbf{u} \mid q(\mathbf{u}) \neq 0\}$ and $Z(\xi) = \bigcup_{\mathbf{u}\in\operatorname{supp}(\xi)} Z(\mathbf{u})$, and then we define $\|\xi\|$ by

$$\|\xi\| = \operatorname*{Max}_{\mathbf{v}\in Z(\xi)} \|\mathbf{v}\|.$$

We say that ξ is *reduced* if $\|\xi\| = 1$. Hence ξ is reduced if and only if all its submodular symbols are unimodular or have determinant 0. Clearly there are only finitely many reduced k-sharblies modulo Γ for any k.

In general the cohomology groups $H^*(\Gamma;\mathbb{C})$ are *not* spanned by reduced sharblies. However, it is known (cf. [**McC91**]) that for $\Gamma \subset \mathrm{SL}_4(\mathbb{Z})$, the group $H^5(\Gamma;\mathbb{C})$ is spanned by reduced 1-sharbly cycles. The best one can say in general is that for each pair n, k, there is an integer $N = N(n,k)$ such that for $\Gamma \subset \mathrm{SL}_n(\mathbb{Z})$, $H^{\nu-k}(\Gamma;\mathbb{C})$ is spanned by k-sharblies of norm $\leq N$. This set of sharblies is also finite modulo Γ, although it is not known how large N must be for any given pair n, k.

A.5.4. Recall that the cells of the well-rounded retract W are indexed by sets of primitive vectors in $\mathbb{Z}^n$. Since any primitive vector determines a point in $\mathbb{Q}^n \smallsetminus \{0\}$ and since sets of such points index sharblies, it is clear that there is a close relationship between S_* and the chain complex associated to W,

[10]The terminology for S_* is due to Lee Rudolph, in honor of Lee and Szczarba. They introduced a very similar complex in [**LS76**] for $\mathrm{SL}_3(\mathbb{Z})$.

although of course S_* is much bigger. In any case, both complexes compute $H^*(\Gamma;\mathbb{C})$.

The main benefit of using the sharbly complex to compute cohomology is that it admits a Hecke action. Suppose $\xi = \sum q(\mathbf{u})\mathbf{u}$ is a sharbly cycle mod Γ, and consider a Hecke operator T_g. Then we have

$$T_g(\xi) = \sum_{h\in\Omega,\mathbf{u}} n(\mathbf{u})h\cdot\mathbf{u}, \tag{A.5.1}$$

where Ω is a set of coset representatives as in (A.4.3). Since $\Omega \not\subset \mathrm{SL}_n(\mathbb{Z})$ in general, the Hecke image of a reduced sharbly is not usually reduced.

A.5.5. We are now ready to describe our algorithm for the computation of the Hecke operators on $H^{\nu-1}(\Gamma;\mathbb{C})$. It suffices to describe an algorithm that takes as input a 1-sharbly cycle ξ and produces as output a cycle ξ' with

(a) the classes of ξ and ξ' in $H^{\nu-1}(\Gamma;\mathbb{C})$ the same, and
(b) $\|\xi'\| < \|\xi\|$ if $\|\xi\| > 1$.

Below, we will present an algorithm satisfying (a). In [**Gun00a**], we conjectured (and presented evidence) that the algorithm satisfies (b) for $n \leq 4$. Further evidence is provided by the computations in [**AGM02**], which relied on the algorithm to compute the Hecke action on $H^5(\Gamma;\mathbb{C})$, where $\Gamma = \Gamma_0(N) \subset \mathrm{SL}_4(\mathbb{Z})$.

The idea behind the algorithm is simple: given a 1-sharbly cycle ξ that is not reduced, (i) simultaneously apply the modular symbol algorithm (Theorem A.22) to each of its submodular symbols, and then (ii) package the resulting data into a new 1-sharbly cycle. Our experience in presenting this algorithm is that most people find the geometry involved in (ii) daunting. Hence we will give details only for $n = 2$ and will provide a sketch for $n > 2$. Full details are contained in [**Gun00a**]. Note that $n = 2$ is topologically and arithmetically uninteresting, since we are computing the Hecke action on $H^0(\Gamma;\mathbb{C})$; nevertheless, the geometry faithfully represents the situation for all n.

A.5.6. Fix $n = 2$, let $\xi \in S_1$ be a 1-sharbly cycle mod Γ for some $\Gamma \subset \mathrm{SL}_2(\mathbb{Z})$, and suppose ξ is not reduced. Assume Γ is torsion-free to simplify the presentation.

Suppose first that all submodular symbols $\mathbf{v} \in Z(\xi)$ are nonunimodular. Select reducing points for each $\mathbf{v} \in Z(\xi)$ and make these choices Γ-equivariantly. This means the following. Suppose $\mathbf{u}, \mathbf{u}' \in \operatorname{supp}\xi$ and $\mathbf{v} \in \operatorname{supp}(\partial\mathbf{u})$ and $\mathbf{v}' \in \operatorname{supp}(\partial\mathbf{u}')$ are modular symbols such that $\mathbf{v} = \gamma\cdot\mathbf{v}'$ for some $\gamma \in \Gamma$. Then we select reducing points w for $\mathbf{v}$ and w' for $\mathbf{v}'$ such that $w = \gamma\cdot w'$. (Note that since Γ is torsion-free, no modular symbol can be identified to itself by an element of Γ; hence $\mathbf{v} \neq \mathbf{v}'$.) This is possible since if $\mathbf{v}$ is a modular symbol and w is a reducing point for $\mathbf{v}$, then $\gamma \cdot w$ is a reducing point for $\gamma \cdot \mathbf{v}$ for any $\gamma \in \Gamma$. Because there are only finitely many Γ-orbits in $Z(\xi)$, we can choose reducing points Γ-equivariantly by selecting them for some set of orbit representatives.

It is important to note that Γ-equivariance is the only global criterion we use when selecting reducing. In particular, there is a priori no relationship among the three reducing points chosen for any $\mathbf{u} \in \operatorname{supp} \xi$.

A.5.7. Now we want to use the reducing points and the 1-sharblies in ξ to build ξ'. Choose $\mathbf{u} = [v_1, v_2, v_3] \in \operatorname{supp} \xi$, and denote the reducing point for $[v_i, v_j]$ by w_k, where $\{i, j, k\} = \{1, 2, 3\}$. We use the v_i and the w_i to build a 2-sharbly chain $\eta(\mathbf{u})$ as follows.

Let P be an octahedron in $\mathbb{R}^3$. Label the vertices of P with the v_i and w_i such that the vertex labeled v_i is opposite the vertex labeled w_i (Figure A.5.1). Subdivide P into four tetrahedra by connecting two opposite vertices, say v_1 and w_1, with an edge (Figure A.5.2). For each tetrahedron T, take the labels of four vertices and arrange them into a quadruple. If we orient P, then we can use the induced orientation on T to order the four primitive points. In this way, each T determines a 2-sharbly, and $\eta(\mathbf{u})$ is defined to be the sum. For example, if we use the decomposition in Figure A.5.2, we have

(A.5.2)

$$\eta(\mathbf{u}) = [v_1, v_3, v_2, w_1] + [v_1, w_2, v_3, w_1] + [v_1, w_3, w_2, w_1] + [v_1, v_2, w_3, w_1].$$

Repeat this construction for all $\mathbf{u} \in \operatorname{supp} \xi$, and let $\eta = \sum q(\mathbf{u})\eta(\mathbf{u})$. Finally, let $\xi' = \xi + \partial\eta$.

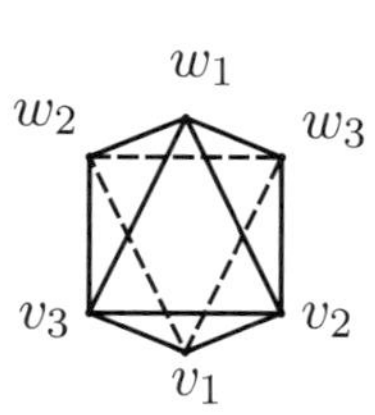

Figure A.5.1.

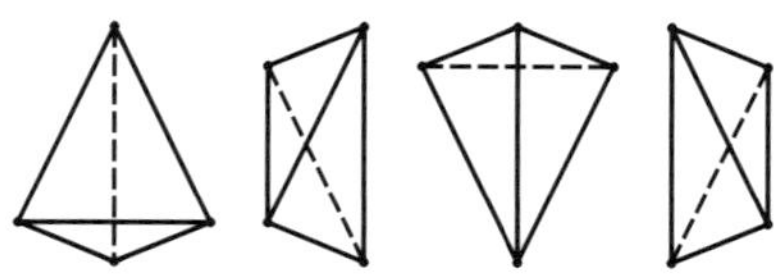

Figure A.5.2.

A.5.8. By construction, ξ' is a cycle mod Γ in the same class as ξ. We claim in addition that no submodular symbol from ξ appears in ξ'. To see this, consider $\partial\eta(\mathbf{u})$. From (A.5.2), we have

$$\begin{aligned}(\text{A.5.3})\quad \partial\eta(\mathbf{u}) = &-[v_1, v_2, v_3] + [v_1, v_2, w_3] + [v_1, w_2, v_3] + [w_1, v_2, v_3]\\ &- [v_1, w_2, w_3] - [w_1, v_2, w_3] - [w_1, w_2, v_3] + [w_1, w_2, w_3].\end{aligned}$$

Note that this is the boundary in S_*, not in $S_{*,\Gamma}$. Furthermore, $\partial\eta(\mathbf{u})$ is independent of which pair of opposite vertices of P we connected to build $\eta(\mathbf{u})$.

From (A.5.3), we see that in $\xi + \partial\eta$ the 1-sharbly $-[v_1, v_2, v_3]$ is canceled by $\mathbf{u} \in \operatorname{supp}\xi$. We also claim that 1-sharblies in (A.5.3) of the form $[v_i, v_j, w_k]$ vanish in $\partial_\Gamma\eta$.

To see this, let $\mathbf{u}$,$\mathbf{u}' \in \operatorname{supp}\xi$, and suppose $\mathbf{v} = [v_1, v_2] \in \operatorname{supp}\partial\mathbf{u}$ equals $\gamma \cdot \mathbf{v}'$ for some $\mathbf{v}' = [v_1', v_2'] \in \operatorname{supp}\partial\mathbf{u}'$. Since the reducing points were chosen Γ-equivariantly, we have $w = \gamma \cdot w'$. This means that the 1-sharbly $[v_1, v_2, w] \in \partial\eta(\mathbf{u})$ will be canceled mod Γ by $[v_1', v_2', w'] \in \partial\eta(\mathbf{u}')$. Hence, in passing from ξ to ξ', the effect in $(S_*)_\Gamma$ is to replace $\mathbf{u}$ with *four* 1-sharblies in $\operatorname{supp}\xi'$:

$$(\text{A.5.4})\quad [v_1, v_2, v_3] \longmapsto -[v_1, w_2, w_3] - [w_1, v_2, w_3] - [w_1, w_2, v_3] + [w_1, w_2, w_3].$$

Note that in (A.5.4), there are no 1-sharblies of the form $[v_i, v_j, w_k]$.

Remark A.27. For implementation purposes, it is not necessary to explicitly construct η. Rather, one may work directly with (A.5.4).

A.5.9. Why do we expect ξ' to satisfy $\|\xi'\| < \|\xi\|$? First of all, in the right hand side of (A.5.4) there are no submodular symbols of the form $[v_i, v_j]$. In fact, any submodular symbol involving a point v_i also includes a reducing point for $[v_i, v_j]$.

On the other hand, consider the submodular symbols in (A.5.4) of the form $[w_i, w_j]$. Since there is no relationship among the w_i, one has no reason to believe that these modular symbols are closer to unimodularity than those in $\mathbf{u}$. Indeed, for certain choices of reducing points it can happen that $\|[w_i, w_j]\| \geq \|\mathbf{u}\|$.

The upshot is that some care must be taken in choosing reducing points. In [**Gun00a**, Conjectures 3.5 and 3.6] we describe two methods for finding reducing points for modular symbols, one using Voronoĭ reduction and one using LLL-reduction. Our experience is that if one selects reducing points using either of these conjectures, then $\|[w_i, w_j]\| < \|\mathbf{u}\|$ for each of the new modular symbols $[w_i, w_j]$. In fact, in practice these symbols are trivial or satisfy $\|[w_i, w_j]\| = 1$.

A.5.10. In the previous discussion we assumed that no submodular symbols of any $\mathbf{u} \in \operatorname{supp} \xi$ were unimodular. Now we say what to do if some are. There are three cases to consider.

First, all submodular symbols of $\mathbf{u}$ may be unimodular. In this case there are no reducing points, and (A.5.4) becomes

$$[v_1, v_2, v_3] \longmapsto [v_1, v_2, v_3]. \tag{A.5.5}$$

Second, one submodular symbol of $\mathbf{u}$ may be nonunimodular, say the symbol $[v_1, v_2]$. In this case, to build η, we use a tetrahedron P' and put $\eta(\mathbf{u}) = [v_1, v_2, v_3, w_3]$ (Figure A.5.3). Since $[v_1, v_2, w_3]$ vanishes in the boundary of η mod Γ, (A.5.4) becomes

$$[v_1, v_2, v_3] \mapsto -[v_1, v_3, w_3] + [v_2, v_3, w_3]. \tag{A.5.6}$$

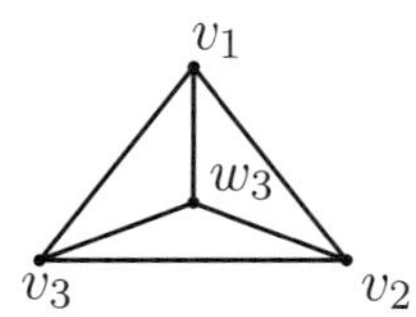

Figure A.5.3.

Finally, two submodular symbols of $\mathbf{u}$ may be nonunimodular, say $[v_1, v_2]$ and $[v_1, v_3]$. In this case we use the cone on a square P'' (Figure A.5.4). To construct $\eta(\mathbf{u})$, we must choose a decomposition of P'' into tetrahedra. Since P'' has a nonsimplicial face, this choice affects ξ' (in contrast to the previous cases). If we subdivide P'' by connecting the vertex labelled v_2 with the vertex labelled w_2, we obtain

$$[v_1, v_2, v_3] \longmapsto [v_2, w_2, w_3] + [v_2, v_3, w_2] + [v_1, v_3, w_2]. \tag{A.5.7}$$

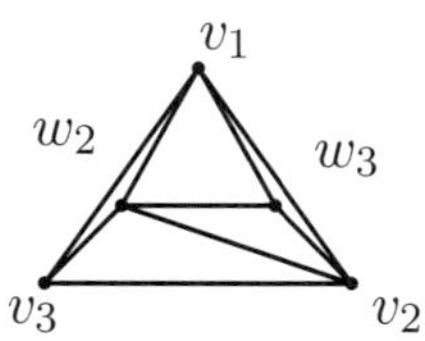

Figure A.5.4.

A.5.11. Now consider general n. The basic technique is the same, but the combinatorics become more complicated. Suppose $\mathbf{u} = [v_1, \ldots, v_{n+1}]$ satisfies $q(\mathbf{u}) \neq 0$ in a 1-sharbly cycle ξ, and for $i = 1, \ldots, n+1$ let $\mathbf{v}_i$ be the submodular symbol $[v_1, \ldots, \widehat{v_i}, \ldots, v_{n+1}]$. Assume that all $\mathbf{v}_i$ are nonunimodular, and for each i let w_i be a reducing point for $\mathbf{v}_i$.

For any subset $I \subset \{1, \dots, n+1\}$, let $\mathbf{u}_I$ be the 1-sharbly $[u_1, \dots, u_{n+1}]$, where $u_i = w_i$ if $i \in I$, and $u_i = v_i$ otherwise. The polytope P used to build $\eta(\mathbf{u})$ is the *cross polytope*, which is the higher-dimensional analogue of the octahedron [**Gun00a**, §4.4]. We suppress the details and give the final answer: (A.5.4) becomes

$$\mathbf{u} \longmapsto -\sum_I (-1)^{\#I} \mathbf{u}_I, \tag{A.5.8}$$

where the sum is taken over all subsets $I \subset \{1, \dots, n+1\}$ of cardinality *at least* 2.

More generally, if some $\mathbf{v}_i$ happen to be unimodular, then the polytope used to build η is an iterated cone on a lower-dimensional cross polytope. This is already visible for $n = 2$:

- The 2-dimensional cross polytope is a square, and the polytope P'' is a cone on a square.
- The 1-dimensional cross polytope is an interval, and the polytope P' is a double cone on an interval.

Altogether there are $n+1$ relations generalizing (A.5.5)–(A.5.7).

A.5.12. Now we describe how these computations are carried out in practice, focusing on $\Gamma = \Gamma_0(N) \subset \mathrm{SL}_4(\mathbb{Z})$ and $H^5(\Gamma; \mathbb{C})$. Besides discussing technical details, we also have to slightly modify some aspects of the construction in Section A.5.6, since Γ is not torsion-free.

Let W be the well-rounded retract. We can represent a cohomology class $\beta \in H^5(\Gamma; \mathbb{C})$ as $\beta = \sum q(\sigma)\sigma$, where σ denotes a codimension 1 cell in W. In this case there are three types of codimension 1 cells in W. Under the bijection $W \leftrightarrow \mathscr{C}$, these cells correspond to the Voronoĭ cells indexed by the columns of the matrices

$$\begin{pmatrix} 1&0&0&0&1\\0&1&0&0&1\\0&0&1&0&1\\0&0&0&1&1 \end{pmatrix}, \quad \begin{pmatrix} 1&0&0&0&1\\0&1&0&0&1\\0&0&1&0&1\\0&0&0&1&0 \end{pmatrix}, \quad \begin{pmatrix} 1&0&0&0&1\\0&1&0&0&1\\0&0&1&0&0\\0&0&0&1&0 \end{pmatrix}. \tag{A.5.9}$$

Thus each σ in W modulo Γ corresponds to an $\mathrm{SL}_4(\mathbb{Z})$-translate of one of the matrices in (A.5.9). These translates determine basis 1-sharblies $\mathbf{u}$ (by taking the points u_i to be the columns), and hence we can represent β by a 1-sharbly chain $\xi = \sum q(\mathbf{u})\mathbf{u} \in S_1$ that is a cycle in the complex of coinvariants $(S_{*,\Gamma}, \partial_\Gamma)$.

To make later computations more efficient, we precompute more data attached to ξ. Given a 1-sharbly $\mathbf{u} = [u_1, \dots, u_{n+1}]$, a *lift* $M(\mathbf{u})$ of $\mathbf{u}$ is defined to be an integral matrix with primitive columns M_i such that $\mathbf{u} =$

$[M_1, \dots, M_{n+1}]$. Then we encode ξ, once and for all, by a finite collection Φ of 4-tuples

$$(\mathbf{u}, n(\mathbf{u}), \{\mathbf{v}\}, \{M(\mathbf{v})\}),$$

where

(1) $\mathbf{u}$ ranges over the support of ξ,

(2) $n(\mathbf{u}) \in \mathbb{C}$ is the coefficient of $\mathbf{u}$ in ξ,

(3) $\{\mathbf{v}\}$ is the set of submodular symbols appearing in the bounday of $\mathbf{u}$, and

(4) $\{M(\mathbf{v})\}$ is a set of lifts for $\{\mathbf{v}\}$.

Moreover, the lifts in (4) are chosen to satisfy the following Γ-equivariance condition. Suppose that for $\mathbf{u}, \mathbf{u}' \in \operatorname{supp} \xi$ we have $\mathbf{v} \in \operatorname{supp}(\partial \mathbf{u})$ and $\mathbf{v}' \in \operatorname{supp}(\partial \mathbf{u}')$ satisfying $\mathbf{v} = \gamma \cdot \mathbf{v}'$ for some $\gamma \in \Gamma$. Then we require $M(\mathbf{v}) = \gamma M(\mathbf{v}')$. This is possible since ξ is a cycle modulo Γ, although there is one complication since Γ has torsion: it can happen that some submodular symbol $\mathbf{v}$ of a 1-sharbly $\mathbf{u}$ is identified to *itself* by an element of Γ. This means that in constructing $\{M(\mathbf{v})\}$ for $\mathbf{u}$, we must somehow choose more than one lift for $\mathbf{v}$. To deal with this, let $M(\mathbf{v})$ be any lift of $\mathbf{v}$, and let $\Gamma(\mathbf{v}) \subset \Gamma$ be the stabilizer of $\mathbf{v}$. Then in ξ, we replace $q(\mathbf{u})\mathbf{u}$ by

$$\frac{1}{\#\Gamma(\mathbf{v})} \sum_{\gamma \in \Gamma(\mathbf{v})} q(\mathbf{u})\mathbf{u}_\gamma,$$

where $\mathbf{u}_\gamma$ has the same data as $\mathbf{u}$, except[11] that we give $\mathbf{v}$ the lift $\gamma M(\mathbf{v})$.

Next we compute and store the 1-sharbly transformation laws generalizing (A.5.5)–(A.5.7). As a part of this we fix triangulations of certain cross polytopes as in (A.5.7).

We are now ready to begin the actual reduction algorithm. We take a Hecke operator $T(l, k)$ and build the coset representatives Ω as in (A.5.1). For each $h \in \Omega$ and each 1-sharbly $\mathbf{u}$ in the support of ξ, we obtain a non-reduced 1-sharbly $\mathbf{u}_h := h \cdot \mathbf{u}$. Here h acts on all the data attached to $\mathbf{u}$ in the list Φ. In particular, we replace each lift $M(\mathbf{v})$ with $h \cdot M(\mathbf{v})$, where the dot means matrix multiplication.

Now we check the submodular symbols of $\mathbf{u}_h$ and choose reducing points for the nonunimodular symbols. This is where the lifts come in handy. Recall that reduction points must be chosen Γ-equivariantly over the entire cycle. Instead of explicitly keeping track of the identifications between modular symbols, we do the following trick:

[11] In fact, we can be slightly more clever than this and only introduce denominators that are powers of 2.

(1) Construct the *Hermite normal form* $M_{\mathrm{her}}(\mathbf{v})$ of the lift $M(\mathbf{v})$ (see [**Coh93**, §2.4] and Exercise 7.5). Record the transformation matrix $U \in \mathrm{GL}_4(\mathbb{Z})$ such that $UM(\mathbf{v}) = M_{\mathrm{her}}(\mathbf{v})$.

(2) Choose a reducing point u for $M_{\mathrm{her}}(\mathbf{v})$.

(3) Then the reducing point for $M(\mathbf{v})$ is $U^{-1}u$.

This guarantees Γ-equivariance: if $\mathbf{v}$, $\mathbf{v}'$ are submodular symbols of ξ with $\gamma\cdot\mathbf{v} = \mathbf{v}'$ and with reducing points u, u', we have $\gamma u = u'$. The reason is that the Hermite normal form $M_{\mathrm{her}}(\mathbf{v})$ is a *uniquely determined* representative of the $\mathrm{GL}_4(\mathbb{Z})$-orbit of $M(\mathbf{v})$ [**Coh93**]. Hence if $\gamma M(\mathbf{v}) = M(\mathbf{v}')$, then $M_{\mathrm{her}}(\mathbf{v}) = M_{\mathrm{her}}(\mathbf{v}')$.

After computing all reducing points, we apply the appropriate transformation law. The result will be a chain of 1-sharblies, each of which has (conjecturally) smaller norm than the original 1-sharbly $\mathbf{u}$. We output these 1-sharblies if they are reduced; otherwise they are fed into the reduction algorithm again. Eventually we obtain a reduced 1-sharbly cycle ξ' homologous to the original cycle ξ.

The final step of the algorithm is to rewrite ξ' as a cocycle on W. This is easy to do since the relevant cells of W are in bijection with the reduced 1-sharblies. There are some nuisances in keeping orientations straight, but the computation is not difficult. We refer to [**AGM02**] for details.

A.5.13. We now give some examples, taken from [**AGM02**], of Hecke eigenclasses in $H^5(\Gamma_0(N);\mathbb{C})$ for various levels N. Instead of giving a table of eigenvalues, we give the *Hecke polynomials*. If β is an eigenclass with $T(l,k)(\beta) = a(l,k)\beta$, then we define

$$H(\beta,l) = \sum_k (-1)^k l^{k(k-1)/2} a(l,k) X^k \in \mathbb{C}[X].$$

For almost all l, after putting $X = l^{-s}$ where s is a complex variable, the function $H(\beta,s)$ is the inverse of the local factor at l of the automorphic representation attached to β.

Example A.28. Suppose $N = 11$. Then the cohomology $H^5(\Gamma_0(11);\mathbb{C})$ is 2-dimensional. There are two Hecke eigenclasses u_1, u_2, each with rational Hecke eigenvalues.

u_1	T_2	$(1-4X)(1-8X)(1+2X+2X^2)$
	T_3	$(1-9X)(1-27X)(1+X+3X^2)$
	T_5	$(1-25X)(1-125X)(1-X+5X^2)$
	T_7	$(1-49X)(1-343X)(1+2X+7X^2)$
u_2	T_2	$(1-X)(1-2X)(1+8X+32X^2)$
	T_3	$(1-X)(1-3X)(1+9X+243X^2)$
	T_5	$(1-X)(1-5X)(1-25X+3125X^2)$
	T_7	$(1-X)(1-7X)(1+98X+16807X^2)$

Example A.29. Suppose $N = 19$. Then the cohomology $H^5(\Gamma_0(19);\mathbb{C})$ is 3-dimensional. There are three Hecke eigenclasses u_1, u_2, u_3, each with rational Hecke eigenvalues.

u_1	T_2	$(1-4X)(1-8X)(1+2X^2)$
	T_3	$(1-9X)(1-27X)(1+2X+3X^2)$
	T_5	$(1-25X)(1-125X)(1-3X+5X^2)$
u_2	T_2	$(1-X)(1-2X)(1+32X^2)$
	T_3	$(1-X)(1-3X)(1+18X+243X^2)$
	T_5	$(1-X)(1-5X)(1-75X+3125X^2)$
u_3	T_2	$(1-2X)(1-4X)(1+3X+8X^2)$
	T_3	$(1-3X)(1-9X)(1+5X+27X^2)$
	T_5	$(1-5X)(1-25X)(1+12X+125X^2)$

In these examples, the cohomology is completely accounted for by the Eisenstein summand of (A.2.8). In fact, let $\Gamma_0'(N) \subset \mathrm{SL}_2(\mathbb{Z})$ be the usual Hecke congruence subgroup of matrices upper-triangular modulo N. Then the cohomology classes above actually come from classes in $H^1(\Gamma_0'(N))$, that is from holomorphic modular forms of level N.

For $N = 11$, the space of weight two cusp forms $S_2(11)$ is 1-dimensional. This cusp form f lifts in two different ways to $H^5(\Gamma_0(11);\mathbb{C})$, which can be seen from the quadratic part of the Hecke polynomials for the u_i. Indeed, for u_i the quadratic part is exactly the inverse of the local factor for the L-function attached to f, after the substitution $X = l^{-s}$. For u_2, we see that the lift is also twisted by the square of the cyclotomic character. (In fact the linear terms of the Hecke polynomials come from powers of the cyclotomic character.)

For $N = 19$, the space of weight two cusp forms $S_2(19)$ is again 1-dimensional. The classes u_1 and u_2 are lifts of this form, exactly as for $N = 11$. The class u_3, on the other hand, comes from $S_4(19)$, the space of weight 4 cusp forms on $\Gamma_0'(19)$. In fact, $\dim S_4(19) = 4$, with one Hecke eigenform defined over $\mathbb{Q}$ and another defined over a totally real cubic extension of $\mathbb{Q}$. Only the rational weight four eigenform contributes to $H^5(\Gamma_0(19);\mathbb{C})$. One can show that whether or not a weight four cuspidal eigenform f contributes to the cohomology of $\Gamma_0(N)$ depends only on the sign of the functional equation of $L(f, s)$ [**Wes**]. This phenomenon is typical of what one encounters when studying Eisenstein cohomology.

In addition to the lifts of weight 2 and weight 4 cusp forms, for other levels one finds lifts of Eisenstein series of weights 2 and 4 and lifts of cuspidal cohomology classes from subgroups of $\mathrm{SL}_3(\mathbb{Z})$. For some levels one finds cuspidal classes that appear to be lifts from the group of symplectic similitudes $\mathrm{GSp}(4)$. More details can be found in [**AGM02, AGM**].

A.5.14. Here are some notes on the reduction algorithm and its implementation:

- Some additional care must be taken when selecting reducing points for the submodular symbols of $\mathbf{u}$. In particular, in practice one should choose w for $\mathbf{v}$ such that $\sum \|\mathbf{v}_i(w)\|$ is minimized. Similar remarks apply when choosing a subdivision of the crosspolytopes in Section A.5.10.
- In practice, the reduction algorithm has *always* terminated with a reduced 1-sharbly cycle ξ' homologous to ξ. However, at the moment we cannot prove that this will always happen.
- Experimentally, the efficiency of the reduction step appears to be comparable to that of Theorem A.22. In other words the depth of the "reduction tree" associated to a given 1-sharbly $\mathbf{u}$ seems to be bounded by a polynomial in $\log\log \|\mathbf{u}\|$. Hence computing the Hecke action using this algorithm is extremely efficient.

  On the other hand, computing Hecke operators on SL_4 is still a much bigger computation—relative to the level—than on SL_2 and SL_3. For example, the size of the full retract W modulo $\Gamma_0(p)$ is roughly $O(p^6)$, which grows rapidly with p. The portion of the retract corresponding to H^5 is much smaller, around $p^3/10$, but this still grows quite quickly. This makes computing with $p > 100$ out of reach at the moment.

  The number of Hecke cosets grows rapidly as well, e.g., the number of coset representatives of $T(l, 2)$ is $l^4 + l^3 + 2l^2 + l + 1$. Hence it is only feasible to compute Hecke operators for small l; for large levels only $l = 2$ is possible.

  Here are some numbers to give an idea of the size of these computations. For level 73, the rank of H^5 is 20. There are 39504 cells of codimension 1 and 4128 top-dimensional cells in W modulo $\Gamma_0(73)$. The computational techniques in [**AGM02**] used at this level (a Lanczos scheme over a large finite field) tend to produce sharbly cycles supported on almost all the cells. Computing $T(2, 1)$

requires a reduction tree of depth 1 and produces as many as 26 reduced 1-sharblies for each of the 15 nonreduced Hecke images. Thus one cycle produces a cycle supported on as many as 15406560 1-sharblies, all of which must be converted to an appropriate cell of W modulo Γ. Also this is just what needs to be done for *one* cycle; do not forget that the rank of H^5 is 20.

In practice the numbers are slightly better, since the reduction step produces fewer 1-sharblies on average and since the support of the initial cycle has size less than 39504. Nevertheless the orders of magnitude are correct.

- Using lifts is a convenient way to encode the global Γ-identifications in the cycle ξ, since it means we do not have to maintain a big data structure keeping track of the identifications on $\partial\xi$. However, there is a certain expense in computing the Hermite normal form. This is balanced by the benefit that working with the data Φ associated to ξ allows us to reduce the supporting 1-sharblies $\mathbf{u}$ *independently*. This means we can cheaply parallelize our computation: each 1-sharbly, encoded as a 4-tuple $(\mathbf{u}, n(\mathbf{u}), \{\mathbf{v}\}, \{M(\mathbf{v})\})$, can be handled by a separate computer. The results of all these individual computations can then be collated at the end, when producing a W-cocycle.

A.6. Complements and Open Problems

A.6.1. We conclude this appendix by giving some complements and describing some possible directions for future work, both theoretical and computational. Since a full explanation of the material in this section would involve many more pages, we will be brief and will provide many references.

A.6.2. Perfect Quadratic Forms over Number Fields and Retracts. Since Voronoï's pioneering work [**Vor08**], it has been the goal of many to extend his results from $\mathbb{Q}$ to a general algebraic number field F. Recently Coulangeon [**Cou01**], building on work of Icaza and Baeza [**Ica97, BI97**], has found a good notion of *perfection* for quadratic forms over number fields[12]. One of the key ideas in [**Cou01**] is that the correct notion of equivalence between Humbert forms involves not only the action of $\mathrm{GL}_n(\mathscr{O}_F)$, where $\mathscr{O}_F$ is the ring of integers of F, but also the action of a certain continuous group U related to the units $\mathscr{O}_F^\times$. One of Coulangeon's basic results is that there are finitely many equivalence classes of perfect Humbert forms modulo these actions.

---

[12]Such forms are called *Humbert forms* in the literature.

On the other hand, Ash's original construction of retracts [**Ash77**] introduces a geometric notion of perfection. Namely he generalizes the Voronoĭ polyhedron Π and defines a quadratic form to be perfect if it naturally indexes a facet of Π. What is the connection between these two notions? Can one use Coulangeon's results to construct cell complexes to be used in cohomology computations? One tempting possibility is to try to use the group U to collapse the Voronoĭ cells of [**Ash77**] into a cell decomposition of the symmetric space associated to $\mathrm{SL}_n(F)$.

A.6.3. The Modular Complex. In his study of multiple ζ-values, Goncharov has recently defined the *modular complex* M^* [**Gon97, Gon98**]. This is an n-step complex of $\mathrm{GL}_n(\mathbb{Z})$-modules closely related both to the properties of multiple polylogarithms evaluated at μ_N, the Nth roots of unity, and to the action of $G_{\mathbb{Q}}$ on $\pi_{1,N} = \pi_1^l(\mathbb{P}^1 \smallsetminus \{0, \infty, \mu_N\})$, the pro-$l$ completion of the algebraic fundamental group of $\mathbb{P}^1 \smallsetminus \{0, \infty, \mu_N\}$.

Remarkably, the modular complex is very closely related to the Voronoĭ decomposition $\mathscr{V}$. In fact, one can succinctly describe the modular complex by saying that it is the chain complex of the cells coming from the top-dimensional Voronoĭ cone of type A_n. This is all of the Voronoĭ decomposition for $n = 2, 3$, and Goncharov showed that the modular complex is quasi-isomorphic to the full Voronoĭ complex for $n = 4$. Hence there is a precise relationship among multiple polylogarithms, the Galois action on $\pi_{1,N}$, and the cohomology of level N congruence subgroups of $\mathrm{SL}_n(\mathbb{Z})$.

The question then arises, how much of the cohomology of congruence subgroups is captured by the modular complex for all n? Table A.3.2 indicates that asymptotically very little of the Voronoĭ decomposition comes from the A_n cone, but this says nothing about the cohomology. The first interesting case to consider is $n = 5$.

A.6.4. Retracts for Other Groups. The most general construction of retracts W known [**Ash84**] applies only to *linear* symmetric spaces. The most familiar example of such a space is $\mathrm{SL}_n(\mathbb{R})/\mathrm{SO}(n)$; other examples are the symmetric spaces associated to SL_n over number fields and division algebras.

Now let $\Gamma \subset \mathbf{G}(\mathbb{Q})$ be an arithmetic group, and let $X = G/K$ be the associated symmetric space. What can one say about cell complexes that can be used to compute $H^*(\Gamma; \mathscr{M})$? The theorem of Borel–Serre mentioned in Section A.3.3 implies the vanishing of $H^k(\Gamma; \mathscr{M})$ for $k > \nu := \dim X - q$, where q is the $\mathbb{Q}$-*rank* of Γ. For example, for the split form of SL_n, the $\mathbb{Q}$-rank is $n - 1$. For the split symplectic group Sp_{2n}, the $\mathbb{Q}$-rank is n. Moreover, this bound is sharp: there will be coefficient modules $\mathscr{M}$ for which $H^\nu(\Gamma; \mathscr{M}) \neq 0$. Hence any minimal cell complex used to compute the cohomology of Γ should have dimension ν.

Ideally one would like to see such a complex realized as a subspace of X and would like to be able to treat all finite index subgroups of Γ simultaneously. This leads to the following question: is there a Γ-equivariant deformation retraction of X onto a regular cell complex W of dimension ν?

For $\mathbf{G} = \mathrm{Sp}_4$, McConnell and MacPherson showed that the answer is yes. Their construction begins by realizing the symplectic symmetric space X_{Sp} as a subspace of the special linear symmetric space X_{SL}. They then construct subsets of X_{Sp} by intersecting the Voronoĭ cells in X_{SL} with X_{Sp}. Through explicit computations in coordinates they prove that these intersections are cells and give a cell decomposition of X_{Sp}. By taking an appropriate dual complex (as suggested by Figures A.3.2 and A.3.3 and as done in [**Ash77**]), they construct the desired cell complex W.

Other progress has been recently made by Bullock [**Bul00**], Bullock and Connell [**BC06**], and Yasaki [**Yas05b, Yas05a**] in the case of groups of $\mathbb{Q}$-rank 1. In particular, Yasaki uses the *tilings* of Saper [**Sap97**] to construct an explicit retract for the unitary group $\mathrm{SU}(2,1)$ over the Gaussian integers. His method also works for Hilbert modular groups, although further refinement may be needed to produce a regular cell complex. Can one generalize these techniques to construct retracts for groups of arbitrary $\mathbb{Q}$-rank? Is there an analogue of the Voronoĭ decomposition for these retracts (i.e., a dual cell decomposition of the symmetric space)? If so, can one generalize ideas in Sections A.4–A.5 and use that generalization to compute the action of the Hecke operators on the cohomology?

A.6.5. Deeper Cohomology Groups. The algorithm in Section A.5 can be used to compute the Hecke action on $H^{\nu-1}(\Gamma)$. For $n > 4$, this group no longer contains cuspidal cohomology classes. Can one generalize this algorithm to compute the Hecke action on deeper cohomology groups? The first practical case is $n = 5$. Here $\nu = 10$, and the highest degree in which cuspidal cohomology can live is 8. This case is also interesting since the cohomology of full level has been studied [**EVGS02**].

Here are some indications of what one can expect. The general strategy is the same: for a k-sharbly ξ representing a class in $H^{\nu-k}(\Gamma)$, begin by Γ-equivariantly choosing reducing points for the nonunimodular submodular symbols of ξ. This data can be packaged into a new k-sharbly cycle as in Section A.5.7ff, but the crosspolytopes must be replaced with *hypersimplices*. By definition, the hypersimplex $\Delta(n,k)$ is the convex hull in $\mathbb{R}^n$ of the points $\{\sum_{i\in I} e_i\}$, where I ranges over all order k subsets of $\{1,\ldots,n\}$ and $e_1,\ldots,e_n$ denotes the standard basis of $\mathbb{R}^n$.

The simplest example is $n = 2$, $k = 2$. From the point of view of cohomology, this is even less interesting than $n = 2$, $k = 1$, since now we are computing the Hecke action on $H^{-1}(\Gamma)$! Nevertheless, the geometry here illustrates what one can expect in general.

Each 2-sharbly in the support of ξ can be written as $[v_1, v_2, v_3, v_4]$ and determines six submodular symbols, of the form $[v_i, v_j]$, $i \neq j$. Assume for simplicity that all these submodular symbols are nonunimodular. Let w_{ij} be the reducing point for $[v_i, v_j]$. Then use the ten points v_i, w_{ij} to label the vertices of the hypersimplex $\Delta(5, 2)$ as in Figure A.6.1 (note that $\Delta(5, 2)$ is 4-dimensional).

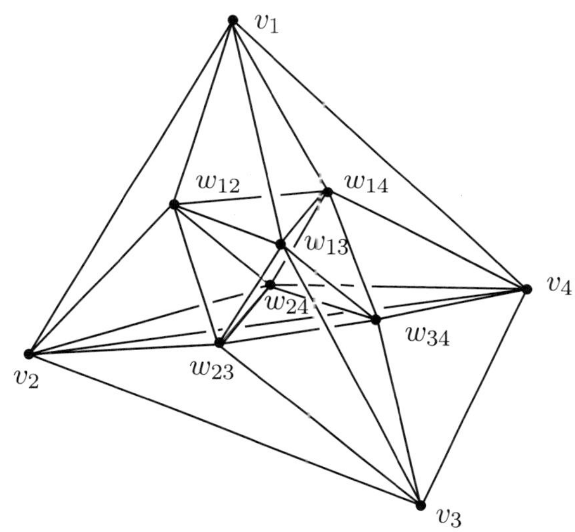

Figure A.6.1.

The boundary of this hypersimplex gives the analogue of (A.5.4). Which 2-sharblies will appear in ξ'? The boundary $\partial\Delta(5, 2)$ is a union of five tetrahedra and five octahedra. The outer tetrahedron will not appear in ξ', since that is the analogue of the left side of (A.5.4). The four octahedra sharing a triangular face with the outer tetrahedron also will not appear, since they disappear when considering ξ' modulo Γ. The remaining four tetrahedra and the central octahedron survive to ξ' and constitute the right side of the analogue of (A.5.4). Note that we must choose a simplicial subdivision of the central octahedron to write the result as a 2-sharbly cycle and that this must be done with care since it introduces a new submodular symbol.

If some submodular symbols are unimodular, then again one must consider iterated cones on hypersimplices, just as in Section A.5.10. The analogues of these steps become more complicated, since there are now many simplicial subdivisions of a hypersimplex[13]. There is one final complication: in general we cannot use reduced k-sharblies alone to represent cohomology classes. Thus one must terminate the algorithm when $\|\xi\|$ is less than some predetermined bound.

A.6.6. Other Linear Groups. Let F be a number field, and let $\mathbf{G} = \mathbf{R}_{F/\mathbb{Q}}(\mathrm{SL}_n)$ (Example A.2). Let $\Gamma \subset \mathbf{G}(\mathbb{Q})$ be an arithmetic subgroup. Can one compute the action of the Hecke operators on $H^*(\Gamma)$?

There are two completely different approaches to this problem. The first involves the generalization of the modular symbols method. One can define the analogue of the sharbly complex, and can try to extend the techniques of Sections A.4–A.5.

This technique has been extensively used when F is *imaginary quadratic* and $n = 2$. We have $X = \mathrm{SL}_2(\mathbb{C})/\,\mathrm{SU}(2)$, which is isomorphic to 3-dimensional hyperbolic space $\mathfrak{h}_3$. The arithmetic groups $\Gamma \subset \mathrm{SL}_2(\mathcal{O}_F)$ are known as *Bianchi groups*. The retracts and cohomology of these groups have been well studied; as a representative sample of works we mention [**Men79, EGM98, Vog85, GS81**].

Such groups have $\mathbb{Q}$-rank 1 and thus have cohomological dimension 2. One can show that the cuspidal classes live in degrees 1 and 2. This means that we can use modular symbols to investigate the Hecke action on cuspidal cohomology. This was done by Cremona [**Cre84**] for *euclidean* fields F. In that case Theorem A.22 works with no trouble (the euclidean algorithm is needed to construct reducing points). For noneuclidean fields further work has been done by Whitley [**Whi90**], Cremona and Whitely [**CW94**] (both for principal ideal domains), Bygott [**Byg99**] (for $F = \mathbb{Q}(\sqrt{-5})$ and any field with class group an elementary abelian 2-group), and Lingham [**Lin05**] (any field with odd class number). Putting all these ideas together allows one to generalize the modular symbols method to *any* imaginary quadratic field [**Cre**].

For F imaginary quadratic and $n > 2$, very little has been studied. The only related work to the best of our knowledge is that of Staffeldt [**Sta79**]. He determined the structure of the Voronoĭ polyhedron in detail for $\mathbf{R}_{F/\mathbb{Q}}(\mathrm{SL}_3)$, where $F = \mathbb{Q}(\sqrt{-1})$. We have $\dim X = 8$ and $\nu = 6$. The cuspidal cohomology appears in degrees $3, 4, 5$, so one could try to use the techniques of Section A.5 to investigate it.

Similar remarks apply to F *real quadratic* and $n = 2$. The symmetric space $X \simeq \mathfrak{h} \times \mathfrak{h}$ has dimension 4 and the $\mathbb{Q}$-rank is 1, which means $\nu = 3$. Unfortunately the cuspidal cohomology appears only in degree 2, which

[13]Indeed, computing all simplicial subdivisions of $\Delta(n, k)$ is a difficult problem in convex geometry.

means modular symbols cannot see it. On the other hand, 1-sharblies can see it, and so one can try to use ideas in Section A.5 here to compute the Hecke operators. The data needed to build the retract W already (essentially) appears in the literature for certain fields; see for example [**Ong86**].

The second approach shifts the emphasis from modular symbols and the sharbly complex to the Voronoĭ fan and its cones. For this approach we must assume that the group Γ is associated to a *self-adjoint homogeneous cone* over $\mathbb{Q}$. (cf. [**Ash77**]). This class of groups includes arithmetic subgroups of $\mathbf{R}_{F/\mathbb{Q}}(\mathrm{SL}_n)$, where F is a totally real or CM field. Such groups have all the nice structures in Section A.3.2. For example, we have a cone C with a G-action. We also have an analogue of the Voronoĭ polyhedron Π. There is a natural compactification $\tilde{C}$ of C obtained by adjoining certain self-adjoint homogeneous cones of lower rank. The quotient $\Gamma\backslash\tilde{C}$ is singular in general, but it can still be used to compute $H^*(\Gamma;\mathbb{C})$. The polyhedron Π can be used to construct a fan $\mathscr{V}$ that gives a Γ-equivariant decomposition of all of $\tilde{C}$. But the most important structure we have is the Voronoĭ reduction algorithm: given any point $x \in \tilde{C}$, we can determine the unique Voronoĭ cone containing x.

Here is how this setup can be used to compute the Hecke action. Full details are in [**Gun99, GM03**]. We define two chain complexes $\mathbf{C}_*^V$ and $\mathbf{C}_*^R$. The latter is essentially the chain complex generated by all simplicial rational polyhedral cones in $\tilde{C}$; the former is the subcomplex generated by the Voronoĭ cones. These are the analogues of the sharbly complex and the chain complex associated to the retract W, and one can show that either can be used to compute $H^*(\Gamma;\mathbb{C})$. Take a cycle $\xi \in \mathbf{C}_*^V$ representing a cohomology class in $H^*(\Gamma;\mathbb{C})$ and act on it by a Hecke operator T. We have $T(\xi) \in \mathbf{C}_*^R$, and we must push $T(\xi)$ back to $\mathbf{C}_*^V$.

To do this, we use the linear structure on $\tilde{C}$ to subdivide $T(\xi)$ very finely into a chain ξ'. For each 1-cone τ in $\operatorname{supp}\xi'$, we choose a 1-cone $\rho_\tau \in \tilde{C} \smallsetminus C$ and assemble them using the combinatorics of ξ' into a polyhedral chain ξ'' homologous to ξ'. Under certain conditions involved in the construction of ξ', this chain ξ'' will lie in $\mathbf{C}_*^V$.

We illustrate this process for the split group SL_2; more details can be found in [**Gun99**]. We work modulo homotheties, so that the three-dimensional cone $\tilde{C}$ becomes the extended upper half plane $\mathfrak{h}^* := \mathfrak{h}\cup\mathbb{Q}\cup\{\infty\}$, with $\partial\tilde{C}$ passing to the cusps $\mathfrak{h}^* \smallsetminus \mathfrak{h}$. As usual top-dimensional Voronoĭ cones become the triangles of the Farey tessellation, and the cones ρ_τ become cusps. Given any $x \in \mathfrak{h}$, let $R(x)$ be the set of cusps of the unique triangle or edge containing x (this can be computed using the Voronoĭ reduction algorithm). Extend R to a function on $\mathfrak{h}^*$ by putting $R(u) = \{u\}$ for any cusp u.

In $\mathfrak{h}$, the support of $T(\xi)$ becomes a geodesic μ between two cusps u, u', in other words the support of a modular symbol $[u, u']$ (Figure A.6.2). Subdivide μ by choosing points $x_0, \ldots, x_n$ such that $x_0 = u$, $x_n = u'$, and $R(x_i) \cap R(x_{i+1}) \neq \varnothing$. (This is easily done, for example by repeatedly barycentrically subdividing μ.) For each $i < n$ choose a cusp $q_i \in R(x_i) \cap R(x_{i+1})$, and put $q_n = u'$. Then we have a relation in H^1:

$$[u, u'] = [q_0, q_1] + \cdots + [q_{n-1}, q_n]. \tag{A.6.1}$$

Moreover, each $[q_i, q_{i+1}]$ is unimodular, since q_i and q_{i+1} are both vertices of a triangle containing x_{i+1}. Upon lifting (A.6.1) back to $\mathbf{C}_*^R$, the cusps q_i become the 1-cones ρ_τ and give us a relation $T(\xi) = \xi'' \in \mathbf{C}_*^V$.

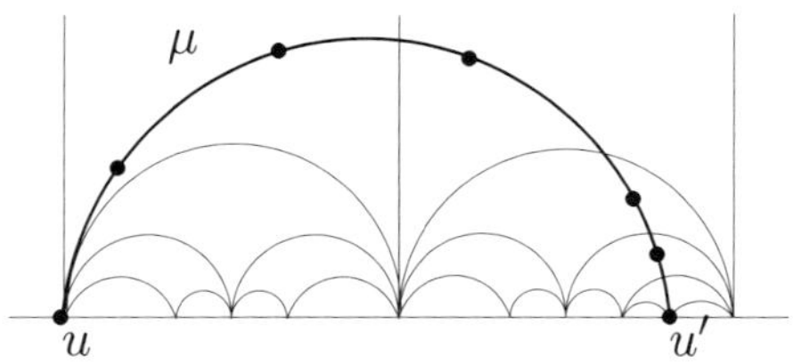

Figure A.6.2. A subdivision of μ; the solid dots are the x_i. Since the x_i lie in the same or adjacent Voronoï cells, we can assign cusps to them to construct a homology to a cycle in $\mathbf{C}_*^V$.

A.6.7. The Sharbly Complex for General Groups. In [**Gun00b**] we generalized Theorem A.22 (without the complexity statement) to the symplectic group Sp_{2n}. Using this algorithm and the symplectic retract [**MM93, MM89**], one can compute the action of the Hecke operators on the top-degree cohomology of subgroups of $\mathrm{Sp}_4(\mathbb{Z})$.

More recently, Toth has investigated modular symbols for other groups. He showed that the unimodular symbols generate the top-degree cohomology groups for Γ an arithmetic subgroup of a split classical group or a split group of type E_6 or E_7 [**Tot05**]. His technique of proof is completely different from that of [**Gun00b**]. In particular he does not give an analogue of the Manin trick. Can one extract an algorithm from Toth's proof that can be used to explicitly compute the action of the Hecke operators on cohomology?

The proof of the main result of [**Gun00b**] uses a description of the relations among the modular symbols. These relations were motivated by the structure of the cell complex in [**MM93, MM89**]. The modular symbols and these relations are analogues of the groups S_0 and S_1 in the sharbly complex. Can one extend these combinatorial constructions to form a *symplectic sharbly complex*? What about for general groups $\mathbf{G}$?

Already for Sp_4, resolution of this question would have immediate arithmetic applications. Indeed, Harder has a beautiful conjecture about certain congruences between holomorphic modular forms and Siegel modular forms of full level [**Hara**]. Examples of these congruences were checked numerically in [**Hara**] using techniques of [**FvdG**] to compute the Hecke action.

However, to investigate higher levels, one needs a different technique. The relevant cohomology classes live in $H^{\nu-1}(\Gamma; \mathscr{M})$, so one only needs to understand the first three terms of the complex $S_0 \leftarrow S_1 \leftarrow S_2$. We understand S_0, S_1 from [**Gun00b**]; the key is understanding S_2, which should encode relations among elements of S_1. If one could do this and then could generalize the techniques of [**Gun00a**], one would have a way to investigate Harder's conjecture.

A.6.8. Generalized Modular Symbols. We conclude this appendix by discussing a geometric approach to modular symbols. This complements the algebraic approaches presented in this book and leads to many new interesting phenomena and problems.

Suppose $\mathbf{H}$ and $\mathbf{G}$ are connected semisimple algebraic groups over $\mathbb{Q}$ with an injective map $f\colon \mathbf{H} \to \mathbf{G}$. Let K_H be a maximal compact subgroup of $H = \mathbf{H}(\mathbb{R})$, and suppose $K \subset G$ is a maximal compact subgroup containing $f(K_H)$. Let $X = G/K$ and $Y = H/K_H$.

Now let $\Gamma \subset \mathbf{G}(\mathbb{Q})$ be a torsion-free arithmetic subgroup. Let $\Gamma_H = f^{-1}(\Gamma)$. We get a map $\Gamma_H\backslash Y \to \Gamma\backslash X$, and we denote the image by $S(H,\Gamma)$. Any compactly supported cohomology class $\xi \in H_c^{\dim Y}(\Gamma\backslash X;\mathbb{C})$ can be pulled back via f to $\Gamma_H\backslash Y$ and integrated to obtain a complex number. Hence $S(H,\Gamma)$ defines a linear form on $H_c^{\dim Y}(\Gamma\backslash X;\mathbb{C})$. By Poincaré duality, this linear form determines a class $[S(H,\Gamma)] \in H^{\dim X - \dim Y}(\Gamma\backslash X;\mathbb{C})$, called a *generalized modular symbol*. Such classes have been considered by many authors, for example [**AB90, SV03, Har05, AGR93**].

As an example, we can take $\mathbf{G}$ to be the split form of SL_2, and we can take $f\colon \mathbf{H} \to \mathbf{G}$ to be the inclusion of connected component of the diagonal subgroup. Hence $H \simeq \mathbb{R}_{>0}$. In this case K_H is trivial. The image of Y in X is the ideal geodesic from 0 to ∞. One way to vary f is by taking an $\mathrm{SL}_2(\mathbb{Q})$-translate of this geodesic, which gives a geodesic between two cusps. Hence we can obtain the support of any modular symbol this way. This example generalizes to SL_n to yield the modular symbols in Section A.4. Here $H \simeq (\mathbb{R} > 0)^{n-1}$. Note that $\dim Y = n-1$, so the cohomology classes we have constructed live in the top degree $H^{\nu}(\Gamma\backslash X;\mathbb{C})$.

Another family of examples is provided by taking $\mathbf{H}$ to be a Levi factor of a parabolic subgroup; these are the modular symbols studied in [**AB90**].

There are many natural questions to study for such objects. Here are two:

- Under what conditions on $\mathbf{G}, \mathbf{H}, \Gamma$ is $[S(H, \Gamma)]$ nonzero? This question is connected to relations between periods of automorphic forms and functoriality lifting. There are a variety of partial results known; see for example [**SV03, AGR93**].
- We know the usual modular symbols span the top-degree cohomology for any arithmetic group Γ. Fix a class of generalized modular symbols by fixing the pair $\mathbf{G}, \mathbf{H}$ and fixing some class of maps f. How much of the cohomology can one span for a general arithmetic group $\Gamma \subset \mathbf{G}(\mathbb{Q})$?

  A simple example is given by the Ash–Borel construction for $\mathbf{G} = \mathrm{SL}_3$ and $\mathbf{H}$ a Levi factor of a rational parabolic subgroup $\mathbf{P}$ of type $(2,1)$. In this case $H \simeq \mathrm{SL}_2(\mathbb{R}) \times \mathbb{R}_{>0}$ and sits inside G via

$$g \begin{pmatrix} \alpha^{-1}M & 0 \\ 0 & \alpha \end{pmatrix} g^{-1}, \quad M \in \mathrm{SL}_2(\mathbb{R}), \quad \alpha \in \mathbb{R}_{>0}, \quad g \in \mathrm{SL}_3(\mathbb{Q}).$$

  For $\Gamma \subset \mathrm{SL}_3(\mathbb{Z})$ these symbols define a subspace

$$S_{(2,1)} \subset H^2(\Gamma \backslash X; \mathbb{C}).$$

  Are there Γ for which $S_{(2,1)}$ equals the full cohomology space? For general Γ how much is captured? Is there a nice combinatorial way to write down the relations among these classes? Can one cook up a generalization of Theorem A.22 for these classes and use it to compute Hecke eigenvalues?

Bibliography

[AB90] A. Ash and A. Borel, *Generalized modular symbols*, Cohomology of arithmetic groups and automorphic forms (Luminy-Marseille, 1989), Springer, Berlin, 1990, pp. 57–75.

[ADT04] Nadia Ben Atti and Gema M. Díaz-Toca, `http://hlombardi.free.fr/publis/ABMAvar.html` (2004).

[AG00] Avner Ash and Robert Gross, *Generalized non-abelian reciprocity laws: a context for Wiles' proof*, Bull. London Math. Soc. **32** (2000), no. 4, 385–397. MR 1760802 (2001h:11142)

[Aga00] A. Agashe, *The Birch and Swinnerton-Dyer formula for modular abelian varieties of analytic rank* 0, Ph.D. thesis, University of California, Berkeley (2000).

[AGG84] Avner Ash, Daniel Grayson, and Philip Green, *Computations of cuspidal cohomology of congruence subgroups of* $\mathrm{SL}(3, \mathbf{Z})$, J. Number Theory **19** (1984), no. 3, 412–436. MR 769792 (86g:11032)

[AGM] Avner Ash, Paul E. Gunnells, and Mark McConnell, *Cohomology of congruence subgroups of* $\mathrm{SL}_4(\mathbb{Z})$ *II*, in preparation.

[AGM02] ———, *Cohomology of congruence subgroups of* $\mathrm{SL}_4(\mathbb{Z})$, J. Number Theory **94** (2002), no. 1, 181–212. MR 1904968 (2003f:11072)

[AGR93] Avner Ash, David Ginzburg, and Steven Rallis, *Vanishing periods of cusp forms over modular symbols*, Math. Ann. **296** (1993), no. 4, 709–723. MR 1233493 (94f:11044)

[Ahl78] Lars V. Ahlfors, *Complex analysis*, third ed., McGraw-Hill Book Co., New York, 1978, An introduction to the theory of analytic functions of one complex variable, International Series in Pure and Applied Mathematics. MR 510197 (80c:30001)

[AL70] A. O. L. Atkin and J. Lehner, *Hecke operators on* $\Gamma_0(m)$, Math. Ann. **185** (1970), 134–160.

[AO01] Scott Ahlgren and Ken Ono, *Addition and counting: the arithmetic of partitions*, Notices Amer. Math. Soc. **48** (2001), no. 9, 978–984. MR 1854533 (2002e:11136)

[AR79] Avner Ash and Lee Rudolph, *The modular symbol and continued fractions in higher dimensions*, Invent. Math. **55** (1979), no. 3, 241–250. MR 553998 (82g:12011)

[Art79] James Arthur, *Eisenstein series and the trace formula*, Automorphic forms, representations and L-functions (Proc. Sympos. Pure Math., Oregon State Univ., Corvallis, Ore., 1977), Part 1, Proc. Sympos. Pure Math., XXXIII, Amer. Math. Soc., Providence, R.I., 1979, pp. 253–274. MR 546601 (81b:10020)

[Ash77] Avner Ash, *Deformation retracts with lowest possible dimension of arithmetic quotients of self-adjoint homogeneous cones*, Math. Ann. **225** (1977), no. 1, 69–76. MR 0427490 (55 #522)

[Ash80] ———, *Cohomology of congruence subgroups* $\mathrm{SL}(n, \mathbb{Z})$, Math. Ann. **249** (1980), no. 1, 55–73. MR 82f:22010

[Ash84] ———, *Small-dimensional classifying spaces for arithmetic subgroups of general linear groups*, Duke Math. J. **51** (1984), no. 2, 459–468. MR 747876 (85k:22027)

[Ash86] ———, *A note on minimal modular symbols*, Proc. Amer. Math. Soc. **96** (1986), no. 3, 394–396. MR 822426 (87e:22024)

[Ash94] ———, *Unstable cohomology of* $\mathrm{SL}(n, \mathcal{O})$, J. Algebra **167** (1994), no. 2, 330–342. MR 1283290 (95g:20050)

[Bar57] E. S. Barnes, *The perfect and extreme senary forms*, Canad. J. Math. **9** (1957), 235–242. MR 0086834 (19,251e)

[Bar94] A. Barvinok, *A polynomial time algorithm for counting integral points in polyhedra when the dimension is fixed*, Math. Oper. Res. **19** (1994), no. 4, 769–779.

[Bas96] Jacques Basmaji, *Ein Algorithmus zur Berechnung von Hecke-Operatoren und Anwendungen auf modulare Kurven,* `http://modular.math.washington.edu/scans/papers/basmaji/`, 1996.

[BC06] S. S. Bullock and C. Connell, *Equivariant retracts of geometrically finite discrete groups acting on negatively pinched Hadamard manifolds*, in preparation, 2006.

[BCDT01] C. Breuil, B. Conrad, Fred Diamond, and R. Taylor, *On the modularity of elliptic curves over* $\mathbf{Q}$*: wild 3-adic exercises*, J. Amer. Math. Soc. **14** (2001), no. 4, 843–939 (electronic). MR 2002d:11058

[BCP97] W. Bosma, J. Cannon, and C. Playoust, *The Magma algebra system. I. The user language*, J. Symbolic Comput. **24** (1997), no. 3–4, 235–265, Computational algebra and number theory (London, 1993). MR 1 484 478

[BCS92] J. P. Buhler, R. E. Crandall, and R. W. Sompolski, *Irregular primes to one million*, Math. Comp. **59** (1992), no. 200, 717–722. MR 1134717 (93a:11106)

[BHKS06] K. Belebas, M. Van Hoeij, J. Klüners, and A. Steel, *Factoring polynomials over global fields*, preprint at `http://www.math.fsu.edu/~hoeij/papers.html` (2006).

[BI97] R. Baeza and M. I. Icaza, *On Humbert-Minkowski's constant for a number field*, Proc. Amer. Math. Soc. **125** (1997), no. 11, 3195–3202. MR 1403112 (97m:11092)

[Bir71] B. J. Birch, *Elliptic curves over* $\mathbf{Q}$*: A progress report*, 1969 Number Theory Institute (Proc. Sympos. Pure Math., Vol. XX, State Univ. New York,

Stony Brook, N.Y., 1969), Amer. Math. Soc., Providence, R.I., 1971, pp. 396–400.

[BK90] S. Bloch and K. Kato, *L-functions and Tamagawa numbers of motives*, The Grothendieck Festschrift, Vol. I, Birkhäuser Boston, Boston, MA, 1990, pp. 333–400.

[BMS06] Yann Bugeaud, Maurice Mignotte, and Samir Siksek, *Classical and modular approaches to exponential Diophantine equations. II. The Lebesgue-Nagell equation*, Compos. Math. **142** (2006), no. 1, 31–62. MR 2196761

[Bro94] Kenneth S. Brown, *Cohomology of groups*, Graduate Texts in Mathematics, vol. 87, Springer-Verlag, New York, 1994, corrected reprint of the 1982 original. MR 1324339 (96a:20072)

[BS73] A. Borel and J.-P. Serre, *Corners and arithmetic groups*, Comment. Math. Helv. **48** (1973), 436–491, avec un appendice: Arrondissement des variétés à coins, par A. Douady et L. Hérault. MR 0387495 (52 #8337)

[BS02] K. Buzzard and W. A. Stein, *A mod five approach to modularity of icosahedral Galois representations*, Pacific J. Math. **203** (2002), no. 2, 265–282. MR 2003c:11052

[BT82] Raoul Bott and Loring W. Tu, *Differential forms in algebraic topology*, Graduate Texts in Mathematics, vol. 82, Springer-Verlag, New York, 1982. MR 658304 (83i:57016)

[Bul00] S. S. Bullock, *Well-rounded retracts of rank one symmetric spaces*, preprint, 2000.

[Bum84] Daniel Bump, *Automorphic forms on* $\mathrm{GL}(3, \mathbf{R})$, Lecture Notes in Mathematics, vol. 1083, Springer-Verlag, Berlin, 1984. MR 765698 (86g:11028)

[Bum97] ———, *Automorphic forms and representations*, Cambridge Studies in Advanced Mathematics, vol. 55, Cambridge University Press, Cambridge, 1997. MR 1431508 (97k:11080)

[Buz96] Kevin Buzzard, *On the eigenvalues of the Hecke operator* T_2, J. Number Theory **57** (1996), no. 1, 130–132. MR 96m:11033

[BW00] A. Borel and N. Wallach, *Continuous cohomology, discrete subgroups, and representations of reductive groups*, second ed., Mathematical Surveys and Monographs, vol. 67, American Mathematical Society, Providence, RI, 2000. MR 1721403 (2000j:22015)

[Byg99] J. Bygott, *Modular forms and modular symbols over imaginary quadratic fields*, Ph.D. thesis, Exeter University, 1999.

[Car59a] L. Carlitz, *Arithmetic properties of generalized Bernoulli numbers*, J. Reine Angew. Math. **202** (1959), 174–182. MR 0109132 (22 #20)

[Car59b] ———, *Some arithmetic properties of generalized Bernoulli numbers*, Bull. Amer. Math. Soc. **65** (1959), 68–69. MR 0104630 (21 #3383)

[CDT99] Brian Conrad, Fred Diamond, and Richard Taylor, *Modularity of certain potentially Barsotti-Tate Galois representations*, J. Amer. Math. Soc. **12** (1999), no. 2, 521–567. MR 1639612 (99i:11037)

[CF67] George E. Cooke and Ross L. Finney, *Homology of cell complexes*, Based on lectures by Norman E. Steenrod, Princeton University Press, Princeton, N.J., 1967. MR 0219059 (36 #2142)

[Che05] Imin Chen, *A Diophantine equation associated to* $X_0(5)$, LMS J. Comput. Math. **8** (2005), 116–121 (electronic). MR 2153792 (2006b:11052)

[CL04] J. Cremona and M. P. Lingham, *Finding all elliptic curves with good reduction outside a given set of primes*, in progress (2004).

[CO77] H. Cohen and J. Oesterlé, *Dimensions des espaces de formes modulaires*, 69–78. Lecture Notes in Math., Vol. 627. MR 57 #12396

[Coh93] H. Cohen, *A course in computational algebraic number theory*, Springer-Verlag, Berlin, 1993. MR 94i:11105

[Cou01] Renaud Coulangeon, *Voronoï theory over algebraic number fields*, Réseaux euclidiens, designs sphériques et formes modulaires, Monogr. Enseign. Math., vol. 37, Enseignement Math., Geneva, 2001, pp. 147–162. MR 1878749 (2002m:11064)

[Cre] J. E. Cremona, personal communication.

[Cre84] ———, *Hyperbolic tessellations, modular symbols, and elliptic curves over complex quadratic fields*, Compositio Math. **51** (1984), no. 3, 275–324.

[Cre92] ———, *Modular symbols for $\Gamma_1(N)$ and elliptic curves with everywhere good reduction*, Math. Proc. Cambridge Philos. Soc. **111** (1992), no. 2, 199–218.

[Cre97a] ———, *Algorithms for modular elliptic curves*, second ed., Cambridge University Press, Cambridge, 1997, `http://www.maths.nott.ac.uk/personal/jec/book/`.

[Cre97b] ———, *Computing periods of cusp forms and modular elliptic curves*, Experiment. Math. **6** (1997), no. 2, 97–107.

[Cre06] ———, Proceedings of the 7th International Symposium (ANTS-VII) (2006).

[CS88] J. H. Conway and N. J. A. Sloane, *Low-dimensional lattices. III. Perfect forms*, Proc. Roy. Soc. London Ser. A **418** (1988), no. 1854, 43–80. MR 953277 (90a:11073)

[CW94] J. E. Cremona and E. Whitley, *Periods of cusp forms and elliptic curves over imaginary quadratic fields*, Math. Comp. **62** (1994), no. 205, 407–429.

[CWZ01] Janos A. Csirik, Joseph L. Wetherell, and Michael E. Zieve, *On the genera of $X_0(N)$*, `http://www.csirik.net/papers.html` (2001).

[Dar97] H. Darmon, *Faltings plus epsilon, Wiles plus epsilon, and the generalized Fermat equation*, C. R. Math. Rep. Acad. Sci. Canada **19** (1997), no. 1, 3–14. MR 1479291 (98h:11034a)

[Dem04] L. Dembélé, *Quaternionic Manin symbols, Brandt matrices and Hilbert modular forms*, preprint, 2004.

[Dem05] L. Dembélé, *Explicit computations of Hilbert modular forms on $\mathbb{Q}(\sqrt{5})$*, Experiment. Math. **14** (2005), no. 4, 457–466. MR 2193808

[DI95] F. Diamond and J. Im, *Modular forms and modular curves*, Seminar on Fermat's Last Theorem, Providence, RI, 1995, pp. 39–133.

[Dia96] F. Diamond, *On deformation rings and Hecke rings*, Ann. of Math. (2) **144** (1996), no. 1, 137–166. MR 1405946 (97d:11172)

[Dix82] John D. Dixon, *Exact solution of linear equations using p-adic expansions*, Numer. Math. **40** (1982), no. 1, 137–141. MR 681819 (83m:65025)

[Dok04] Tim Dokchitser, *Computing special values of motivic L-functions*, Experiment. Math. **13** (2004), no. 2, 137–149.

[DP04] H. Darmon and R. Pollack, *The efficient calculation of Stark-Heegner points via overconvergent modular symbols*.

[DS05] Fred Diamond and Jerry Shurman, *A first course in modular forms*, Graduate Texts in Mathematics, vol. 228, Springer-Verlag, New York, 2005.

[DVS05] M. Dutour, F. Vallentin, and A. Schürmann, *Classification of perfect forms in dimension* 8, talk at Oberwolfach meeting *Sphere packings: Exceptional structures and relations to other fields*, November 2005.

[Ebe02] Wolfgang Ebeling, *Lattices and codes*, revised ed., Advanced Lectures in Mathematics, Friedr. Vieweg & Sohn, Braunschweig, 2002, a course partially based on lectures by F. Hirzebruch.

[ECdJ+06] Bas Edixhoven, Jean-Marc Couveignes, Robin de Jong, Franz Merkl, and Johan Bosman, *On the computation of coefficients of modular form*, http://www.arxiv.org/abs/math.NT/0605244 (2006).

[EGM98] J. Elstrodt, F. Grunewald, and J. Mennicke, *Groups acting on hyperbolic space*, Springer Monographs in Mathematics, Springer-Verlag, Berlin, 1998, Harmonic analysis and number theory. MR 1483315 (98g:11058)

[Eil47] Samuel Eilenberg, *Homology of spaces with operators. I*, Trans. Amer. Math. Soc. **61** (1947), 378–417; errata, 62, 548 (1947). MR 0021313 (9,52b)

[Elk98] Noam D. Elkies, *Elliptic and modular curves over finite fields and related computational issues*, Computational perspectives on number theory (Chicago, IL, 1995), AMS/IP Stud. Adv. Math., vol. 7, Amer. Math. Soc., Providence, RI, 1998, pp. 21–76. MR 1486831 (99a:11078)

[EVGS02] Philippe Elbaz-Vincent, Herbert Gangl, and Christophe Soulé, *Quelques calculs de la cohomologie de* $\mathrm{GL}_N(\mathbb{Z})$ *et de la K-théorie de* $\mathbb{Z}$, C. R. Math. Acad. Sci. Paris **335** (2002), no. 4, 321–324. MR 1931508 (2003h:19002)

[FH91] William Fulton and Joe Harris, *Representation theory*, Graduate Texts in Mathematics, vol. 129, Springer-Verlag, New York, 1991, A first course, Readings in Mathematics. MR 1153249 (93a:20069)

[FJ02] D.W. Farmer and K. James, *The irreducibility of some level 1 Hecke polynomials*, Math. Comp. **71** (2002), no. 239, 1263–1270 (electronic). MR 2003e:11046

[FL] D.W. Farmer and Stefan Lemurell, *Maass forms and their L-functions*, AIM 2005-15, arXiv:math.NT/0506102.

[FM99] G. Frey and M. Müller, *Arithmetic of modular curves and applications*, Algorithmic algebra and number theory (Heidelberg, 1997), Springer, Berlin, 1999, pp. 11–48.

[Fra98] J. Franke, *Harmonic analysis in weighted* L_2*-spaces*, Ann. Sci. École Norm. Sup. (4) **31** (1998), no. 2, 181–279.

[FT93] A. Fröhlich and M. J. Taylor, *Algebraic number theory*, Cambridge University Press, Cambridge, 1993.

[FvdG] C. Faber and G. van der Geer, *Sur la cohomologie des Systémes Locaux sur les Espaces des Modules des Courbes de Genus 2 and des Surfaces Abéliennes*, arXiv:math.AG/0305094.

[Gel75] Stephen S. Gelbart, *Automorphic forms on adèle groups*, Princeton University Press, Princeton, N.J., 1975, Annals of Mathematics Studies, No. 83. MR 0379375 (52 #280)

[GH81] M.J. Greenberg and J.R. Harper, *Algebraic topology*, Benjamin/Cummings Publishing Co. Inc. Advanced Book Program, Reading, Mass., 1981, A first course. MR 83b:55001

[GLQ04] Josep González, Joan-Carles Lario, and Jordi Quer, *Arithmetic of $\mathbb{Q}$-curves*, Modular curves and abelian varieties, Progr. Math., vol. 224, Birkhäuser, Basel, 2004, pp. 125–139. MR 2058647 (2005c:11068)

[GM03] P. E. Gunnells and M. McConnell, *Hecke operators and $\mathbb{Q}$-groups associated to self-adjoint homogeneous cones*, J. Number Theory **100** (2003), no. 1, 46–71.

[Gol05] Dorian Goldfeld, *Automorphic forms and L-functions on the general linear group*, to appear, 2005.

[Gon97] A. B. Goncharov, *The double logarithm and Manin's complex for modular curves*, Math. Res. Lett. **4** (1997), no. 5, 617–636.

[Gon98] ______, *Multiple polylogarithms, cyclotomy and modular complexes*, Math. Res. Lett. **5** (1998), no. 4, 497–516.

[Gor93] D. Gordon, *Discrete logarithms in* $\mathrm{GF}(p)$ *using the number field sieve*, SIAM J. Discrete Math. **6** (1993), no. 1, 124–138. MR 94d:11104

[Gor04] ______, *Discrete logarithm problem,* `http://www.win.tue.nl/~henkvt/content.html`.

[GP05] Benedict H. Gross and David Pollack, *On the Euler characteristic of the discrete spectrum*, J. Number Theory **110** (2005), no. 1, 136–163. MR 2114678 (2005k:11100)

[Gre83] Ralph Greenberg, *On the Birch and Swinnerton-Dyer conjecture*, Invent. Math. **72** (1983), no. 2, 241–265. MR 700770 (85c:11052)

[Gri05] G. Grigorov, *Kato's Euler System and the Main Conjecture*, Harvard Ph.D. Thesis (2005).

[Gro98] Benedict H. Gross, *On the Satake isomorphism*, Galois representations in arithmetic algebraic geometry (Durham, 1996), London Math. Soc. Lecture Note Ser., vol. 254, Cambridge Univ. Press, Cambridge, 1998, pp. 223–237. MR 1696481 (2000e:22008)

[GS81] F. Grunewald and J. Schwermer, *A nonvanishing theorem for the cuspidal cohomology of SL_2 over imaginary quadratic integers*, Math. Ann. **258** (1981), 183–200.

[GS02] Mark Giesbrecht and Arne Storjohann, *Computing rational forms of integer matrices*, J. Symbolic Comput. **34** (2002), no. 3, 157–172. MR 1935075 (2003j:15016)

[Gun99] P. E. Gunnells, *Modular symbols for $\mathbb{Q}$-rank one groups and Voronoĭ reduction*, J. Number Theory **75** (1999), no. 2, 198–219.

[Gun00a] ______, *Computing Hecke eigenvalues below the cohomological dimension*, Experiment. Math. **9** (2000), no. 3, 351–367. MR 1 795 307

[Gun00b] ______, *Symplectic modular symbols*, Duke Math. J. **102** (2000), no. 2, 329–350.

[Hara] G. Harder, *Congruences between modular forms of genus 1 and of genus 2*, Arbeitstagung.

[Harb] ______, *Kohomologie arithmetischer Gruppen*, lecture notes, Universität Bonn, 1987–1988.

[Har87] ______, *Eisenstein cohomology of arithmetic groups. The case* GL_2, Invent. Math. **89** (1987), no. 1, 37–118. MR 892187 (89b:22018)

[Har91] ———, *Eisenstein cohomology of arithmetic groups and its applications to number theory*, Proceedings of the International Congress of Mathematicians, Vol. I, II (Kyoto, 1990) (Tokyo), Math. Soc. Japan, 1991, pp. 779–790. MR 1159264 (93b:11057)

[Har05] ———, *Modular symbols and special values of automorphic L-functions*, preprint, 2005.

[HC68] Harish-Chandra, *Automorphic forms on semisimple Lie groups*, Notes by J. G. M. Mars. Lecture Notes in Mathematics, No. 62, Springer-Verlag, Berlin, 1968. MR 0232893 (38 #1216)

[Hel01] Sigurdur Helgason, *Differential geometry, Lie groups, and symmetric spaces*, Graduate Studies in Mathematics, vol. 34, American Mathematical Society, Providence, RI, 2001, corrected reprint of the 1978 original. MR 1834454 (2002b:53081)

[Hij74] H. Hijikata, *Explicit formula of the traces of Hecke operators for* $\Gamma_0(N)$, J. Math. Soc. Japan **26** (1974), no. 1, 56–82.

[Hsu96] Tim Hsu, *Identifying congruence subgroups of the modular group*, Proc. Amer. Math. Soc. **124** (1996), no. 5, 1351–1359. MR 1343700 (96k:20100)

[HT01] Michael Harris and Richard Taylor, *The geometry and cohomology of some simple Shimura varieties*, Annals of Mathematics Studies, vol. 151, Princeton University Press, Princeton, NJ, 2001, with an appendix by Vladimir G. Berkovich. MR 1876802 (2002m:11050)

[Hum80] James E. Humphreys, *Arithmetic groups*, Lecture Notes in Mathematics, vol. 789, Springer, Berlin, 1980. MR 584623 (82j:10041)

[Ica97] M. I. Icaza, *Hermite constant and extreme forms for algebraic number fields*, J. London Math. Soc. (2) **55** (1997), no. 1, 11–22. MR 1423282 (97j:11034)

[Jaq91] David-Olivier Jaquet, *Classification des réseaux dans* $\mathbf{R}^7$ *(via la notion de formes parfaites)*, Astérisque (1991), no. 198-200, 7–8, 177–185 (1992), Journées Arithmétiques, 1989 (Luminy, 1989). MR 1144322 (93g:11071)

[JBS03] A. Jorza, J. Balakrishna, and W. Stein, *The Smallest Conductor for an Elliptic Curve of Rank Four is Composite*, `http://modular.math.washington.edu/rank4/`.

[JC93] David-Olivier Jaquet-Chiffelle, *Énumération complète des classes de formes parfaites en dimension* 7, Ann. Inst. Fourier (Grenoble) **43** (1993), no. 1, 21–55. MR 1209694 (94d:11048)

[Kan00] Masanobu Kaneko, *The Akiyama-Tanigawa algorithm for Bernoulli numbers*, J. Integer Seq. **3** (2000), no. 2, Article 00.2.9, 6 pp. (electronic). MR 1800883 (2001k:11026)

[Kel06] Bernd C. Kellner, *Bernoulli numbers*, `http://www.bernoulli.org` (2006).

[Kna92] A.W. Knapp, *Elliptic curves*, Princeton University Press, Princeton, NJ, 1992.

[Knu] Donald E. Knuth, *The art of computer programming. Vol. 2*, third ed., Addison-Wesley Publishing Co., Reading, Mass., Seminumerical algorithms, Addison-Wesley Series in Computer Science and Information Processing.

[Kob84] N. Koblitz, *Introduction to elliptic curves and modular forms*, Graduate Texts in Mathematics, vol. 97, Springer-Verlag, New York, 1984. MR 86c:11040

[Kri90] Aloys Krieg, *Hecke algebras*, Mem. Amer. Math. Soc. **87** (1990), no. 435, x+158. MR 1027069 (90m:16024)

[Laf02] Laurent Lafforgue, *Chtoucas de Drinfeld et correspondance de Langlands*, Invent. Math. **147** (2002), no. 1, 1–241. MR 1875184 (2002m:11039)

[Lan66] R. P. Langlands, *Eisenstein series*, Algebraic Groups and Discontinuous Subgroups (Proc. Sympos. Pure Math., Boulder, Colo., 1965), Amer. Math. Soc., Providence, R.I., 1966, pp. 235–252. MR 0249539 (40 #2784)

[Lan76] Robert P. Langlands, *On the functional equations satisfied by Eisenstein series*, Springer-Verlag, Berlin, 1976, Lecture Notes in Mathematics, Vol. 544. MR 0579181 (58 #28319)

[Lan95] S. Lang, *Introduction to modular forms*, Springer-Verlag, Berlin, 1995, with appendixes by D. Zagier and W. Feit, corrected reprint of the 1976 original.

[Lem01] Dominic Lemelin, *Mazur-tate type conjectures for elliptic curves defined over quadratic imaginary fields.*

[Leo58] Heinrich-Wolfgang Leopoldt, *Eine Verallgemeinerung der Bernoullischen Zahlen*, Abh. Math. Sem. Univ. Hamburg **22** (1958), 131–140. MR 0092812 (19,1161e)

[Li75] W-C. Li, *Newforms and functional equations*, Math. Ann. **212** (1975), 285–315.

[Lin05] M. Lingham, *Modular forms and elliptic curves over imaginary quadratic fields*, Ph.D. thesis, Nottingham, 2005.

[LLL82] A. K. Lenstra, H. W. Lenstra, Jr., and L. Lovász, *Factoring polynomials with rational coefficients*, Math. Ann. **261** (1982), no. 4, 515–534. MR 682664 (84a:12002)

[LS76] Ronnie Lee and R. H. Szczarba, *On the homology and cohomology of congruence subgroups*, Invent. Math. **33** (1976), no. 1, 15–53. MR 0422498 (54 #10485)

[LS90] J.-P. Labesse and J. Schwermer (eds.), *Cohomology of arithmetic groups and automorphic forms*, Lecture Notes in Mathematics, vol. 1447, Berlin, Springer-Verlag, 1990. MR 1082959 (91h:11033)

[LS02] Joan-C. Lario and René Schoof, *Some computations with Hecke rings and deformation rings*, Experiment. Math. **11** (2002), no. 2, 303–311, with an appendix by Amod Agashe and William Stein. MR 1959271 (2004b:11072)

[LS04] Jian-Shu Li and Joachim Schwermer, *On the Eisenstein cohomology of arithmetic groups*, Duke Math. J. **123** (2004), no. 1, 141–169. MR 2060025 (2005h:11108)

[Lub94] A. Lubotzky, *Discrete groups, expanding graphs and invariant measures*, Progress in Mathematics, vol. 125, Birkhäuser Verlag, Basel, 1994, with an appendix by Jonathan D. Rogawski.

[Man72] J.I. Manin, *Parabolic points and zeta functions of modular curves*, Izv. Akad. Nauk SSSR Ser. Mat. **36** (1972), 19–66. MR 47 #3396

[Mar01] François Martin, *Périodes de formes modulaires de poids 1.*

[Mar03] Jacques Martinet, *Perfect lattices in Euclidean spaces*, Grundlehren der Mathematischen Wissenschaften [Fundamental Principles of Mathematical Sciences], vol. 327, Springer-Verlag, Berlin, 2003. MR 1957723 (2003m:11099)

[Mar05] Greg Martin, *Dimensions of the spaces of cusp forms and newforms on $\Gamma_0(N)$ and $\Gamma_1(N)$*, J. Number Theory **112** (2005), no. 2, 298–331. MR 2141534 (2005m:11069)

[Maz73] B. Mazur, *Courbes elliptiques et symboles modulaires*, Séminaire Bourbaki, 24ème année (1971/1972), Exp. No. 414, Springer, Berlin, 1973, pp. 277–294. Lecture Notes in Math., Vol. 317. MR 55 #2930

[McC91] M. McConnell, *Classical projective geometry and arithmetic groups*, Math. Ann. **290** (1991), no. 3, 441–462 MR 92k:22020

[Men79] Eduardo R. Mendoza, *Cohomology of* PGL_2 *over imaginary quadratic integers*, Bonner Mathematische Schriften [Bonn Mathematical Publications], 128, Universität Bonn Mathematisches Institut, Bonn, 1979, Dissertation, Rheinische Friedrich-Wilhelms-Universität, Bonn, 1979. MR 611515 (82g:22012)

[Mer94] L. Merel, *Universal Fourier expansions of modular forms*, On Artin's conjecture for odd 2-dimensional representations, Springer, 1994, pp. 59–94.

[Mer99] ———, *Arithmetic of elliptic curves and Diophantine equations*, J. Théor. Nombres Bordeaux **11** (1999), no. 1, 173–200, Les XXèmes Journées Arithmétiques (Limoges, 1997). MR 1730439 (2000j:11084)

[Mes86] J.-F. Mestre, *La méthode des graphes. Exemples et applications*, Proceedings of the international conference on class numbers and fundamental units of algebraic number fields (Katata) (1986), 217–242.

[Miy89] T. Miyake, *Modular forms*, Springer-Verlag, Berlin, 1989, translated from the Japanese by Yoshitaka Maeda.

[MM89] R. MacPherson and M. McConnell, *Classical projective geometry and modular varieties*, Algebraic analysis, geometry, and number theory (Baltimore, MD, 1988), Johns Hopkins Univ. Press, Baltimore, MD, 1989, pp. 237–290. MR 98k:14076

[MM93] ———, *Explicit reduction theory for Siegel modular threefolds*, Invent. Math. **111** (1993), no. 3, 575–625. MR 94a:32052

[MTT86] B. Mazur, J. Tate, and J. Teitelbaum, *On p-adic analogues of the conjectures of Birch and Swinnerton-Dyer*, Invent. Math. **84** (1986), no. 1, 1–48.

[MW94] Colette Mœglin and Jean-Loup Waldspurger, *Décomposition spectrale et séries d'Eisenstein*, Progress in Mathematics, vol. 113, Birkhäuser Verlag, Basel, 1994, Une paraphrase de l'Écriture [A paraphrase of Scripture]. MR 1261867 (95d:11067)

[Nec94] V. I. Nechaev, *On the complexity of a deterministic algorithm for a discrete logarithm*, Mat. Zametki **55** (1994), no. 2, 91–101, 189. MR 96a:11145

[Ong86] Heidrun E. Ong, *Perfect quadratic forms over real-quadratic number fields*, Geom. Dedicata **20** (1986), no. 1, 51–77. MR 823160 (87f:11023)

[PR94] Vladimir Platonov and Andrei Rapinchuk, *Algebraic groups and number theory*, Pure and Applied Mathematics, vol. 139, Academic Press Inc., Boston, MA, 1994, translated from the 1991 Russian original by Rachel Rowen. MR 1278263 (95b:11039)

[Que06] J. Quer, *Dimensions of spaces of modular forms for* $\Gamma_H(N)$, Preprint.

[Rib92] K. A. Ribet, *Abelian varieties over* $\mathbf{Q}$ *and modular forms*, Algebra and topology 1992 (Taejŏn), Korea Adv. Inst. Sci. Tech., Taejŏn, 1992, pp. 53–79. MR 94g:11042

[Ros86] M. Rosen, *Abelian varieties over* **C**, Arithmetic geometry (Storrs, Conn., 1984), Springer, New York, 1986, pp. 79–101.

[RS01] K. A. Ribet and W. A. Stein, *Lectures on Serre's conjectures*, Arithmetic algebraic geometry (Park City, UT, 1999), IAS/Park City Math. Ser., vol. 9, Amer. Math. Soc., Providence, RI, 2001, pp. 143–232. MR 2002h:11047

[Sap97] Leslie Saper, *Tilings and finite energy retractions of locally symmetric spaces*, Comment. Math. Helv. **72** (1997), no. 2, 167–202. MR 1470087 (99a:22019)

[Sar03] Peter Sarnak, *Spectra of hyperbolic surfaces*, Bull. Amer. Math. Soc. (N.S.) **40** (2003), no. 4, 441–478 (electronic). MR 1997348 (2004f:11107)

[SC03] Samir Siksek and John E. Cremona, *On the Diophantine equation* $x^2+7=y^m$, Acta Arith. **109** (2003), no. 2, 143–149. MR 1980642 (2004c:11109)

[Sch86] Joachim Schwermer, *Holomorphy of Eisenstein series at special points and cohomology of arithmetic subgroups of* $\mathrm{SL}_n(\mathbf{Q})$, J. Reine Angew. Math. **364** (1986), 193–220. MR 817646 (87h:11048)

[Sch90] A. J. Scholl, *Motives for modular forms*, Invent. Math. **100** (1990), no. 2, 419–430.

[Sch95] R. Schoof, *Counting points on elliptic curves over finite fields*, J. Théor. Nombres Bordeaux **7** (1995), no. 1, 219–254, Les Dix-huitièmes Journées Arithmétiques (Bordeaux, 1993). MR 1413578 (97i:11070)

[Ser73] J-P. Serre, *A Course in Arithmetic*, Springer-Verlag, New York, 1973, Translated from the French, Graduate Texts in Mathematics, No. 7.

[Ser87] ———, *Sur les représentations modulaires de degré* 2 *de* $\mathrm{Gal}(\overline{\mathbf{Q}}/\mathbf{Q})$, Duke Math. J. **54** (1987), no. 1, 179–230.

[Shi59] G. Shimura, *Sur les intégrales attachées aux formes automorphes*, J. Math. Soc. Japan **11** (1959), 291–311.

[Shi94] ———, *Introduction to the arithmetic theory of automorphic functions*, Princeton University Press, Princeton, NJ, 1994, reprint of the 1971 original, Kan Memorial Lectures, 1.

[Sho80a] V. V. Shokurov, *Shimura integrals of cusp forms*, Izv. Akad. Nauk SSSR Ser. Mat. **44** (1980), no. 3, 670–718, 720. MR 582162 (82b:10029)

[Sho80b] ———, *A study of the homology of Kuga varieties*, Izv. Akad. Nauk SSSR Ser. Mat. **44** (1980), no. 2, 443–464, 480. MR 571104 (82f:14023)

[Sho97] Victor Shoup, *Lower bounds for discrete logarithms and related problems*, Advances in cryptology—EUROCRYPT '97 (Konstanz), Lecture Notes in Comput. Sci., vol. 1233, Springer, Berlin, 1997, pp. 256–266. MR 98j:94023

[Sil92] J. H. Silverman, *The arithmetic of elliptic curves*, Springer-Verlag, New York, 1992, corrected reprint of the 1986 original.

[Sou75] Christophe Soulé, *Cohomologie de* $SL_3(Z)$, C. R. Acad. Sci. Paris Sér. A-B **280** (1975), no. 5, Ai, A251–A254. MR 0396849 (53 #709)

[Sta79] R. E. Staffeldt, *Reduction theory and* K_3 *of the Gaussian integers*, Duke Math. J. **46** (1979), no. 4, 773–798. MR 552526 (80m:22014)

[Ste] Allan Steel, *Advanced matrix algorithms*, Seminar Talk at Harvard University.

[Ste97] ———, *A new algorithm for the computation of canonical forms of matrices over fields*, J. Symbolic Comput. **24** (1997), no. 3-4, 409–432, Computational algebra and number theory (London, 1993). MR 1484489 (98m:65070)

[Ste99a] Norman Steenrod, *The topology of fibre bundles*, Princeton Landmarks in Mathematics, Princeton University Press, Princeton, NJ, 1999, reprint of the 1957 edition, Princeton Paperbacks. MR 1688579 (2000a:55001)

[Ste99b] W. A. Stein, HECKE*: The Modular Symbols Calculator*, software (available online) (1999).

[Ste00] ———, *Explicit approaches to modular abelian varieties*, Ph.D. thesis, University of California, Berkeley (2000).

[Ste06] ———, SAGE: *Software for Algebra and Geometry Experimentation,* `http://sage.scipy.org/sage`.

[Str69] Volker Strassen, *Gaussian elimination is not optimal*, Numerische Mathematik **13** (1969), 354–356.

[Stu87] J. Sturm, *On the congruence of modular forms*, Number theory (New York, 1984–1985), Springer, Berlin, 1987, pp. 275–280.

[SV01] W. A. Stein and H. A. Verrill, *Cuspidal modular symbols are transportable*, LMS J. Comput. Math. **4** (2001), 170–181 (electronic). MR 1 901 355

[SV03] B. Speh and T. N. Venkataramana, *Construction of some generalised modular symbols*, preprint, 2003.

[SW02] William A. Stein and Mark Watkins, *A database of elliptic curves—first report*, Algorithmic number theory (Sydney, 2002), Lecture Notes in Comput. Sci., vol. 2369, Springer, Berlin, 2002, pp. 267–275. MR 2041090 (2005h:11113)

[SW05] Jude Socrates and David Whitehouse, *Unramified Hilbert modular forms, with examples relating to elliptic curves*, Pacific J. Math. **219** (2005), no. 2, 333–364. MR 2175121

[Tat75] J. Tate, *Algorithm for determining the type of a singular fiber in an elliptic pencil*, Modular functions of one variable, IV (Proc. Internat. Summer School, Univ. Antwerp, Antwerp, 1972), Springer, Berlin, 1975, pp. 33–52. Lecture Notes in Math., Vol. 476. MR 52 #13850

[Tho89] J. G. Thompson, *Hecke operators and noncongruence subgroups*, Group theory (Singapore, 1987), de Gruyter, Berlin, 1989, including a letter from J.-P. Serre, pp. 215–224. MR 981844 (90a:20105)

[Tot05] A. Toth, *On the Steinberg module of Chevalley groups*, Manuscripta Math. **116** (2005), no. 3, 277–295.

[TW95] R. Taylor and A. J. Wiles, *Ring-theoretic properties of certain Hecke algebras*, Ann. of Math. (2) **141** (1995), no. 3, 553–572.

[vdG] Gerard van der Geer, *Siegel Modular Forms*, arXiv:math.AG/0605346.

[vGvdKTV97] Bert van Geemen, Wilberd van der Kallen, Jaap Top, and Alain Verberkmoes, *Hecke eigenforms in the cohomology of congruence subgroups of* $\mathrm{SL}(3,\mathbf{Z})$, Experiment. Math. **6** (1997), no. 2, 163–174. MR 1474576 (99a:11059)

[Vig77] Marie-France Vignéras, *Séries thêta des formes quadratiques indéfinies*, Séminaire Delange-Pisot-Poitou, 17e année (1975/76), Théorie des nombres: Fasc. 1, Exp. No. 20, Secrétariat Math., Paris, 1977, p. 3. MR 0480352 (58 #521)

[Vog85] K. Vogtmann, *Rational homology of Bianchi groups*, Math. Ann. **272** (1985), no. 3, 399–419.

[Vog97] David A. Vogan, Jr., *Cohomology and group representations*, Representation theory and automorphic forms (Edinburgh, 1996), Proc. Sympos. Pure Math., vol. 61, Amer. Math. Soc., Providence, RI, 1997, pp. 219–243. MR 1476500 (98k:22064)

[Vor08] G. Voronoĭ, *Nouvelles applications des paramétres continus à la théorie des formes quadratiques, I. Sur quelques propriétés des formes quadratiques positives parfaites*, J. Reine Angew. Math. **133** (1908), 97–178.

[VZ84] David A. Vogan, Jr. and Gregg J. Zuckerman, *Unitary representations with nonzero cohomology*, Compositio Math. **53** (1984), no. 1, 51–90. MR 762307 (86k:22040)

[Wan82] Kai Wang, *A proof of an identity of the Dirichlet L-function*, Bull. Inst. Math. Acad. Sinica **10** (1982), no. 3, 317–321. MR 679019 (84c:10040)

[Wan95] Xiang Dong Wang, *2-dimensional simple factors of $J_0(N)$*, Manuscripta Math. **87** (1995), no. 2, 179–197. MR 1334940 (96h:11059)

[Wes] U. Weselman, personal communication.

[Whi90] E. Whitley, *Modular symbols and elliptic curves over imaginary quadratic number fields*, Ph.D. thesis, Exeter University, 1990.

[Wie05] Gabor Wiese, *Modular Forms of Weight One Over Finite Fields*, Ph.D. thesis (2005).

[Wil95] A. J. Wiles, *Modular elliptic curves and Fermat's last theorem*, Ann. of Math. (2) **141** (1995), no. 3, 443–551. MR 1333035 (96d:11071)

[Wil00] ______, *The Birch and Swinnerton-Dyer Conjecture*, `http://www.claymath.org/prize_problems/birchsd.htm`.

[Yas05a] D. Yasaki, *On the cohomology of* SU(2, 1) *over the Gaussian integers*, preprint, 2005.

[Yas05b] ______, *On the existence of spines for* **Q**-*rank 1 groups*, preprint, 2005.

Index

Symbol Index

Algorithm Index

Definition Index

SAGE Index

General Index

Titles in This Series

79 **William Stein,** Modular forms, a computational approach (with an appendix by Paul E. Gunnells), 2007

78 **Harry Dym,** Linear algebra in action, 2007

77 **Bennett Chow, Peng Lu, and Lei Ni,** Hamilton's Ricci flow, 2006

76 **Michael E. Taylor,** Measure theory and integration, 2006

75 **Peter D. Miller,** Applied asymptotic analysis, 2006

74 **V. V. Prasolov,** Elements of combinatorial and differential topology, 2006

73 **Louis Halle Rowen,** Graduate algebra: Commutative view, 2006

72 **R. J. Williams,** Introduction the the mathematics of finance, 2006

71 **S. P. Novikov and I. A. Taimanov,** Modern geometric structures and fields, 2006

70 **Seán Dineen,** Probability theory in finance, 2005

69 **Sebastián Montiel and Antonio Ros,** Curves and surfaces, 2005

68 **Luis Caffarelli and Sandro Salsa,** A geometric approach to free boundary problems, 2005

67 **T.Y. Lam,** Introduction to quadratic forms over fields, 2004

66 **Yuli Eidelman, Vitali Milman, and Antonis Tsolomitis,** Functional analysis, An introduction, 2004

65 **S. Ramanan,** Global calculus, 2004

64 **A. A. Kirillov,** Lectures on the orbit method, 2004

63 **Steven Dale Cutkosky,** Resolution of singularities, 2004

62 **T. W. Körner,** A companion to analysis: A second first and first second course in analysis, 2004

61 **Thomas A. Ivey and J. M. Landsberg,** Cartan for beginners: Differential geometry via moving frames and exterior differential systems, 2003

60 **Alberto Candel and Lawrence Conlon,** Foliations II, 2003

59 **Steven H. Weintraub,** Representation theory of finite groups: algebra and arithmetic, 2003

58 **Cédric Villani,** Topics in optimal transportation, 2003

57 **Robert Plato,** Concise numerical mathematics, 2003

56 **E. B. Vinberg,** A course in algebra, 2003

55 **C. Herbert Clemens,** A scrapbook of complex curve theory, second edition, 2003

54 **Alexander Barvinok,** A course in convexity, 2002

53 **Henryk Iwaniec,** Spectral methods of automorphic forms, 2002

52 **Ilka Agricola and Thomas Friedrich,** Global analysis: Differential forms in analysis, geometry and physics, 2002

51 **Y. A. Abramovich and C. D. Aliprantis,** Problems in operator theory, 2002

50 **Y. A. Abramovich and C. D. Aliprantis,** An invitation to operator theory, 2002

49 **John R. Harper,** Secondary cohomology operations, 2002

48 **Y. Eliashberg and N. Mishachev,** Introduction to the h-principle, 2002

47 **A. Yu. Kitaev, A. H. Shen, and M. N. Vyalyi,** Classical and quantum computation, 2002

46 **Joseph L. Taylor,** Several complex variables with connections to algebraic geometry and Lie groups, 2002

45 **Inder K. Rana,** An introduction to measure and integration, second edition, 2002

44 **Jim Agler and John E. McCarthy,** Pick interpolation and Hilbert function spaces, 2002

43 **N. V. Krylov,** Introduction to the theory of random processes, 2002

42 **Jin Hong and Seok-Jin Kang,** Introduction to quantum groups and crystal bases, 2002

41 **Georgi V. Smirnov,** Introduction to the theory of differential inclusions, 2002

40 **Robert E. Greene and Steven G. Krantz,** Function theory of one complex variable, third edition, 2006

39 **Larry C. Grove,** Classical groups and geometric algebra, 2002

38 **Elton P. Hsu,** Stochastic analysis on manifolds, 2002

37 **Hershel M. Farkas and Irwin Kra,** Theta constants, Riemann surfaces and the modular group, 2001

36 **Martin Schechter,** Principles of functional analysis, second edition, 2002

35 **James F. Davis and Paul Kirk,** Lecture notes in algebraic topology, 2001

34 **Sigurdur Helgason,** Differential geometry, Lie groups, and symmetric spaces, 2001

33 **Dmitri Burago, Yuri Burago, and Sergei Ivanov,** A course in metric geometry, 2001

32 **Robert G. Bartle,** A modern theory of integration, 2001

31 **Ralf Korn and Elke Korn,** Option pricing and portfolio optimization: Modern methods of financial mathematics, 2001

30 **J. C. McConnell and J. C. Robson,** Noncommutative Noetherian rings, 2001

29 **Javier Duoandikoetxea,** Fourier analysis, 2001

28 **Liviu I. Nicolaescu,** Notes on Seiberg-Witten theory, 2000

27 **Thierry Aubin,** A course in differential geometry, 2001

26 **Rolf Berndt,** An introduction to symplectic geometry, 2001

25 **Thomas Friedrich,** Dirac operators in Riemannian geometry, 2000

24 **Helmut Koch,** Number theory: Algebraic numbers and functions, 2000

23 **Alberto Candel and Lawrence Conlon,** Foliations I, 2000

22 **Günter R. Krause and Thomas H. Lenagan,** Growth of algebras and Gelfand-Kirillov dimension, 2000

21 **John B. Conway,** A course in operator theory, 2000

20 **Robert E. Gompf and András I. Stipsicz,** 4-manifolds and Kirby calculus, 1999

19 **Lawrence C. Evans,** Partial differential equations, 1998

18 **Winfried Just and Martin Weese,** Discovering modern set theory. II: Set-theoretic tools for every mathematician, 1997

17 **Henryk Iwaniec,** Topics in classical automorphic forms, 1997

16 **Richard V. Kadison and John R. Ringrose,** Fundamentals of the theory of operator algebras. Volume II: Advanced theory, 1997

15 **Richard V. Kadison and John R. Ringrose,** Fundamentals of the theory of operator algebras. Volume I: Elementary theory, 1997

14 **Elliott H. Lieb and Michael Loss,** Analysis, 1997

13 **Paul C. Shields,** The ergodic theory of discrete sample paths, 1996

12 **N. V. Krylov,** Lectures on elliptic and parabolic equations in Hölder spaces, 1996

11 **Jacques Dixmier,** Enveloping algebras, 1996 Printing

10 **Barry Simon,** Representations of finite and compact groups, 1996

9 **Dino Lorenzini,** An invitation to arithmetic geometry, 1996

8 **Winfried Just and Martin Weese,** Discovering modern set theory. I: The basics, 1996

7 **Gerald J. Janusz,** Algebraic number fields, second edition, 1996

6 **Jens Carsten Jantzen,** Lectures on quantum groups, 1996

5 **Rick Miranda,** Algebraic curves and Riemann surfaces, 1995

For a complete list of titles in this series, visit the AMS Bookstore at **www.ams.org/bookstore/**.